CLUSTER ANALYSIS

Cluster Analysis

Robert C. Tryon

late professor of psychology
university of california

Daniel E. Bailey

associate professor of psychology and computing science
university of colorado
president, tryon-bailey associates, inc.
boulder, colorado

•

McGRAW-HILL BOOK COMPANY

new york   st. louis   san francisco   düsseldorf   london   mexico   panama
sydney   toronto

This book was set in News Gothic by The Maple Press
Company, and printed on permanent paper and bound
by The Maple Press Company. The designer was Paula
Tuerk; the drawings were done by John Cordes, J. & R.
Technical Services, Inc. The editors were Walter
Maytham and Anne Marie Horowitz. Matt Martino
supervised the production.

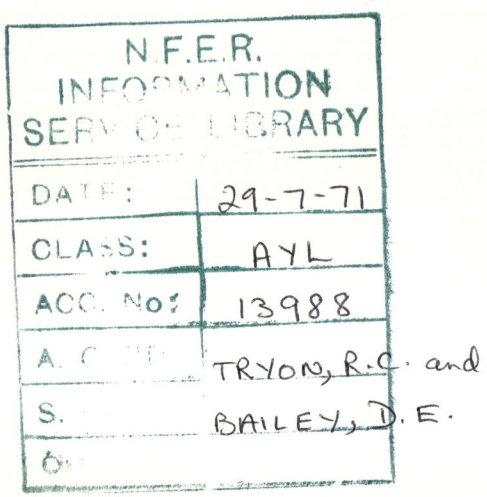
CLUSTER ANALYSIS

Library of Congress Catalog Card Number 71-115152
65325

1234567890 MAMM 79876543210

foreword

This book distills the multivariate thinking of Robert Tryon and his students over a period of more than 30 years. Tryon's initial statement of cluster analysis was presented in his 1939 monograph entitled "Cluster Analysis." At that time, all computations had to be done by hand; Tryon was later to speak of his misspent youth, because too much of his time had been spent with a desk calculator. In the 1950s the practice of cluster analysis was restated in computer terms to enable the investigator to escape from hand calculations. Tryon and Bailey therefore planned this book to be the definitive account of postcomputer cluster analysis. The manuscript was almost finished when Tryon died suddenly in 1967. Bailey has perforce had to accept sole responsibility for the final revisions.

Since the book is appearing posthumously, the reader may find it helpful to know about Robert Tryon's psychological career and his intellectual objectives in formulating cluster analysis. His academic record was a simple one. He received both bachelor's and doctor's degrees from the University of California, he was then invited to join the faculty of the university, and he stayed at Berkeley in the Department of Psychology for the rest of his life. His only lengthy absence was his war service for the Office of Strategic Services from 1941 to 1945 in Washington, D.C.

His intellectual history was more complicated. His interests were broad, but they were linked in a unique way by his personal philosophy and his intellectual objectives. His first research contributions were in experimental psychology, but he was already an experimenter with a difference. Most experimentalists of the period focused on the construction of general theories of perception, learning, and so on, and viewed individual differences in behavior as a nuisance. Tryon, however, like Galton and Binet before him, was interested in individual differences for their own sake. He thought that science should concern itself with the differences as well as the similarities in behavior. In particular, he wanted to know how far the differences could be explained by genetic inheritance and how far by the physical and social environment. Thus his research and writing span an area from behavioral genetics through abilities and personality characteristics to social psychology and social ecology. Three examples of cluster analysis are repeatedly used in this book. The first is concerned with abilities; the second, with personality measures; and the third, with social areas. These represent the range of his interests fairly well, except that his genetic interests are not included. Tryon seems to have been the founder of the field we now call behavioral genetics, and this was the area in which he first became widely known. He carried out extensive selective breeding experiments in rats. The differences in rats' skills in finding their way through mazes to food became the central phenomenon for analysis. After about three generations, bright and dull strains of maze learners had been established. Although Tryon did not return to laboratory experiments after his departure for war service, he continued to retain a lively interest in behavioral genetics throughout his life, and students of his have since attained prominence in the field.

An interest in individual differences is not, of course, unique to the twentieth century. Classification is usually one of the earliest forms of scientific activity, and it depends upon differences in individuals. As everyone knows, Aristotle delighted in classificatory systems. So did Linnaeus, who, without the aid of any measurement, devised a biological taxonomy in which membership of one class rather than another usually depended upon a single key attribute.

Perhaps the main change in classification in the last 100 years has been a new emphasis upon measurement. The post-Darwinians began a search for a reliable method of quantifying association in measures of biological variability. Galton's discovery of the correlation coefficient in 1883 brought this quest to fruition. Tables of correlations began to appear in scientific journals by the turn of the century. The way was then clear for the multivariate methods, such as factor analysis and cluster analysis, which classify by the analysis of correlations.

When Tryon finished his doctorate in 1928, the most generally accepted view of human ability was Spearman's two-factor theory. Spearman devised an early form of factor analysis (the analysis of tetrad differences) in an attempt to demonstrate in human beings the existence of an all-encompassing ability which he called intelligence. He theorized that this would be present in differing degree in a wide variety of mental tests but that in each test there would also be represented an ability which was specific to that test. Spearman's two-factor theory and his methods of analysis had come under increasing criticism by the 1920s. Godfrey Thomson had shown that any correlational proportionality which accorded with the two-factor theory could also be explained in other than two-factor terms. Cyril Burt found how to extract further factors than the general one and, by so doing, demonstrated the need for a verbal factor as well as the factor of general intelligence if the observed correlations were to be explained. Tryon arrived on the professional scene just in time to add his criticisms of Spearman's theory. He postulated many abilities of low generality, because abilities were the product of multiple genes and diverse social environments, and he recognized the need for new methods to identify these lower-level abilities. Cluster analysis was his answer.

In the early 1930s, Thurstone was the leading innovator in factor analysis. His method of multiple factor analysis was a much more satisfactory procedure for identifying abilities than Spearman's method had been. Tryon spent a sabbatical year at the University of Chicago in the later 1930s. He grasped the point, as many others at Chicago must have done, that similar tests would have high correlations between them and that clusters of related tests could therefore be identified without the labor of a centroid factor analysis by direct search of the correlations. Thus cluster analysis, as originally conceived by Tryon, was a poor man's factor analysis. Practicality was always a guiding motive for him. In his animal studies, he had found a way of getting automatic maze readings to lighten the experimenter's load. He now wanted to save the immense labor that Thurstone and his associates were devoting to factorial studies. Tryon also wanted methods which relied as much as possible upon logic and as little as possible upon mathematics. This preference continues to be displayed in this book. Actually cluster analysis has many equations, but this book is written with rather few. The logic and the practice occupy the center of the stage.

When Tryon worked mathematically, his usual choice was for geometry. He seems to have seen clusters in his space rather as astronomers see galaxies in theirs. There has been some discussion in the psychological literature about who first thought of the multiple-group method of factor analysis, in which axes were put through the centers of clusters. Guttman, Holzinger, and Thurstone all seem to have discovered variants of the method at about the same time. Since Tryon had been at the University of Chicago shortly before then, my guess—and it is only a guess—is that Tryon's cluster approach was one stimulant to the multiple-group method. When computer programs later came to be written for cluster analysis, Tryon required the output to include configurations as well as numbers. Many illustrations will be found in this book. Inflatable balloons became a standard part of Tryon's baggage when he went to professional meetings in his later years.

Whenever psychology and mathematics came into conflict, Tryon almost always preferred a realistic psychological solution to a neat mathematical one. One example is his insistence (in agreement with Thurstone) that there is no compelling psychological reason why factors or clusters should be at right angles to one another. If they are oblique, then we must use oblique axes despite the mathematical unpalatibility of doing so. Another example appears in Tryon's method for getting a score for each individual on each cluster. He was not patient with the elaborate regressional techniques by which even the variables outside the cluster played their part in the complicated weighting scheme and through which the scores on the different clusters were uncorrelated (or very nearly so). As Tryon saw it, the price of such mathematical elegance was one's not knowing exactly what one had measured. For the sake of psychological understanding, he decided to obtain the cluster scores by straightforward addition, even though they were then correlated with one another.

After Tryon's service in World War II he was chairman of his department for several years, so that one might have wondered in 1950 whether he would again contribute appreciably to multivariate methods. At the most, so it might have seemed, there would be need only for a revised version of the 1939 monograph. Then a computer became available to Tryon, and he entered upon another period of sustained effort in cluster analysis which ended only with his death. Tryon found that a computer made it possible for cluster analyses to be carried out as the logic demanded, without too many concessions to computational convenience. He realized, perhaps more firmly than any other psychologist of his time, that his quantitative methods must be translated into a package of computer programs. Psychologists are indebted to the National Institutes of Health for their generous financial support for Tryon's ambitious programming effort. This enabled him to collect a group of associates (often his former students) who knew cluster analysis very well, were able to express its ideas in convenient computer terms, and then to write the individual programs or to supervise their writing. The package, under the strange name of BC TRY (Bailey explains the origin of this name in the Preface), forms a model of its kind, and it is now widely used. In designing the package, Tryon was eclectic. He provided various options so that each investigator could use his own judgment in selecting whatever method and computer output he needed. Hence BC TRY is essentially a package in factor analysis as well as cluster analysis. Indeed, the distinction between the two has become somewhat blurred. Tryon's eclecticism reflects his tolerance, which was one of his strongest personality characteristics.

Tryon and Bailey assume, correctly enough, that any reader of this book will have access to a computer. The account will be of only limited value to anyone who continues to use desk calculators. The text bristles with the new computer words—DAP, SPAN, OMARK and dozens of others—to which quantitative behavioral scientists must now become accustomed. Chapter 3 gives an overview of the computer system, and Chap. 12 describes its statistical and logical basis. (Incidentally, equations are more abundant in this later chapter.) Details of the computer control cards are given in Chap. 13, where a User's Manual supplements the book.

Years of labor went into BC TRY. We are reminded how much effort is needed to get our computer programs into good order. The task was periodically delayed by changes in the computer hardware. Bailey points out in the Preface that the cost to researchers of the frequent changes of computers made by almost all universities has been very great. I agree strongly with him. I do not think that computer administrators have always paid sufficient regard to the need for continuity in users' research. Programs have had to be rewritten too many times. It has become clear that computers should have been changed less often and that bigger leaps should have been taken in computer power whenever a change became inevitable. Furthermore, the outgoing machine should almost always have been left to work for longer periods of time alongside the incoming one which was scheduled to replace it. It is pleasing to know that Bailey has grappled firmly with this problem of programming continuity. He continues to update programs and the Users' Manual as computers change. Furthermore, he has planned financial arrangements to enable BC TRY to be modified to fit the computers of various manufacturers. Many of us have complained about the wastefulness of the present haphazard system by which each university duplicates the efforts of the others in building its independent library of basic computer programs. Interchange of programs has been erratic, quality of those being exchanged has often been unsatisfactory, and far too much programming has been duplicated. A national library of behavioral science computer programs is clearly needed. Bailey has pioneered the way to this by his system for continuous maintenance and national distribution of BC TRY.

The BC TRY package not only gives the user many options. Tryon and Bailey repeatedly urge the cluster analyst, after studying his initial results, to redefine his conditions and then make a further analysis. In desk calculator days, no one ever wanted to repeat his analysis. With a computer, it has become easy for any researcher to do his work a second time, profiting from what he has learned in the initial analysis. Critics sometimes allege (with some justification) that the computer will dehumanize behavioral research by eliminating the opportunity for judgment in the statistical analysis. Tryon and Bailey demonstrate unequivocally that this need not be so.

I should like to conclude this foreword by paying public tribute to Robert Choate Tryon as a man. I found him almost invariably courteous, kind, humorous, and humane. He set for himself the ideals of a gentleman and a scholar, and he was wise as a counselor and of loyal and generous spirit as a friend. *I believe these were the generally accepted impressions of the man.*

Charles Wrigley

Preface

This book is designed to present an integrated theoretical and applied description of cluster analysis as conceived of by the late Robert C. Tryon and his colleagues over the span of more than 30 years. The material presented is the current state of the science, its methods, and research findings in a variety of application fields. The reader is given enough information to be able to comprehend and use the most sophisticated and elaborate cluster-analysis procedures and to understand the complex results of cluster-analysis applications. Consequently, the book can be used as a text in both undergraduate and more advanced courses. The book is also intended as a guide for the more seasoned researcher just beginning to feel the need for cluster analysis in his own research. To the sophisticated scientist the book offers many new ideas in cluster analysis methodology and research findings.

The range of topics discussed extends over the most basic concepts of observation of natural phenomena, the logical foundations for the cluster analysis of variables, the mathematical formulation of cluster analysis, the logical and operational bases of cluster analysis of objects, and the technology of cluster analysis computer systems. The intent is as much substantive as methodological. The substantive content of the book is drawn from the behaviorial and social sciences. However, in the last few years, the science of cluster analysis has been discovered to be a valuable tool in the physical, economic, and biological sciences. We have witnessed the growth of the use of the methods of cluster analysis presented in this book to include such diverse scientific disciplines as chemistry, genetics, hydrology, zoology, glaciology, medicine, and soil morphology, in addition to the more expected disciplines such as psychology, education, sociology, anthropology, and economics.

The finished product presented here was conceived by Robert C. Tryon in the middle 1930s and pushed by him to its present form through many stages of sophistication and technology. In 1939, Tryon published a monograph, "Cluster Analysis," containing the broad outlines of this book. The methodology and theoretical principles of cluster analysis stated in that monograph were expressed in "oxcart" desk calculator "programs" that taxed the patience and skill of the most ardent cluster analyst. That period culminated in the first laboriously worked-out example in the form of a monograph, "Identification of Social Areas by Cluster Analysis," by Tryon in 1955. Almost all the methodological principles in the current form of cluster analysis were anticipated or already had been achieved in that publication. The advent of the computer (an IBM 701 at the University of California in the fall of 1956) and the presence of Charles Wrigley at Berkeley at the same time stimulated a period of great activity and inventiveness in cluster analysis on the part of Tryon that ended only with his death in September, 1967. Tryon was able to capitalize on the stimulation from others working in allied areas at that time in Berkeley, principally Charles Wrigley, Henry Kaiser, Louis Guttman, and John Neuhaus. Out of these associations Tryon found tremendous inspirations and sources of motivation that led eventually to the modern concepts of cluster analysis presented in this book. I had the good fortune to be a graduate student at the initiation of this productive, exciting period and wrote, with the guidance and assistance of John Neuhaus, the first computer program to execute the central analysis of cluster analysis, a key-cluster analysis of variables from a correlation matrix (Tryon, 1958a).

In 1959 we conceived of a general computer system to execute all the main forms of cluster and factor analysis as mere special cases or options within the general system. This effort was encouraged by Dr. Philip Sapir, of the National Institutes of Mental Health, and funds were awarded for Project CAP, Cluster Analysis Programs, in 1960 by the National Institutes of Mental Health to Professor Tryon. During subsequent years the staff of the project met regularly, exchanging ideas and developing the theoretical, statistical, and computer aspects of the cluster analysis system. Early in the

development it became necessary to attach a name to the system of computer programs. I suggested the name TRYON in recognition of the source
of the entire development of cluster analysis. Professor Tryon, in modesty,
demurred. However, the other staff members and I were able to convince
him that the name TRY would be appropriate because of the experimental
nature of much of the work we were doing; we were indeed going to *try* many
ideas. Because of the computing center participation in the IBM consortium
SHARE, the letters BC (standing for the Berkeley chapter of SHARE) were
tacked onto the front of the name, and the somewhat appalling designation
of the "BC TRY System" has become so widespread that it probably cannot
be changed.

In the intervening 10 years we have seen many changes in computer
technology, computers, and computer languages. All these changes have
been painful to a certain extent, and we are not completely convinced that
the overall result has been beneficial. We went from the IBM 701 to the IBM
704, to the IBM 7090, the IBM 7094, to the CDC 6400 and at Colorado from
the 7090 to the IBM 709. In the process, changes in computer operating
systems and programming languages have been more frequent than
changes in computer. As a consequence, the productive work is somewhat
less than half of what would have been accomplished under stable computer conditions. Perhaps that is progress, but it reminds one of the
pioneers who struck out due west only to have to keep changing their
route because of impassable mountain ranges and deserts.

Many hands and minds have built the BC TRY System over the years
since 1959. We are particularly indebted to the programmers and the programming supervisors (in which category I worked on project CAP for two
years). Also critical in the development of the System and the production
of the research leading to this book are the research assistants and secretarial staff of project CAP and its successors. It is impossible here to do jus-
tice to the contributions. John Vinsonhaler played a key role in supervision
of the programming on the 704 in its final stages and in the 7090 programming. Robert Russell was the main innovator of many special programming
features of the System and wrote many of its programs, especially on the
704 and the 7090.   Many others made signal contributions to the programming, especially John Vinsonhaler, David Matula, and John Bauer (who is a
collaborator with the authors in Chap. 13), Tom Kibler, and Eleanor Krasnow.
In formulating the programming of BC TRY those most centrally involved
were Professor Tryon, myself, John Vinsonhaler, Robert Russell, William
Meredith, and John Bauer. Chen-lin Chu and Jin-Yu Yen (also a collaborator
in Chap. 13) assisted Professor Tryon in performing many methodological
studies that led to decisions incorporated in the programs. Others who have

made important contributions in development are Richard Burack, James Cameron, Mike Davidson, Don Flory, Robert Menhenett, Kent Mitchell, and John Wolfe. On the secretarial side we were valuably served by Valerie Siebert, Louis Ruhland, Sara Bailey, and Nancy Geidel.

Dr. Harley B. Messinger read the completed manuscript and offered many valuable suggestions that led to improvements in the finished product. His assistance is deeply appreciated.

I wish to thank the Editor of the journal *Multivariate Behavioral Research* for permission to reproduce copious material from the papers Professor Tryon and I published in the journal. Many of the chapters of this book are reworked versions of those papers. The Abridged User's Manual of the BC TRY System appears here as Chap. 13 by courtesy of Tryon-Bailey Associates, Inc.

The support of the National Institutes of Mental Health in grants MH 0811 and MH 08314 is gratefully acknowledged.

Our wives, Freida Tryon and Sara Bailey, deserve a special expression of gratitude. Their encouragement and interest played an important role in this book. Mrs. Tryon's active interest and participation was an important contribution to Professor Tryon's work.

It is fitting that the close of this preface and the dedication of this book be the statement appearing in the October, 1967, edition of *Multivariate Behavioral Research:*

Robert C. Tryon, great scientist, true friend, died on the 27th day of September, 1967. The lives of his colleagues and friends will never again be quite so full. The field of multivariate experimental psychology will never again be quite so rich.

Daniel E. Bailey

Contents

CLUSTER ANALYSIS

Chapter 1
INTRODUCTION

U nderstanding our world requires conceptualizing the similarities and differences between the entities that compose it. By entities we mean either the objects or the properties of objects that constitute our world. In a study of the mental abilities of children, for example, the objects are the children, and their properties are the abilities in which they vary. Objects may be persons or animals, or subgroups of them, or any identifiable physical or psychological structures. Their properties are the characteristics with respect to which the objects vary from each other and on the basis of which we differentiate them from each other; thus, properties are generally known as "variables" or "attributes" or in biology as "characters." Variables may be measurements on a continuum, or they may be discontinuous qualities.

Cluster analysis is the general logic, formulated as a procedure, by which we objectively group together entities on the basis of their similarities and differences. When the entities compared are variables, such as mental abilities, the procedure is called "the cluster analysis of variables" or, more simply, "V-analysis." V-analysis has its historical roots in the initial work of two Englishmen, Karl Pearson (1901) and Charles Spearman (1904),[1]

[1] References in text are found alphabetized and dated in reference section at end of book.

and the later developments of Godfrey Thomson (1916, 1951) and Cyril Burt (1915, 1941). Around the 1930s V-analysis took a special form of mathematical dimensional analysis as the result of the work of Truman Kelley (1928), Karl Holzinger (1930, 1941), and Leon Thurstone (1931, 1947); Thurstone labeled the new approach "factor analysis," a term still used (Harman, 1967) though strictly speaking it refers only to those mathematical procedures of V-analysis called "factoring" (see Chap. 6).

The "factors" derived from variables by the process of factoring are often interpreted as "underlying" the observed variables—as if they represent genetic or psychological dispositions of persons. In the 1930s Tryon opposed this trend to reify factors (Tryon, 1932, 1935) and devised the procedures now called "cluster analysis" (Tryon, 1939); this term was chosen to stress the fact that one can discover the general properties of objects by an objective clustering procedure of grouping variables without imputing causative underlying dynamics to the properties.

The term cluster analysis applies equally well to entities that are objects, such as persons. The procedure of grouping together objects that have similar patterns of characteristics is called the "cluster analysis of objects" or simply, "O-analysis." This is the field of typology, much older than V-analysis, having its scientific roots in the taxonomy of Linnaeus, in the diagnostic syndromes of physicians, and in genotype analysis of geneticists. O-analysis is new as a quantitative field in biology, where it is known as "numerical taxonomy" (Sokal and Sneath, 1963). By factor analysts it is called "inverse factor analysis" or "Q-technique" (Stephenson, 1935; Burt, 1937; Cattell, 1952, chap. 7). Factor analysts treat O-analysis merely as the application of dimensional factoring to objects treated as variables. So conceived, it plays a minor role in factor analysis and is even ignored in some systematic treatments of the field (e.g., Harman, 1967).

This book presents O-analysis as a highly developed form of cluster analysis in which the dimensional factoring approach is handled as only one method and, indeed, one that does not resemble orthodox Q-technique factor analysis. We can anticipate rapid development of O-analysis, as indicated by the cross-disciplinary Conference on Cluster Analysis held in New Orleans in 1966.

From the above discussion it is clear that the term cluster analysis embraces both V-analysis and O-analysis. The specialized term factor analysis is retained in cluster analysis, where it is properly allocated to its important, though subordinate, role as a dimensional procedure. Cluster analysis should not be called factor analysis, because factoring is only one of its subordinate methods. It should not be thought of as referring only to the clustering of objects because it embraces both V-analysis and O-analysis.

Since assessment of similarities and differences among entities is a universal conceptual problem, we can anticipate a rapidly expanding use of the procedures of cluster analysis in nearly all fields of human thought and scientific study. The real possibilities of its objective methods could not be capitalized on until the advent of modern digital computers because the observable properties of entities can be very numerous and the objects having such properties usually also occur in vast numbers; hence only the modern computer can manage V-analysis or O-analysis. Throughout this book, therefore, all analyses are executed by a computer. The computer programs employed are those of the BC TRY System of cluster and factor analysis, especially designed to handle all facets of V-analysis and O-analysis. When the first modern computer (IBM 701) was introduced at the University of California in 1956, we immediately undertook the programming of the BC TRY System, and it has taken a full 10 years to think out and program the complete logic of the system, the developmental process being much delayed by changing programming languages and machines. Much of this book is a description of this computer system and its application.

The intent of this book is as much substantive as methodological. Complete cluster-analytic procedures have been applied to three separate problems. The first is in the cognitive domain of the intellectual abilities, being the famous Holzinger study of the scores of 301 schoolchildren on 24 diverse tests of verbal, spatial, speed, memory, and mathematical abilities. The second is in the well-known personality domain of self-conception, being the responses to the 566 items of the MMPI (Minnesota Multiphasic Personality Inventory) of 310 patients and normal adult subjects. The third is in the field of the ecology of metropolitan social areas, in which the objects are groups instead of individuals; they are the several hundred neighborhoods of the San Francisco Bay Area, observed both on demographic and voting-attitude characteristics before and after World War II.

In order to give a suggestion of the full scope of a cluster analysis of the data of a given problem, in this chapter we first review the main analytic steps and findings of the Holzinger problem and then do the same, except more succinctly, for the MMPI and the social-area problems. The chapters of this book generally follow the summary analyses presented in this introduction.

The Holzinger study of intellectual abilities

Cluster analysis of the variables (V-analysis)

When one observes 24 different intellectual abilities of many children, is it necessary to preserve all 24, or can a reduced number of composites of them fully account for all that is general among the 24 abilities? This is the question posed by V-analysis, and it is answered by applying the inherent cognitive processes by means of which we generally organize entities in terms of their similarities and differences.

The degree of generality of individual differences in each of the 24 abilities is revealed by the degree to which individual differences in it correspond to, i.e., correlate with, differences in the other 23 abilities. By the process of comparing each ability with every other ability, with respect to the similarity in which they order individual differences, a detailed statement of the relationships of each ability with the others is summarized in a "correlation matrix." This is taken up, with other basic measurement problems, in Chap. 2.

One type of V-analysis is to form "rational composites" out of the 24 variables, grouping them by inspection into a priori content categories, such as verbal, speed, spatial, memory, and mathematical composites. But this logical grouping procedure takes no account of the correlations among the 24 abilities. Instead, we prefer to employ the process of empirical grouping of the 24 variables, casting together into the same group those which correlate positively with each other, and especially those whose patterns of correlations with the other abilities are similar. Variables that form such like-patterned groups are called "collinear" clusters. They are the composites we seek in V-analysis, as is fully explained in Chap. 4. The contrast between the rational- and the empirical-grouping approach is illustrated in the Holzinger problem, where we find that though there appear to be five logical groupings of the 24 variables based on obvious content similarity, there are in fact only four collinear clusters among the 24 tests. It appears that the five tests of mathematical abilities are heterogeneous and redundant.

By what process do we discover that we can properly reduce the 24 abilities to the specific number of only four empirical clusters? We do so by the process of concentrating information, for which the procedures of "dimensional analysis" or "factoring" have been devised. These processes enable us to determine the number of dimensions, or clusters, that are sufficient to account for all the correlations among the 24 tests. The procedure begins by deciding ahead of time the amount of variance among individuals that we want to account for by the reduced number of composites. For each test this variance is called its "communality," an index of such critical importance that we devote Chap. 5 to it. The factoring procedure itself, developed in Chap. 6, reveals that only four salient cluster defined dimensions are sufficient to account for all the communalities of all 24 variables.

We must finally choose the defining variable of each of the four composites. We do so by the process of examining the total configuration of the relationships among the 24 variables. There are three main selection criteria: each cluster of variables should be (1) as "tight," i.e., collinear, as possible, (2) as nearly independent of the others as possible, and (3) able to account for as much general variability as possible. In Chap. 7, which considers this problem, it will be seen that there are simple computerized aids, both graphical and metric, that enable the analyst to observe the cluster structure among the 24 tests and to make the best selection of four composites. The composites turn out to be subsets of variables that define the four basic abilities of V (verbal), S (speed), F (form or space), and M (memory).

Cluster analysis of objects (O-analysis or typology)

The purpose of O analysis in the Holzinger problem is to differentiate the 301 children of the study into a set of salient ability types. Each type consists of children who have the same profile of scores on the four basic composite abilities found in the V-analysis, namely, on V, F, S, and M. In solving this problem we use the same four logical processes employed in discovering the clusters of variables in V-analysis, i.e., the processes of comparing, grouping, concentrating, and inspecting structure.

By the process of comparing each child with every other child in their profiles on V, S, F, and M, we develop a similarity matrix among the children (analogous to the correlation matrix among variables in V-analysis). To secure this matrix we direct the computer to represent each child as a point in the four-dimensional score space of V, S, F, and M. In this space, the degree of similarity of any two children is defined by the distance between their two points, an index called the "euclidean distance D."

The process of grouping is implemented by directing the computer to search in the score space and find those concentrations of points which define distinctive types of children.

The process of concentrating this information consists of directing the computer to find, by an iterative process, a minimally sufficient set of types that represents the whole configuration of children and to set aside any child whose score pattern is too unique to be included in any type. In the Holzinger data, our computer program locates 15 ability types, into which all but 16 unique children are cast.

The final process of observing the whole structured configuration of children in relation to each other is most critical. The precise description of the whole configuration is the total matrix of D values between all the children, but since it is too complex to comprehend as such, various other metric and graphical aids (fully developed and illustrated in Chap. 8) are employed to help decide on the final set of interrelated types to be used in classifying the children. Among these aids is a hierarchical ordering of the types, such as the genus-species classification system of biology.

Comparative cluster analysis of variables (VCOMP) and of objects (OCOMP)

Just as in biology, where it is well known that different organic species and varieties live in different ecological settings and are thus selected for different characteristics, so ecologically different groups of human beings are likely to differ in their patterns of abilities. This point is demonstrated by a comparison of the cluster structure of the abilities of two different groups of children, one from a suburban school and the other from a school in a factory area. In Chap. 9, comparative procedures, called "COMP analysis," are used to make indirect and direct comparisons of the cluster structure of the 24 abilities in the two school groups. They turn out, in fact, to be quite similar, except that the test clusters of the factory children are more independent of each other than those of the suburban children, a fact with both social and genetic implications.

Though the structures of the abilities of factory and suburban children are fairly similar, it does not follow that when the typological structures of these groups are compared, they, too, will be similar. OCOMP procedures, also presented in Chap. 9, are designed to find out. When we compare the two groups of children on a common array of ability types, the frequency of cases in the different types varies significantly; furthermore, when we perform independent typologies on the two groups and project the two arrays of types into the same analysis, we find some marked differences in the typological structures of the two groups. Such findings as these have obvious educational, occupational, and social implications, hitherto ignored for want of methods capable of revealing such salient differences. As a further illustration of comparative analysis of two typologies, a comparison between the ability typologies of boys and girls, long a matter of speculation in the psychology of sex differences, is also given in Chap. 9.

Prediction from clusters of variables and from object clusters

Suppose one wished to predict a child's mathematical ability from knowledge of his scores on the verbal, speed, form, and memory tests. One way is univariate prediction, in which the predictor variable is a single score, either from a single test or from a composite of tests. In Chap. 10, on prediction in cluster analysis, we will see that individual tests are generally poorer predictors than the cluster composites discovered in V-analysis. It is also found that multivariate prediction, in which the predictors are all four cluster composites, V, S, F, and M, weighted by the multiple correlation methods, is generally better than univariate. But the superior form is differential typological prediction. This is the prediction of mathematical ability from knowledge of the distinctive "O-type profile pattern" of a child's scores on the four predictor cluster composites. For children having one distinctive type of pattern of scores on V, S, F, and M, we can predict mathematical ability with considerable accuracy, whereas for those of another distinctive pattern type, our prediction is no better than chance. Thanks to the computer, we can also now realistically assess the role of sampling error in such predictions by means of Monte Carlo runs, from which we can determine with accuracy which O-type predictions are significant and do so without making the usual normal-curve assumptions that have made such estimations of error rather questionable in the past. One special merit of differential O-type prediction is that we can now cumulate knowledge about a given O-type from a discovery of those "outside" attributes which are predictable from it. This form of prediction is a new approach in science. It means that each O-type is likely to have a highly predictable but distinctive pattern of characteristics differing from that of other O-types.

The personality domain of self-conception: the MMPI

The second substantive problem treated throughout this book deals with the personality domain of reporting on emotional and social reactions in oneself that might be maladaptive. The data set consists of the responses to the 566 items of the MMPI (Minnesota Multiphasic Personality Inventory) of 90 normal subjects and 220 psychiatric patients.

V-analysis and O-analysis

This MMPI study nicely illustrates the perennial problem of reducing by V-analysis the large number of item-variables that comprise some tests, questionnaires, or other surveys. The BC TRY System includes programs that can perform such an analysis on as many as 2,000 variables. The special procedure for such a large-scale solution is called BIGNV analysis, the logic of which is presented in Chap. 11 and illustrated there on the MMPI problem. In principle, there is no limit to the number of variables that can be tackled by BIGNV analysis. In a nutshell, here are the steps. The computer calculates the communalities of all item-variables, orders them by magnitude, and casts out all items of trivial generality, since they can play no general role in differentiating persons from each other. Thus, about two-thirds of the 566 MMPI item pool go out at the very beginning as sheer dross. The cluster structure of the remaining pool of good items is then computed in manageable subsamples and then merged to give, finally, the salient cluster structure of the total supply of all item-variables. The logic of this approach is as old as statistical theory: cumulate the results on successive subsamples until they converge. The converged results describe the cluster structure in the full pool.

The special power of V-analysis to reveal the basic domains sampled by such a large pool of items as that of the MMPI inventory is clearly evident in this study. The empirical cluster search procedure finds seven item-clusters. The four most salient V-clusters are the disturbing conditions of I (introversion), B (body malfunctions), S (suspicion and mistrust), and T (tenson, worry, and fears). The four scores of subjects on the I, B, S, T item-clusters largely measure the common variance of the whole MMPI item pool, though three additional clusters, D (depression), R (resentment), and A (autism) are also clearly evident; however, the scores on these three are largely redundant, i.e., predictable from scores on I, B, S, and T. Each of these seven item-clusters forms an objective scale, each with reliability coefficients approaching .90.

When the inclusive group of 310 subjects is scored on the basic four I, B, S, T item-clusters and the O-analysis procedures are applied to them, a clear, meaningful typology of 14 types of person appears (Chap. 8), with no unique persons.

Comparative analysis

Perhaps the most dramatic discovery in the MMPI problem, and one that has a potent moral for cluster and factor analysis generally, is the finding (Chap. 9) from comparative V-analysis and O-analysis of the two subgroups of normals and patients that make up the subject sample. The results on these two groups are radically different from each other. This finding illustrates how a factor analysis of an inclusive group may differ from factor analyses performed on the subgroups that compose it. This finding further proves that when the MMPI is cluster analyzed by the methods described in this book, this questionnaire turns out to have very potent validity in differentiating normal subjects from psychiatric patients.

Prediction

The power of prediction from V-clusters and O-clusters is sharply clear in this MMPI analysis (Chap. 10). When the subjects' scores on the three item-clusters D (depression), R (resentment), and A (autism) are predicted from their scores on the basic four, I, B, S, and T, we find, just as in the Holzinger study, that as one moves from univariate, to multivariate, to differential O-type prediction, the ability to predict sharply increases. Indeed, for some subjects having a distinctive profile on I, B, S, and T, we find their scores on D, R, and A are almost perfectly predictable. For others, of course, the prediction is poor.

The social-area study

The third substantive study analyzed in this book moves to a different type of object, namely, groups rather than individuals. In our own case, the groups are neighborhoods (census tracts) making up the metropolitan San Francisco Bay Area in prewar 1940 and in the postwar period circa 1950. No attempt is made in this introduction to review the findings of this social-area study in any detail. Suffice it to say here that in the following chapters the same types of analyses are made on the neighborhood data as on the individuals in the Holzinger and MMPI problems; but here are some highlights. The focal variables that describe the properties of these neighborhoods are 34 demographic characteristics of the neighborhoods published by the Census Bureau in the prewar and postwar decades of 1940 and 1950. The findings of the V-analysis show that only three basic domains are sampled by the 34 characteristics, namely, F (family life), A (assimilation), and S (socioeconomic). Thus, if one knows the F, A, S scores of a neighborhood, one knows all there is to know generally about its demographic character (as reported in the census).

In addition to the demographic features of the neighborhoods, data on the voting of the neighborhoods are also carried along in the analyses. Perhaps the most surprising discoveries are the findings (1) that the demographic and voting attitudes of neighborhoods have an identical tridimensional cluster structure, (2) that this structure remains constant over a decade despite the social upheavals of a great war, and (3) that, in typological prediction, if one knows the demographic pattern of a neighborhood, one can predict how it will vote on some issues before the election takes place—in some cases 15 years before the vote is taken—and in spite of the fact that the people voting are largely different people from those on whom the demographic data were collected 15 years before!

Chapter 2

GENERALITY OF
INDIVIDUAL DIFFERENCES

Since the purpose of cluster and factor analysis is to discover the general properties of objects (V-analysis) and the general types into which the objects can be classed (O-analysis), we should agree at the start on what "generality" means. A general property, variable, or attribute of objects, e.g., an intellectual ability of children, is defined here as a composite of two or more variables that similarly order the subjects observed with respect to that property. The word "similarly" appears consistently here in the definitions of generality; hence we need a clear understanding of what we mean objectively by similarity.

This chapter is a preliminary exposition of how the similarities of properties and of objects are described and measured. In describing the similarity of two variables, the focus is on the variation among objects in each of the two variables, i.e., on individual differences. To be clear on what is meant by similarity we must be clear on what is meant by individual differences among objects. We therefore present some basic concepts on the observation and description of individual differences in a single attribute, or property. First we discuss the observation of individuals and then the means by which individual differences in an attribute are described.

Methods of describing the generality of individual differences are then discussed.

In this treatment we illustrate the concepts mainly on the intellectual abilities of the Holzinger study, in particular on tests of the so-called "verbal" and "mathematical" abilities. This particular problem is interesting since it bears on the important question of whether there are two such general abilities or factors and, if so, the degree to which they are similar. Only the briefest account of the science of psychological measurement of individual differences can be given here, and only those psychometric principles which have direct relevance to the problem of assessing similarity of individual differences in two or more characteristics will be introduced.

With the psychometric concepts behind us, we can address the basic designs by which cluster analysis and factor analysis reveal general traits and types of individuals. First we deal with methods of discovering and describing the degree of similarity of variables and of objects. Finally we discuss the graphical representation of similarity among variables and objects.

The observation of individual differences in an attribute

When we use the word "observation," we generally are interested in the principles by which some conceptualized property of objects comes to be represented by scores. For example, in the Holzinger study, the 24 different abilities are represented by scores obtained by observing the behavior of a group of children, 145 children in the instance described (see Table 4.1). The process may be a simple tabulation of the number of "correct" responses made to a task, such as the sentence completion task representing a specific verbal ability. It is instructive to look carefully but in an elementary fashion at this problem.

Abstract and operational definitions

The investigator first forms an abstract concept of what property or attribute of objects he wishes to observe. Thus, in the example of verbal ability, he decides that he wishes to observe a child's ability to complete, verbally, an idea that is only partially presented to him in the test situation. How this ability is expressed in a real situation requires a specification of the verbal conditions under which the abstractly defined ability is believed to emerge in a child in the observation situation. The investigator might think up some incomplete sentence that partially expresses an idea and give a child a choice of words to complete the sentence, only one of which correctly does so. An example:

At home she does not understand how to make
1) difficulties, 2) quarrels, 3) arrangements, 4) labor.

The format of the abstract and operational definition with regard to a "number" ability is much the same except that the incompleted idea is quantitative, as shown in this example:

1   3   2   4   3   ——.

The child completes the series by supplying the number, "out of his head," that correctly continues the series. All 24 variables in the Holzinger study are defined in a similar manner.

Knowing the abstract and operational definitions of each variable in a study is necessary for an understanding of the meaning of general clusters and types derived by V-analysis and O-analysis. For example, imagine we discover empirically that five of the six memory tests form a cluster, i.e., that they tend to order the subjects in the same way. The meaning of a single composite score on the five tests becomes clear only by carefully studying the abstract and operational definitions of the items.

Scoring (coding)

The operationally defined property of an object, like the responses of a person to a stimulus item, usually is assigned a number according to some principle like the degree of intensity or the "goodness" of the responses. The number-coding of responses is usually based on a key, sometimes arbitrarily decided upon by the investigator. Even more arbitrariness is introduced when stimulus items are cast in "objective" test form, like the multiple-choice designs of the illustrative vocabulary and mathematics items cited above. In such cases, responses are a priori prejudged "right" or "wrong" and are scored accordingly.

With modern computer facilities this arbitrariness can be reduced. All the bit responses of a subject in any experimental test situation can be marked by the subject on scoring sheets, which can then go to an automatic photoreader that converts the responses to records to be input to a computer. Then, applying methods described in this book, those responses which are found empirically to cluster together can be composited together to form not an arbitrarily defined but an empirically discovered composite or variable. Thus, cluster and factor analysis procedures can be used to discover how to score the many seemingly chaotic responses of subjects.

The procedures designate what composite each item belongs to and the manner in which it should be scored.

Describing individual differences in specific and general attributes

The first steps in observing each separate attribute are themselves forms of condensing, reducing, or compositing of many entities, either by grouping them together by definition or by some empirical clustering procedure. The procedures of V-analysis continue this initial process by condensing or compositing the scores on the observed variables into even larger composites. Larger composites can also be made by grouping the 24 tests into rational categories or by performing a V-analysis of all 24 tests and forming composites on empirically derived clusters.

The observed raw scores of the individuals on each ability must be described on (or transformed to) a common scale, because the raw scores in the different tests may be scored in noncomparable units. In some studies, the different variables may be in such diverse metrics as inches, pounds, items correct, ratings, elapsed time, decibels, and so on. We need a common scaling that informs us directly of individual differences in the attribute. The scale generally accepted is the standard score, written

$$x = \frac{X - \bar{X}}{\sigma_X} \tag{2.1}$$

where X is the raw observed score of an individual on a given attribute, $\bar{X}$ is the mean score on it, and σ_X is the standard deviation. For example, the first child in the suburban group has a score on the sentence completion test of 17 items correct. The mean and standard deviation of all the children's scores on this test are 19 and 5, respectively, whence the first child's standard score is

$$x = \frac{17 - 19}{5} = -.40$$

The standard score has a universal meaning: it tells where an individual stands in his group relative to the standard deviation of all the scores. Note that for an individual with a score equal to the mean, i.e., whose X is the mean $\bar{X}$, his standard score is zero, and for an individual at 1 standard deviation below the mean, i.e., whose $X = \bar{X} - 1\sigma$, his standard score is -1.00. One knows from these reference points that the first child with $x = -.40$ is below the mean by nearly $\frac{1}{2}$ a standard deviation.

   With the raw scores of all individuals on the 24 tests converted to standard scores on each test, we are in a position to form any general composites we wish of the 24 scores merely by deciding what shall be the defining variables of each composite and then for each individual adding up his standard scores on those definers.

   Suppose we wish to reduce the total number of observed test scores to a smaller number of rational composites, e.g., the verbal composite consisting of the standard scores on five verbal tests, $X_5, X_6, \ldots, X_9$. The raw score on this composite would be

$$Y = x_5 + x_6 + x_7 + x_8 + x_9 \tag{2.2}$$

This raw composite score may not be comparable to other composite scores, e.g., a raw composite score on four different (say vocabulary) tests, since the verbal composite has five scores added together whereas the vocabulary has only four, and the relationships among the scores in the composites may be different for the two composites. We convert the raw composite score in expression (2.2) to a standard score (calling it z for now)

$$z = \frac{Y - \bar{Y}}{\sigma_Y} \tag{2.3}$$

In this form, the standard scores on all composites are comparable, each having a mean of .00 and a standard deviation of 1.00. We could therefore tell a child's relative standing in all the rational composites in the Holzinger study by his standard scores.

   There is a final convention. To avoid carrying plus and minus signs on a subject's score we convert the standard score to a new scale score that has a mean of 50 and a standard deviation of 10, thus,

$$Z = 10z + 50 \tag{2.4}$$

   This standard score Z is the final form in which an object's standing in any given composite is described in cluster and factor analysis, whether the composites are rational groupings of variables or empirical cluster (or factor) scores. When we plot profiles of individuals in O-analysis, these Z scores are used. It is clear from Eqs. (2.1), (2.3), and (2.4) that standard scores, whether in z or Z form, are linear transformations of the raw scores and therefore take exactly the same form of distribution as the raw scores. Thus, if an investigator wishes to make a nonlinear transformation of the raw scores to another metric, such as to logarithms or normalized scores, the conversion of the raw scores is made before performing the z or Z transformation.

Describing the generality of individual differences in an attribute by its correlations

Generality and correlation

Variation among individuals in scores on a variable is said to be general if the rankings of objects on that variable correspond to a noticeable degree with their rankings on other variables. The traditional index of similarity of a variable with another variable is the well-known correlation coefficient r, which takes the value of $+1.00$ if the correspondence is perfect, .00 if there is no generality, and -1.00 if perfect but inverse (high scores in one attribute matching lows in the other). For example, in the Holzinger sample of 145 children the generality of the sentence completion variable is highest with the other four verbal tests, its correlations with them being .65, .73, .63, and .68. However, variation among children in sentence completion does not correspond as well with mathematics ability as indicated by the number completion test since the correlation between these two abilities is only .41. Perhaps if we formed a single composite score from scores on all five verbal tests and correlated that score with an analogous composite of all five mathematics tests, the variation among children in these two broadened composites would correlate more, i.e., be more general. This is true: the correlation between scores on the verbal composite and on the mathematics composite is .65.

What would be the correlation between the V, for verbal, and N, for numerical or mathematical, composites if we could increase the number of tests in each composite to a very large number covering the whole domain of children's verbal abilities and the entire domain of their mathematical abilities? This correlation between the verbal and mathematical domains (or factors) can be estimated to a close approximation by methods to be discussed later. The correlation between the "extended" composites is .76. This ceiling means that there would be considerable generality of individual difference across the verbal and mathematical abilities if we could secure extensive domain scores on these two abilities, but since the correlation would not be 1.00, it also means that the generality would not be perfect. Children equal in a domain score on verbal ability could still vary quite a bit in mathematical ability and vice versa.

One difficulty with expressing a relationship as a single index number, like the correlation coefficient, is that it really does not communicate fully the complex meaning of the relationship between two variables. A graphical display tells the story much better. To illustrate, Fig. 2.1 is the "scattergram" of the 145 scores of the suburban children on the composite Z_V

ORDINATE (Y) = CLST05 (Mathematical)
ABSCISSA (X) = CLST02 (Verbal)

[12 STRIPS USED FOR ETA COMPUTATIONS]
ETA(YX)= 0.6892   UNBIASSED ETA(YX)= 0.6669
ETA(XY)= 0.6971   UNBIASSED ETA(XY)= 0.6659
PRODUCT MOMENT CORRELATION = 0.6509

			PREDICTION OF Y FROM X						PREDICTION OF X FROM Y				
		MEAN X	PREDICTED Y		SE YX				MEAN Y	PREDICTED X		SE XY		
SLICE	SLICE BOUNDS		LINEAR	CURV.	CURV. (BIAS)	CURV. (UNBIAS)	N	SLICE BOUNDS		LINEAR	CURV.	CURV. BIAS	CURV. (UNBIAS)	N
12		76.22	67.07	70.05	7.34	9.00	3		75.01	66.28	75.48	4.16	5.10	3
	71.25							71.25						
11		69.02	62.38	67.23	4.75	5.31	5		70.48	63.33	63.84	3.45	4.88	2
	67.00							67.00						
10		64.37	59.36	56.69	4.67	5.11	6		64.27	59.29	57.16	10.45	11.08	9
	62.75							62.75						
9		60.23	56.66	56.07	6.12	6.31	17		60.72	56.98	57.89	9.00	9.31	15
	58.50							58.50						
8		55.87	53.82	55.71	6.61	6.86	14		55.82	53.79	49.61	7.60	7.82	18
	54.25							54.25						
7		52.23	51.45	48.55	7.31	7.50	20		52.17	51.41	53.10	7.01	7.16	25
	50.00							50.00						
6		47.63	48.46	49.03	7.09	7.20	32		47.99	48.69	49.11	5.67	5.77	28
	45.75							45.75						
5		44.23	46.24	45.81	6.79	6.93	26		43.84	45.99	45.51	5.21	5.35	19
	41.50							41.50						
4		39.48	43.15	39.94	5.91	6.26	9		39.88	43.41	42.75	5.05	5.32	10
	37.25							37.25						
3		36.27	41.07	44.02	14.18	15.85	5		35.34	40.46	43.13	6.75	7.29	7
	33.00							33.00						
2		31.11	37.70	44.14	11.22	12.29	6		30.59	37.36	37.05	9.60	10.52	6
	28.75							28.75						
1		20.95	31.09	27.38	1.39	2.82	2		26.43	34.66	33.59	11.13	13.63	3

SE YX LINEAR = 7.59   SE XY LINEAR = 7.59
IN LINEAR PREDICTED Y, A= 17.46, B= 0.65.   IN LINEAR PREDICTED X, A= 17.46, B= 0.65.

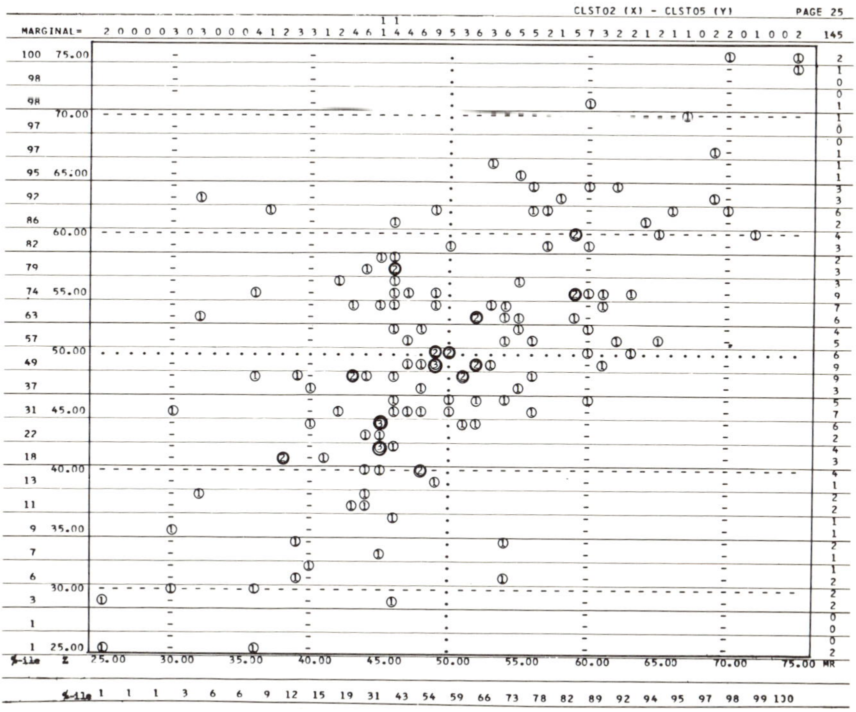

CLST02 (X) - CLST05 (Y)   PAGE 25

MARGINAL= 2 0 0 0 0 3 0 3 0 0 0 4 1 2 3 3 1 2 4 6 1 4 4 6 9 5 3 6 3 6 5 5 2 1 5 7 3 2 2 1 2 1 1 0 2 2 0 1 0 0 2   145

%-ile Σ 1 1 1 3 6 6 9 12 15 19 31 43 54 59 66 73 78 82 89 92 94 95 97 98 99 130

FIGURE 2.1
Correlation scattergram and prediction tables of V (verbal) ability and N (mathematical) ability for 145 suburban children in the Holzinger study.

score on the five verbal tests, and the composite Z_N of the five mathematics test scores. This scatter is a photographic reproduction of the grid produced by the BC TRY component RSCAT. The horizontal scale is the Z_V score axis, the vertical the Z_N, and the circled entries are the number of children at each intersecting score point. This RSCAT scattergram is especially designed to show the standard score sectors in which individuals fall. Note that the vertical and horizontal lines of dots represent Z scores of 50 (at the means) and that vertical and horizontal lines consisting of dashes lie at $\pm 1\sigma$ and $\pm 2\sigma$.

Correlation and prediction

Study of this scattergram, which corresponds to a correlation of .65, reveals that it is fan-shaped, with the wide part of the fan at the lower left. This structure means that one ability can be predicted by the other with greater accuracy in the high ranges of ability than in the middle and low ranges; i.e., very high scores on either ability represent a more general ability than lower scores do. The single index number of $r = .65$ is a gross summary statement of this relationship but tells us nothing about the fan.

A computer program like RSCAT provides both the graphical and the metric facts of correlation and prediction. From study of this output, the meaning of correlation in terms of prediction can be quickly grasped. Above the scattergram to the left can be seen a table headed "Prediction of Y from X." X stands for the Z_V score on the verbal composite, Y for Z_N on the mathematical composite. The program cuts the verbal score X scale into 12 categories or slices and predicts the mathematics Y scores for the children in each slice. For example, for the three children in the top slice, number 12, whose mean X (verbal) score is 76.22 (2.6 sigmas above the mean verbal score of 50), the linear regression predicted Y (mathematics) score is 67.07 (1.7 sigmas above the mathematics mean of 50). Note that the predicted mathematics score of the three children at the extreme has "regressed" toward the mean of 50 compared with their extreme predictor verbal score, i.e., to 67 from 76. A line through the linear predicted Y scores is a "line of regression" of Y on the X scores. The constants of the equation of this line are printed below the prediction table. The slope B is .65, which, it will be recalled, is the correlation coefficient.

This slope gives us one meaning of r, namely, that when the subject's X and Y scores are converted to the same scale (z or Z), the best-fitting straight line through the predicted Y scores from X has a slope r. When the swarm of person points in RSCAT all lies on this regression line, one can predict Y perfectly from X, whence the slope of the line is 1.00. But when the swarm lies randomly over the whole scatter, then whatever the value of an X score, its predicted Y is the mean Y; hence the regression line is through the horizontal dotted mean Y line, which has a slope of $r = .00$. (The other type of prediction is of X from Y: its prediction table is upper right in Fig. 2.1.)

If the relationship of Y to X is not linear, then a smoothed line drawn through a plotting of the column of values headed "Predicted Y, Curv." will reveal the fact. The existence of curvilinear prediction is also summarily revealed by the correlation ratio η printed at the upper right. When η is significantly greater than r, the relation is curvilinear. Since there are two predictions, Y from X and X from Y, there are two values of η. In the illustration, both η's are only slightly higher than r, and so we know that the relation between these two abilities is essentially linear. The "unbiased" η is a corrected value of η (called "Tryon's eta"), which compensates for a tendency for η to be biased upward when the number of cases is small.

The fan-shaped plot, showing that prediction is better in the higher reaches of the two abilities than in the lower, is metrically expressed in the "error of prediction," shown in the column "SE YX," meaning "standard error of predicting Y from X." This value is the standard deviation of the Y values in the X slices, corrected for small sampling in the Unbiased column but uncorrected in the Biased column. The error of predicting the mathematics score is greater when the verbal scores are smaller than when larger, which is the same as saying that the plot is fan-shaped. This difference in the error of prediction at different levels (slices) of X is technically referred to as "heteroscedasticity." If the errors in each slice were the same, the relation would be "homoscedastic."

Generally it is wise always to look at the scattergrams between composite scores derived by cluster or factor analysis. They show in greater detail the nature of the relationships between the composites than the correlation coefficients taken alone do.

Not only are such vitally important matters as curvilinearity and heteroscedasticity revealed in the scattergrams, but simpler statistical features of individual differences in each of the composites are also revealed. Thus in the RSCAT presentation of Fig. 2.1, the histograms of the verbal and mathematics score can be drawn from the frequencies given in the extreme right columns, headed N, of the prediction tables (and in more detail at the top and right margins of the scattergram). Also, the percentile ranks corresponding to every $.20\sigma$ score are printed at the extreme left and bottom margins.

Discovering and describing the degree of similarity of variables and of objects

In V-analysis we form general composites of variables that are similar; in O-analysis we form general types composed of individual objects that are similar. We do so by observing the objects in their relations to a common body of other referent entities. To the degree to which the patterns of their relations to these referents are the same we judge them to be similar.

Similarity of variables in ordering individuals

Take, first, the situation in which this format appears in the V-analysis of the Holzinger problem, namely, discovering and describing the degree of similarity of the 24 separate abilities. The schematized score matrix in Fig. 2.2, in which dashes represent observed scores, illustrates the data from which the rest of the analysis proceeds. In the schema of Fig. 2.2 the columns are the 24 tests set up as comparison entities. The rows are the 145 children taken as common referent entities. The observations in the matrix are raw test scores. The degree of similarity of any two column variables, for example, V7 (sentence completion) and N23 (number completion), is the degree to which the patterns of children's scores in these two columns are similar, or correspond. The index calculated is the correlation coefficient, which is therefore the "index of similarity in ordering individuals." It turns out to be .41 for these two variables in the Holzinger sample of 145 children.

Similarity of variables in sampling domains

The indexes of correlation among all variables are set up in a paired-comparison matrix, schematized in Fig. 2.3.

We can ask the similarity question about any two column variables in this matrix: What is the similarity of V7 and N23 in relation to all the vari-

		Comparison entities (n = 24 abilities)							
		1	2	3	· · ·	V7	· · ·	N23	24
Referent	1	—	—	—	· · ·	—	· · ·	—	—
entities	2	—	—	—	· · ·	—	· · ·	—	—
(N = 145	· ·	· · ·	· · ·	· · ·	· · ·	· · ·	· · ·	· · ·	· · ·
children)	144	—	—	—	· · ·	—	· · ·	—	—
	145	—	—	—	· · ·	—	· · ·	—	—

FIGURE 2.2
Schematized observation or score matrix. Observations are raw scores on each of the column variables.

Comparison entities (n = 24 abilities)

		1	2	3	$\cdots$	V7	$\cdots$	N23	24
Referent	1	—	—	—	$\cdots$	—	$\cdots$	—	—
entities	2	—	—	—	$\cdots$	—	$\cdots$	—	—
(n = 24	$\cdot\cdot$	$\cdots\cdots\cdots\cdots\cdots\cdots\cdots\cdots\cdots\cdots$							
abilities)	24	—	—	—	$\cdots$	—	$\cdots$	—	—

FIGURE 2.3
Paired-comparison matrix, the matrix of correlations of comparison entities. Observations are the raw scores on each of the column variables.

ables as a common body of row referent entities? Since the referent entities are now variables and not objects, we seek an index of the similarity of V7 and N23 in the way the two variables relate to, or sample, the 24 test domains from which the array of test samples is drawn. The similarity is a function of the degree to which the columns of correlations of the two tests follow the same pattern. This is measured by an "index of proportionality" (see Chap. 4), also variously called "the index of similarity of domain sampling," "the Interdomain correlation," and "the common factor correlation." This similarity is expressed graphically as the spatial separation between variables when they are plotted as points in a geometric space, e.g., a sphere.

The important matter here is not the details but the general design by which the similarity of any two entities is determined. The entities are expressed as two comparison entities in their relations to a specified body of common referent entities. The index of their similarity with respect to the referents is the degree to which they have the same pattern of observations on the referents. In the case of the score matrix, the similarity index is their correlation. In the case of the correlation matrix it is the index of proportionality.

Similarity of dimensions in the same and in different groups

The same design is used in determining the degree of similarity of any two dimensions defined by composites of the variables, like the verbal and mathematical composites. In this case the column comparison entities are the composites, the row referent entities are the 24 variables, and the observations are the columns of correlations of the 24 variables with the composites. The measure of similarity of any two dimensions is therefore their index of similarity in domain sampling across the 24 tests. The dimensions can be those determined on different groups. As long as the row referents of 24 variables are the same in the different groups, we can determine the similarity of dimensions expressed as column comparison entities.

Similarity of objects in their profile patterns

The same matrix formulation provides the basis in O-analysis of determining the similarity of objects, such as the 145 suburban children. The data for N separate children are expressed as N column comparison entities. In the Holzinger example the four cluster composite scores verbal (V), speed (S), form (F), and memory (M) are expressed as row referent variables. The observations in the matrix are the standard Z scores on the four composites. The index of similarity of the profile patterns of any two children is the similarity in their two columns of Z scores. The actual index computed is D, the euclidean distance between the two children represented as two points in the four-dimensional score space of V, S, F, and M. It is on the basis of the values of D among all the children in this score space that the typology of the group is formed. In comparative typological analysis the same matrix formulation is employed to discover the degree of similarity of typologies of different groups.

This matrix design has quite universal application and lies at the root of V-analysis and O-analysis. With a little extrapolation it is the broad design by which we assess the degree of similarity and difference between the entities that compose our phenomenal world.

Spherical representation of the relations among variables and objects

From high school geometry most readers are already familiar with the cartesian coordinate system of plotting entities on a set of orthogonal axes. In cluster and factor analysis that method is inefficient as a means of depicting the relations among variables or objects because it does not utilize directly the special knowledge about the relations among entities gained from the processes of factoring and it is of limited practical use when the number of dimensions is three or more.

In the V-analysis of the 24 Holzinger abilities, for example, we find, by factoring the correlation matrix, that the relations among the abilities can be adequately described in four dimensions. The factoring process results in coordinates on four dimensions for the 24 tests. If we restrict our attention to three-dimensional subspaces, we can build up an intuitive understanding of the larger spaces. Since "how to do it" details are given

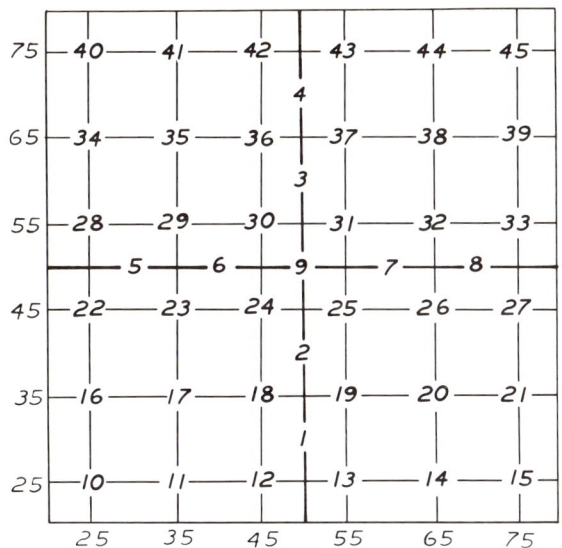

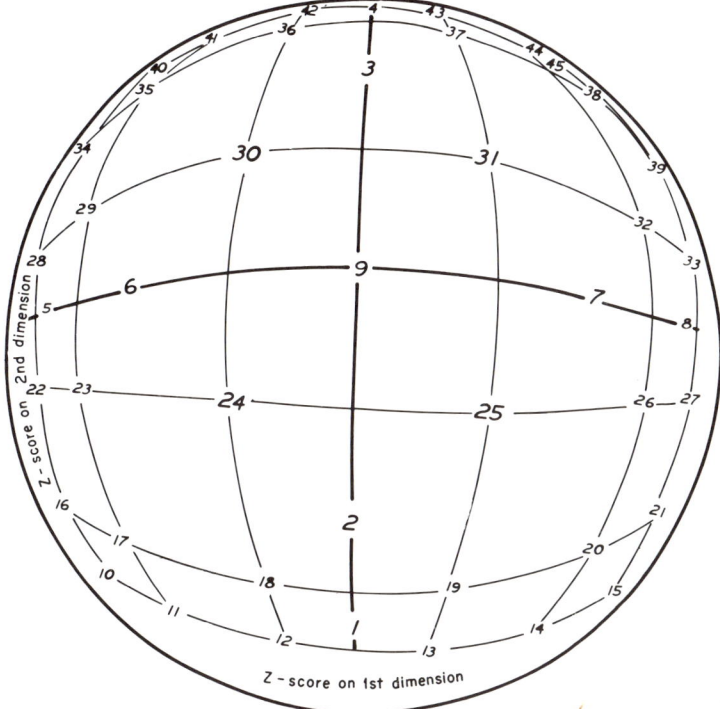

FIGURE 2.4
Configuration of individuals in score space (*upper*) and on the surface of a sphere (*lower*).

in Chap. 7, we skip them here and ask the reader merely to look at the results. Looking forward to Figs. 7.3 and 7.4, it can be seen how the configured relations among the 24 abilities become revealed as a display of points on spheres (three-dimensional). Since, by factoring, the successive dimensions involve progressively less test variance and at a different rate for different variables, it is possible to depict the total configuration on subsets of spheres in such a fashion that the first sphere depicts a maximal part of the configuration, the next one maximally picks up an additional part of the configuration, and so on. Since "marker" variables carry over from one sphere to another, it usually happens that, however many dimensions there may be, the salient features of the configuration can be observed on only several spheres.

When we come to O-analysis, the problem of depicting the relations among all the individual objects in a spherical configuration is more difficult. In V-analysis the spatial surface distance between any two variables on a sphere is a function of the correlation between them.

In O-analysis it is not immediately evident that for any two objects in score space (say any two children plotted as points in Fig. 2.1) the euclidean distance between them is monotonically related to the spatial surface distance between them as points in a spherical configuration. That this relationship does, in fact, hold can be demonstrated mathematically and operationally. The BC TRY procedures of euclidean analysis (EUCO) map objects into score space perfectly. Imagine that the two scores Z_1 and Z_2 on 45 subjects resulted in a subject space like that in the upper part of Fig. 2.4. The two scores on each subject are input to EUCO, which computes a 45 by 45 matrix of intersubject distances in the subject space. Another program transforms this distance matrix to a correlation matrix, which is subjected to a full key-cluster analysis (as described in Chaps. 6 and 7), finally revealing the configuration (in program SPAN) on the surface of a sphere, as shown in the lower part of Fig. 2.4. Except for a systematic distortion in which the points around the origin of the two axes Z_1 and Z_2 are spread out more than at the edges, the configuration given in cartesian score space is preserved in the spherical representation.

Chapter 3

THE BC TRY COMPUTER PROGRAM SYSTEM

When the methods outlined in the first two chapters are applied to large numbers of variables or objects (greater than seven or eight perhaps), the advantage of having a computer to do the actual analysis is clear. The very large number of comparisons and calculations implied in V-analysis and O-analysis is a discouraging prospect without the assistance of a modern digital computer.

Although a current trend toward acceptance of the computer by behavioral scientists is clear, it has been slow in developing. Part of the resistance stems from the myth that the "electronic brain" somehow intrudes into the domain of the scientist in a way harmful to the understanding of data or the freedom of the investigator. The main resistance to learning about how to use computers stems from an equally irrational belief that computers can be used only if the user personally knows how to "program" the computer or at least knows the meaning of a host of fanciful terms and jargon.

These irrational positions are, of course, quite irrelevant. It takes nothing special in intelligence or knowledge to become quite expert in applying computer programs. Learning to use a computer program in an application to data takes less time and effort than learning how to operate a desk calculator. As a user of a computer program one need know nothing about computer programming proper. Anyone who wishes can leave the hardware to the engineers and the programming to programmers.

In this day and age, however, it is not difficult to find ways and means of learning how to program a computer. Several books aimed at teaching behavioral scientists the arts and methods of computer programming are available (e.g., Lehman and Bailey, 1968). Since one of us (DEB) has been teaching computer programming to psychology students (undergraduate as well as graduate) for several years, we now know that in approximately 32 hours of instructional time ordinary students can become reasonably adept at computer programming. Consequently, the fact that one need not know programming to make expert use of computer programs, such as the BC TRY System of cluster and factor analysis, should not inhibit anyone from learning how to program the computer if he finds it useful or interesting to do so.

On computer use and computer programming

To use a computer program with expertise requires only knowledge of a few simple facts. The program must perform the kind of analysis that is wanted on data like the data it is intended to analyze. How to punch the data into computer cards must be known, along with the control data the program requires to proceed with the analysis, e.g., the parameters describing features of the data such as their title, the number of variables, and the number of observations. If the program performs alternative kinds of analyses at some points in its work, control data may have to be provided (again, on punched cards prepared by the user) informing the program of the choices the user wishes to make. In many programs such choices can be left to the judgment of the programmer (expressed as part of the program) by opting to have the program follow standard alternatives— optimal alternatives, or so one might hope. Where the alternatives chosen (either the user's or the programmer's) turn out not to be good choices, a reanalysis of the data by the program often is a matter of repunching a control card (a reanalysis in the days of paper-and-pencil programs would have been a serious matter). The main thing a user must know is the kind of analysis a program can perform and how to use the program, i.e., how to punch the control cards.

Although the developmental stages of the programming process are of secondary interest to the program user, it is useful to have a general understanding of the stages by which the analytical tools are provided. Table 3.1 schematizes the six major stages in the development and use of a program when there are a number of users.

The first stage in developing a program is the conceptual formulation of an objective that needs to be achieved. The conceptualizer must specify, in a step-by-step logical fashion, the successive steps to attain the objective. For example, the objective may be to compute the means of n variables. Almost anyone can lay out a step-by-step method by which this objective might be accomplished. The objective may be more complex, e.g., encompassing a hierarchy of sub-objectives such as "(1) to compute the means and standard deviations of n variables, (2) to read all the scores of all N individuals on the n variables as they are punched on cards, (3) to record these scores on a magnetic tape to be called the Data Storage Tape (DST) so that programs used later can have access to them, and (4) to record the means and standard deviations on another tape, to be called an Intermediate Storage Tape (IST), for a similar use to that of the DST."

The hierarchy of objectives is really an extension of the simple objective of finding the means. The solution of each sub-objective can be laid

TABLE 3.1 STAGES IN THE DEVELOPMENT AND USE OF A MULTIPLE–USER COMPUTER PROGRAM

Stages	Formal Result	Agent
1. Verbal formulation of the purpose of the program and the step-by-step method of achieving it	Program design (prose algorithm)	Conceptualizer
2. Translating the formulation to a general symbolic language, e.g., Fortran), punched on cards	Source program (program listing)	Programmer
3. Translating the symbolic language on the source deck to the specific machine language accepted by the computer	Object program	Compiler (a program)
4. Storing the object deck for ready application by the computer to a particular data set	Entry into program library on tape or disk (system tape)	Computer center depository
5. Printed instructions on the use of the program, giving purpose, method, required control cards, restrictions	User's description	Programmer and conceptualizer
6. Calling the program by required control cards for the execution of it by the computer on a given data set	Computer run	User

out in a step-by-step fashion, and so can the sub-objectives themselves in relation to each other. This complex example is quite real; it represents the initial formulation stage of an actual program called the DAta Processor program, abbreviated DAP from the capitalized letters in its full title.

When the step-by-step method of reaching a goal can be laid out conceptually, each element in each step can be expressed by a symbol. For example, from elementary statistics, "to find the mean of the X scores" is conventionally symbolized as $\Sigma X/N$, where Σ means "add up," X means "the X scores," / means "and divide by," and N is the number of scores. The computer is, in simplest terms, a machine expressly designed to manipulate sequentially the symbols defining the successive stages that reach a desired goal. The symbols can refer to numbers, but they can also stand for other entities, and the manipulations of them, in addition to the usual arithmetic algebraic kinds, can be those of retrieving, ordering, transferring, sorting, storing, rejecting, and many others. Since so many of man's mental and artistic achievements and emotional displays can be symbolically represented, it is clear why computer use has rapidly moved into nearly all fields of man's activities.

A step-by-step procedure of achieving an objective is known as an "algorithm." The job of the conceptualizer of a computer program breaks down, therefore, into specifically framing the algorithms of his intended program into a program design, also known as a prose (or verbal) algorithm.

Stage 2 in the development of a program involves the specialist known as a programmer. He translates the prose algorithm of the conceptualizer into a family of symbols, e.g., Fortran, that, with additional processing (in stage 3 of Table 3.1), can be manipulated in the desired fashion by the computer. As the result of interaction between conceptualizer and programmer in stages 1 and 2, the final program design may take a radically different shape from that originally proposed by the conceptualizer. Also, in the welter of symbols that embody a program, mistakes, or "bugs," can occur. The last operation of a programmer in the second stage is to "debug" the program.

The remaining stages of development are perhaps apparent from Table 3.1. Because the symbols written down and punched on cards (the source program) by the programmer are usually not those actually manipulated by the computer itself, they need to be translated into the machine language that the particular computer can manipulate. This translation is done by the computer itself, using a program called a "compiler." The result of this is expressed as an "object program," which may be punched on cards by the computer or recorded into the "library" of the computer. When the program is compiled, the next step, stage 4, is to store it in a program library (on tape or disk) from which, on call, it can be read for the purpose of executing the algorithm on a particular user's data set.

If the user's description of stage 5, prepared jointly by conceptualizer and programmer, is clearly written, the user is in fact primarily (perhaps exclusively) interested in this description and in his own computer run in stage 6.

The BC TRY System of programs

The BC TRY System consists of about 30 different programs, some combination of which, in a particular sequence, executes an aspect of cluster and factor analysis on a particular data set. To use the System, all that the user needs to know is the particular sequence of programs to use and the control cards necessary for each program. For example, if he wants to have the scattergram between two variables computed, he need only know the sequence of programs necessary to get that scattergram. Looking in the User's Manual, he discovers that the scattergram program RSCAT must be preceded by program DAP but that DAP itself requires no other program preceding it. The system cards for DAP are preceded by a card punched START, which prepares the BC TRY System, located on a disk (or tape), for action. The computer determines that the BC TRY System is the one system among several available that it should select. When the user's job is activated, the computer is under the overall control of a superprogram, called a "monitor," and the user has indicated on his first cards (monitor cards) that the monitor should call the BC TRY System. The sequence needed to compute a scattergram is

    Monitor cards
    START card
    DAP card (followed by its control cards and data deck)
    RSCAT card (followed by its control card)
    END card

More briefly, we write this sequence as

    JOB—START—DAP—RSCAT—END

BC TRY is an integrated system of programs designed on a general conceptual scheme to execute the grand algorithm by which a user discovers the general attributes and general types of individuals in a score matrix (Tryon, 1958b, 1959). A particular computer run accomplishes one of the sub-objectives of the overall objective, executing a special-case solution of the general logic of V-analysis and O-analysis. In a general way the many methods of factor analysis and cluster analysis have the common objective of discovering a minimal number of composites among a collection of variables or objects. The composites can replace the full array of variables or objects without loss of generality, in the sense that the reduced set will reproduce the intercorrelations among the full array.

The BC TRY System provides the user with many options on how to solve for a reduced set of dimensions and objects. For the user who is overwhelmed by the embarrassment of riches, the System offers standard options that have been found in many problems to give satisfactory, perhaps optimal, solutions.

Though all the main forms of factor analysis are available in BC TRY to investigators who have predilections for one rather than another, the main emphasis is on the methods of key-cluster analysis based on the domain-sampling formulation of cluster and factor analysis (Tryon, 1959 and 1958*b*). The emphasized methods will be denoted simply as "cluster analysis."

As a preliminary to the procedures of cluster analysis, the data are first input to the computer in proper form. The initial step of V-analysis is to compute the correlations among all the variables. The proportion of the variances to be reproduced is next set in the diagonal cells of the r matrix (the inserted values are usually estimates of their communalities, though one can insert reliability coefficients or unities). Next, the dimensionality of the matrix is estimated by factoring on the most collinear subsets of variables. Sufficiency of the dimensions is then tested by computing residuals. Finally, an oblique cluster structure solution, called a "direct oblique rotation" in orthodox factor analysis, is described both in statistical and in geometric terms. The investigator may then proceed to an O-analysis of his data, first scoring individuals on the several clusters and then by a series of five integrated steps isolating general types of individuals. If he wishes to test the predictability of other "outside" attributes from these O-types, components for doing so are available.

BC TRY includes many other important supplemental procedures. These are components to perform comparative V-analysis ("matching factors") as well as comparative O-analysis; the latter enables one to assess the forces of multivariate selection in different groups measured on the cluster defined dimensions. Special components permit V-analysis and O-analysis however many variables or subjects one has. For users who wish to perform the traditional kinds of factor analysis, including rotation, these are included in the System, and since these methods are programmed as integral components of the System, one can compare the results of these analyses with those from key-cluster solutions.

The computing system attains its various objectives through the control of component programs by a General Executive Program (GEP). For example, when the System embarks on V-analysis, GEP at the top hierarchical level initiates this analysis by passing control to a component that processes and stores the data. Control then returns to GEP, which in turn passes it to a second component program to compute a paired-comparison (correlation) matrix. After completion of this computation, control again passes back to the executive which then proceeds to pass control successively to further component programs, each of which embodies and completes a logical step of V-analysis.

The System is somewhat more complex than stated above. Some of the components themselves are subhierarchies. Further, the investigator is permitted to chose the form of analysis he wishes the machine to undertake.

Any type of cluster or factor analysis performed by the BC TRY System is executed by separate components linked together in tandem. Each program itself is named by an alphameric (alphabetic and/or numerical) label like GEP, DAP, COR2, and so on. Identified only by these symbols, a sequence may at first seem to be unintelligible to the person learning about BC TRY. However, all parts fit into the general logic by which one reduces many specific attributes of individuals to a smaller number of general attributes, i.e., V-analysis, and in terms of which many particular individuals are reduced to a small number of general types, i.e., O-analysis. Presently the user must communicate his wishes by control cards both to the executive and to the components. In time it is hoped to simplify this interaction by making it necessary for the user to communicate only with the executive, which alone will manage the components without further intervention by the user, who would thus be rid of the onerous task of preparing component control cards.

An illustration of how BC TRY executes the logical analytical procedures in a full-cycle cluster analysis of variables and of individuals will provide a broader acquaintance with some of the components that constitute the System. The successive stages of such an analysis are listed in Table 3.2.

Some procedures are "compounds," consisting of linked programs. There are two stages labeled Stop (steps 6 and 10) where the investigator will normally, though not necessarily, intervene and size up what the System has produced and, if indicated, direct the System to rework some steps.

Early components in a sequence produce results that are utilized by later components. These events are achieved through the mediation

TABLE 3.2 STANDARD CLUSTER ANALYSIS OF VARIABLES AND OF INDIVIDUALS

Stages of analysis	Component
A. Input of raw data	
1. Preparing the n variables on N individuals for processing by later components	
a. Setting the data on the Data Storage Tape (DST)	DAP
b. When n or N are very large, choosing random or forced samples	PICK
B. V-analysis: cluster analysis of variables	
2. Determining generality of variables: correlation matrix (missing data require COR3)	COR2
3. Deciding on variance to be factored	DVP
4. Performing dimensionality analysis (factoring)	
a. Selecting maximally collinear dimension-defining clusters and computing orthogonal factor coefficients on them	CC
b. Computing residuals to test dimensional sufficiency	CC
5. Describing the oblique structure of the dimension-defining clusters (and of dependent clusters, if any)	
a. By statistical quantities (a "direct solution")	CSA
b. By a geometric configuration	SPAN
6. Stop. After study of CSA and SPAN, revising the cluster selection by redefining the subsets (including hierarchical condensation, i.e., higher-order analysis) if necessary and repeating steps 4 or 5	
C. O-analysis: cluster analysis of individuals (objects)	
7. Scoring individuals on the oblique dimension-defining clusters	FACS
8. Linear and nonlinear relating of cluster scores and two-dimensional O-analysis on correlation scattergrams	RSCAT
9. Selecting core O-types from cluster score space	
a. Calculating euclidean distances between individuals	EUCO
b. Introducing marker individuals in the configuration	OMARK
c. Selecting core O-types by dimensionality and structure analysis	NC analysis
10. Stop. Revision of core O-analysis, as in stage 6	
11. Determining O-clusters by identifying each individual with a core O-type, however large N is	OTYPE
12. Describing the cluster score pattern and homogeneity of resulting O-clusters	OSTAT
13. Nonlinear predicting of "outside" attributes from the categorical series of O-clusters	4CAST

of the Intermediate Storage Tape (IST); the components write their own unique results on IST, and from it they read what they need. Just how this works in the full sequence of Table 3.2 down through step 7 is illustrated in Table 3.3.

Under the column headed Component Program, the first component is a data processor called DAP. DAP sets up the raw data on the Data Storage Tape (DST), computes the means (MEANS1) and standard deviations (STDEV1) of all the n variables, and outputs these constants to the storage tape, IST, as shown by the two O's opposite DAP in Table 3.3. In the

TABLE 3.3 HOW COMPONENT PROGRAMS PRODUCE AND UTILIZE FILES[a]

Component Program	Means MEANS1	Sigmas STDEV1	Correlations CORRM1	Diagonal Values DIAGV1	Dimension Definers REFLX1	Orthogonal factor Coefficient UFACT1	Oblique factor Coefficient RFACT1	r between Clusters BASIS1
DAP (for DST)	O	O						
COR2	I	O	O					
DVP			I	O				
CC			I	I,O	O,(I)	O		
CSA			I	I	I		O	O
SPAN						I		
FACS	I	I	I		I	I	I	I

[a] A O = output by the component, I = input to it, (I) = optional input to the component.

MEANS1 column, it is seen that means are later used in calculating correlations (I opposite COR2) and much later in computing cluster or factor scores (I opposite FACS). But, as shown in the STDEV1 column the standard deviations are not used by the correlation program, COR2, since there is no I in the COR2 row under STDEV1. COR2 itself computes standard deviations for the second time (O opposite COR2); these are used much later by FACS. And so the accumulation of results goes on as the operation proceeds down the table, until the time comes to compute scores on the oblique clusters discovered in the V-analysis. FACS leans heavily on results from almost all the prior components.

Table 3.3 gives only a few examples of the complex manipulation of files by BC TRY. Other procedures are listed in Table 3.4, which also gives the symbolic name of the component of BC TRY that does the work.

All the main forms of orthodox factor analysis (Harman, 1967; Thomson, 1951) can be computed by the BC TRY System, as shown in Table 3.5. Actually, these types are merely different forms of factoring (Table 3.2, step 4), which is only one stage in the full-cycle cluster analysis. Except for square root factoring, these orthodox forms of factor analysis define dimensions by different patterns of weights on all n variables. Such dimensions are usually difficult to understand. Indirect, or "derived," rotations (Harman, 1967, pt. III) are therefore resorted to, usually by varimax or quartimax.

The components that execute the procedures of Tables 3.2 to 3.5 are statistical and logical programs, given the symbolic name of STANALOGS in the BC TRY System.

Finally, there arises the problem of making the BC TRY System available to users. Official versions of the BC TRY System are being made available to computer centers at universities and research facilities. However, the problem is complicated not only because the system provides a great wealth of interconnected components but also because the components are being constantly revised and will be programmed on different computers. This problem is being met in several ways.

Accompanying the System is a User's Manual written from punched cards onto magnetic tape and dated to refer to a particular version in a particular language on a particular machine. The cards and tapes can easily be changed as versions change. Multiple printout copies can be made available to local users quickly and cheaply.

To give some idea of the speeds of computation, some time estimates taken from the detailed time charts in the IBM 7090 User's Manual system tape storage are shown below. The estimate for a 25-variable standard cluster analysis by steps 1 to 5 given in Table 3.2 for this direct oblique

TABLE 3.4 SUPPLEMENTAL PROCEDURES OF BC TRY

Procedure	Name
A. Comparative analyses	
1. Comparing cluster defined dimensions discovered by V-analysis of different groups (or of the same group by different factoring methods)	COMP[a]
2. Comparing the O-clusters discovered in different groups, i.e., assessing multivariate selection	OCOMP[a]
B. Large-scale V-analysis and O-analysis	
3. However large n or N, converging in random subsamples on the cluster defined dimensions in the full domain of n variables or on the O-clusters in the full supply of N individuals	BIGNV[a]
C. Miscellaneous	
4. Suppressing designated variables during factoring but reactivating them in a summary analysis	SLEP[a]
5. Dimensional and oblique structure analysis without communalities	NC,NCSA
6. Dimensional analysis on preset dimension-defining clusters including designed and higher-order (hierarchical) analysis	CC, NC
7. Computing communalities in any one of six different ways and utilizing them (or reliability coefficients and unities) as diagonal values in the correlation matrix	DVP
8. Rational nondimensional cluster analysis, e.g., item analysis	CSA, NCSA
9. Multiple correlation and regression of dependent clusters on the minimal set of dimension-defining clusters	SMIS
10. Missing data management	COR3, FACS3

[a] Linked programs.

TABLE 3.5 ORTHODOX FACTOR ANALYSES BY BC TRY

Procedure	Name
A. Simple-sum factoring	
1. Thurstone centroid for $n \leq 20$; salient centroid for $n > 20$	CC(CENT)
2. Bifactor analysis (first dimension a centroid, remaining dimensions key cluster)	CC(CENT+)
3. Square root or diagonal factoring (or pivot-variable analysis)	CC(PV)
B. Least-squares total-set factoring and auxiliaries	
4. Principal-component or principal-axes factoring	FALS(PFA)
5. Canonical factoring	FALS(CFA)
6. Augmented (or α) factoring (also by simple-sum factoring)	FALS(AFA)
7. Residuals from least-squares (or any other) factoring	FAST(RESID)
8. Reproduced correlation matrix from least squares (or any other) factoring	FAST(REPRODUCED)
9. Rotation of total-set factors (any type) varimax or quartimax	GYRO
10. Regression scores of orthogonal or oblique total-set factors	FACS
11. Comparison of total-set factors in different groups with those of a best-fitting population	SIMRO

solution is

DAP—COR2—CC—CSA—SPAN: 1.74 min

Compare this time with that required of two orthodox indirect orthogonal "factor analysis" solutions which the user obtains by simply replacing the CC-CSA segment by distinctive calls that define these two solutions. First are shown the times for the cluster analysis segments, followed by those for the defining segments of a centroid-quartimax and a principal-axes varimax solution (the GYRO components could be interchanged):

Cluster analysis		Centroid-quartimax		Principal-axes varimax	
Segment	Time, min	Segment	Time, min	Segment	Time, min
CC	.45	CC(CENT)	.45	FALS(PFA)	1.42
CSA	.37	GYRO(QRTMAX)	.81	FAST(RESID)	.93
				GYRO(VARMAX)	1.31
	.82		1.26		3.66

The slower principal-axes speed of nearly 4 min. is required despite the use of Householder-Ortega-Wilkinson (HOW) lemmas in the subroutine that computes eigenvalues and eigenvectors with great speed and accuracy on matrices up to 120 variables.

Such time comparisons can be quite misleading due to vitiating "overhead" hardware features. Thus, much of the above slowness of the principal-axes segment is an accidental result of the location of its components at the end of the system tape. When corrections for tape positioning are made (or actually achieved in disk storage), the principal-axes segment takes only about one-third more time than the cluster analysis, which is itself much speeded up.

Other practical restrictions are these: input is restricted to 100 variables, more or less, depending on the installation on no more than 9,999 subjects (5,000 on some installations). Factored dimensions cannot exceed 15. As machines increase in capacity, these limits can, of course, be raised. But even now, when one uses the designs of BIGNV (see Table 3.4), there is really no limit on numbers of variables or subjects.

Systematic computing procedures in BC TRY

This section describes features of the BC TRY System of interest to those engaged in the development and use of computing procedures in the behavioral sciences. The computing science features described here are a varied lot, some simple, others complex in their logic and execution. Although many of the details of the procedures are of interest, treatment here will deal only with general aspects.

Central storage of programs

The entire system of component programs (30 at the time this section was written) is stored on a tape (or disk) in a linkage or overlay mode (depending on the computer) that allows any of the components to be called into execution at any time by any of the components. In this way the user of the System never handles the program decks. The linkage tape is on deposit with the machine operators, who mount it at request of the user; or the linked programs are permanently stored on a disk. The request for the BC TRY System is made via a short program (ACCESS) available to the user as an object deck. Hence no technical computing knowledge is required of users of the System.

The most significant result of this system of program storage and access is integration and linkage of the separate component programs in the System. Because the System is unified as a collection of links, access to the System implies access to each of the components. Executive-program operation features and the data sharing facilities interact with the mode of program storage to produce, in effect, a program of some 500,000 machine words with a practically unlimited amount of common storage and flexibility in determining the pathway through the System at execution time.

Two modes of transition from one component to another in the System are provided. In the most general mode the GEP interrogates the monitor input unit for instructions regarding the component to set into execution. GEP then calls, via the linkage (CHAIN, OVERLAY) subroutine, the designated program from the unit on which the system is resident. Under normal conditions the STANALOG component in operation exits on termination of its execution with a call for the GEP. This is executed via the linkage subroutine. If an error occurs in calculation or control card or there is machine failure (in most cases except errors which cause machine halts directly), each component will call the system component FRGIVE, which punches the entire contents of the IST onto binary cards so that the computations can be restarted with the program that terminated execution in an error.

The general executive program (GEP)

The primary component of BC TRY, as a system, is the General Executive Program (GEP). This program presently operates as a submonitor. Once the monitor initiates the execution of GEP, the GEP performs several preliminary operations and then monitors the sequence of components in their execution as determined by executive control cards on the monitor input unit.

The initiation features of the GEP result in the establishment and checking of the assignment of input-output (I/O) units by unit designations for symbolic functions. The details of this are described below (DYNATAPE). In addition, the positioning and initiation of the storage units are overseen by GEP. Perhaps most important is GEP's role in setting up tables and common storage regions containing information about the location of component programs on the BC TRY System and the files of computed data on IST.

In its current form, the GEP recognizes several specific cards, each with a distinctive form. Correlated with recognition the GEP performs the pertinent operations, the first being the initiation functions discussed above. Second, the GEP prepares for EXIT to monitor by rewinding tapes, unloading the BC TRY System tape, and attending to other minor duties. Third and fourth, the contents of the IST can be punched in binary on cards or be made up from similar cards read by the executive. Fifth, the entire contents of the IST can be printed. Sixth, the GEP can call programs as links from a second chain tape in order to execute programs not on the BC TRY System tape. Seventh, the GEP interprets the name of a component program, e.g., COR2, the correlation program, on the executive control card and calls the indicated link on the BC TRY chain tape. Should an error occur, e.g., an incorrectly punched executive control card or a non-executive control card occurring where an executive control card is expected, the GEP enters an error mode, resulting in the punching of the contents of IST onto binary cards.

An important feature of the System is the ability to modify the System without handling the deck of binary cards for the entire System. This feature also saves a great deal of computer time. Only those components which have been modified need be dealt with. The EDIT option remakes the chain tape, replacing obsolete links with their modified versions.

Data sharing

In the early conceptions of BC TRY a fixed order of program execution was used. Whatever data were calculated by a program and needed by succeeding programs were recorded seriatim by files on a binary tape. When the data were needed for the subsequent calculations, they were simply read from that tape. The location of the data was fixed by the file number. Consequently if program PPP needed the data calculated by program QQQ, PPP had to contain information regarding the location, i.e., the file number, say XXX, of the data. This information was provided by having QQQ always record the data as file XXX. This method required some programs to write dummy files or to restructure the binary storage tape, which was a satisfactory procedure until the full force of the modularity of the master design became apparent. To achieve the grand design and have a truly superprogram with virtually unlimited flexibility in the serial use of the programs we had to abandon the rigid seriatim nature of our file keeping. As often seems the case, the problem and its solution were proposed almost at the same time. We were trying to solve the file-keeping problem at the time Fortran II was introduced. Without Fortran II and its CHAIN and COMMON features the solution might have been delayed indefinitely.

The CHAIN and COMMON features of Fortran II allow a program on

the CHAIN tape to be loaded into core without disrupting selected segments of the COMMON portion of the core. This core area is called ISTCOM
in the BC TRY System and contains a table of the symbolic names of files
currently available on the IST. The System has since been translated to
Fortran IV and uses the corresponding features COMMON and OVERLAY.

The data-sharing system of BC TRY is a method for saving selected
intermediate calculations as files on a binary tape or disk. These files
include logical information (as in variable names, titles, etc.), vectors,
matrices, and lists. A number of standard forms and parameters are
defined. Corresponding subroutines which read or write these forms and
parameters are called the IST subroutines and automatically transmit the
respective quantities to and from the IST.

When a file is transmitted to the IST, a unique character label is
given to it. This label is written on the tape with the rest of the information
of the file and is read when the file is read. The labels are defined by the
programmer at the time the component program is written and are
associated with information generated by the program.

There are six basic elements to the logic and mechanics of IST usage:

1. The information involved, which may be generated by more than one component program
2. The information label
3. Parameters associated with the form of the information
4. The specific IST subroutine used to transmit the information
5. The format of the information to be transmitted
6. The array of file labels and the associated array of file numbers

When information is of potential use in later stages of analysis (as
determined at the time the program was written), it is recorded as a file
on the IST at the end of the string of files already on the tape. In addition,
the symbolic name, or label, of that file and the physical file number of
the file on the IST are recorded in a pair of locations in the ISTCOM area.
When a program, at a later point in the job, requires a file with the label
of a file on IST, it interrogates ISTCOM. The physical file number of the
file containing the information desired is determined by matching the
label of the desired information with a label in ISTCOM and hence with a
physical file number.

Recovering a file of information from IST does not alter the contents
of the ISTCOM or IST. However, when transmitting a file of information to
IST, two possibilities must be considered. If there is no old file with the
same label as the new file, the new file is added to IST and the new label
and physical file number are added to ISTCOM. If a file with the same label
as the new file is already on IST (and hence its label is in ISTCOM), the

new file is added to IST and the physical file number of the new file replaces the physical file number of the old file.

Updating the physical file numbers in ISTCOM implies that the System does not have old files available once a new file with the same label is generated. This is a distinct advantage in that files of the same name and hence confusable are never present simultaneously excepting as dead or nonaccessible information on the IST. However, one disadvantage is inescapable: old information in physical files having the same labels as new information in other physical files is not available to most components of the System. This is vitiated in large part by the data retrieval components of the System discussed below.

An important aspect of this system is the recording in the file itself of the parameters determining the specific character of the information in the file. The number of rows and columns in a matrix, the number of records in the file, the number of elements in a list, the format of the file, etc., are all indicated in the first few words of the file. The advantage of this recording form is that the component needing a certain kind of information generated by another program need not have information regarding the specifics. For example, the rotation component GYRO need not have the number of factors before interrogating IST to obtain the factor coefficients to be rotated.

Data transmission

In addition to providing within-system sharing of data, BC TRY permits a full range of user-system data sharing procedures. Most of these procedures are involved in saving data from calculations in a convenient and usable form. However, the System provides for direct intervention by the user in the data sharing aspects of the system during execution. These features involve an error initiated procedure giving a binary deck reproduction of the IST which can at a later date be used to begin the job where it was terminated by the error. In addition, the user can produce the same binary deck for the IST by calling for it through the GEP. When such a deck is in turn encountered by the GEP, the IST and ISTCOM are reconstructed to correspond to the ISTCOM and the IST files current at the time GIVE was invoked. This option of GEP is especially useful where two sets of calculations based on one set of intermediate data are planned (as in two types of analyses of a single correlation matrix). After calculating the basic data the IST is punched through a call to GIVE. The second analysis avoids recalculation of the basic data by constructing the IST with the binary cards obtained with the earlier call to GIVE.

The user can introduce individual files in IST by the use of a separate component of the System, GIST. Hence if a file is used by a component and it has not been generated in the sequence of calculations already executed, it can be made available. Also if special information is required, e.g., predefined clusters in a cluster analysis, previously calculated factor matrix, it can be put on IST with the appropriate ISTCOM entries.

DYNATAPE

One of the major problems in moving a computing system from one computing installation to another is the lack of consistency in the use of Input/Output units at various installations. The I/O configurations and assignment of units to various functions have not been standardized in the computing profession. An attempt to make the I/O of the BC TRY System as installation-independent as possible is reflected in the system of subroutines called the DYNATAPE system. These subroutines assign the I/O unit functions to I/O units symbolically at execution time. The I/O units which are used at a given installation for systems, input, output, and scratch can be assigned appropriately at a receiving installation without a great deal of technical knowledge and, most important, with the compilation of only one very short subroutine after modification of one block of statements, involving no technical programming knowledge.

Each type of unit function in the BC TRY System has been given a symbolic name. Thus, each I/O statement has one of these symbolic names associated with it. In BC TRY there are nine distinct functions for I/O units. Each unit is assigned to the function by (1) the installation conventions and (2) the needs of BC TRY. The symbolic names and the associated unit functions are:

Symbolic Name:	Unit Function:
MONIN	Monitor input
MONOUT	Monitor output
MONPUN	Monitor punch
MTSYS	BC TRY System
MTIST	Intermediate storage
MTDST	Data storage
MTSC1	First scratch
MTSC2	Second scratch
MTTEST	Program test

The logical number of a unit is set by assigning the desired number as the value of the symbolic name. In all components the values of the symbolic names are set in the main program. References to the units are carried through the calling sequence of each subroutine. The tables of symbolic unit names and their associated values are contained in an area of COMMON shared by all links in the System. The DYNATAPE system establishes the tables, checks them for consistency, redundancy, and undefined logical numbers, and restores the tables should a component program destroy the information in COMMON.

Macro matrix and vector manipulations

In order to provide the ultimate in flexibility of calculation within the System the program SMIS, the symbolic matrix interpretive system, written by E. Wilson (see SHARE F1 BC SMIS) has been adapted to BC TRY. This program provides a complete array of matrix and vector operations with access to the IST through the IST subroutines.

SMIS provides a flexible means of performing matrix operations under the control of a sequence of punched cards. A program run consists of reading the input deck and executing the operation designated by symbols which are selected by the user. The input deck of SMIS is composed of control, data, and remark cards. The control cards designate the operation to be performed on a given matrix. The matrix involved may be input on data cards or called from IST by reference to the file label of the file containing the data.

Some 35 separate commands are included in SMIS. For example, the commands allow data sharing with IST, input and output from and to the monitor, scalar multiplication, finding functions of elements of a matrix, eigenvalue and eigenvector solutions, matrix addition, and matrix multiplication.

Illustration: the program DAP

The principles governing the functioning of the System may be illustrated with the first program needed in a BC TRY System use. This is the DAta Processor (DAP) mentioned above. Generally, the objective of this program is to get the raw score matrix stored in records on the Data Storage Tape (DST), to check it thoroughly, to compute the means and standard deviations of all the variables, and to put these values in files on the IST.

How does the user "control" this program? The Use section of the "User's Description" in the Manual gives all the details of how to do the job. Briefly, one first commands the executive to call DAP by punching an executive control card for DAP. This card has simply "/DAP2" on it. The slash is a control symbol for the executive GEP, and the DAP2 is the signal for the executive to give control over to the DAP program.

Specific directions are given to the DAP program itself by punching certain alphanumeric symbols on the specific component control cards of DAP. If, as with most problems, a V-analysis is planned for not more than 120 variables and not more than a few hundred subjects, the control cards of DAP are quite simple. On the first card is punched the title of the problem, which will be printed on all output. The second card is the parameter card, on which is to be punched in specified columns only four items of control information positively required by DAP: the number of variables, the number of subjects, the number of columns on data cards devoted to the raw scores on each variable, and a number indicating whether there are any missing data. Following these two control cards comes the data deck, consisting of cards containing the scores of all the subjects on all the variables.

It is possible to choose to input more information than this minimal amount, and sometimes it is necessary. For example, if the scores on the cards are not in columns of constant width, one must input a format card telling DAP of the fact. If some variables on the data deck are to be skipped, an input format card is punched to tell DAP to do so. Usually one wants to assign a symbolic name to each variable, a name to be printed on all output; if so, a VNAMS card must be input. There are other options such as inputting names of subjects or reordering the variables and reflecting some of them, but they are all extras; only the first two control cards are necessary in the usual case.

So controlled, DAP executes its algorithms on the data and outputs the results in various forms. It automatically outputs the raw scores onto the DST together with certain necessary identifying information and parameters; these scores are in records on DST so that they can be read later by such programs as COR2, COR3, FACS, and RSCAT. It also outputs some results of its work in files on the IST, the most important of which are the title of the problem, the means and standard deviations of variables, and the names of the variables; these files are extensively used by other programs. The printed output gives a printing of the options that have been taken, and it prints out in detail the values and entries it has stored in the different IST files.

In general, then, DAP illustrates the potentialities of the computer system. For the purposes of cluster and factor analysis the computer does little computing work, but its role in preparing the data for other programs is crucial.

Chapter 4

GENERAL ATTRIBUTES: CONCEPTUALLY DEFINED
GROUPINGS OF VARIABLES VS. EMPIRICAL CLUSTERS

A collection of many variables can be reduced in number by grouping the variables into categories. The individuals observed on the variables can then be compared by their composite scores on the categories rather than by the whole set of variables. What should be the nature of the groupings of the variables? The answer, beyond doubt, is that the variables composing each group, called the "definers" of the category, must be similar in some way. But in what way?

Conceptually defined groups of variables

Traditionally, the standard of similarity has been that each category of variables should be a "rational grouping." This expression means that the definers must all share some common abstract quality or similarity of content. On this standard only those variables which were alike on the basis of some theoretical or social construction would be grouped together.

Table 4.1 shows the five categories in which the 24 test variables of the famous Holzinger-Swineford study (1939), generally known as the

Holzinger-Harman problem (Harman, 1967), were cast by its investigators. The variables were selected so that there are at least four test samples in each of the five groupings. For tests in the first category, the spatial tests, most reasonable persons would agree that the work that the subjects, school children, go through in taking them shares the common property of imaginatively manipulating visual figures and shapes in a spatial way. The tests actually present no real things in space—only representations of them on paper. The five verbal tests, V5 to V9, all require dealing with

TABLE 4.1 DEFINING VARIABLES OF THE RATIONAL
GROUPS IN THE HOLZINGER–SWINEFORD STUDY

Spatial tests

F1	Vis	Visual figure completions
F2	Cub	Cube similarities
F3	Fbd	Paper form board
F4	Loz	Lozenge shape rotations

Verbal tests

V5	Inf	General information
V6	Cmp	Paragraph comprehension
V7	Snt	Sentence completion
V8	Wcl	Word classification
V9	Wmn	Word meaning (vocabulary)

Speed tests

S10	Add	Addition
S11	Cod	Code substitution
S12	Cnt	Counting groups of dots
S13	Scc	Straight or curved capitals discrimination

Memory tests

M14	Wrg	Word recognition
M15	Nrg	Number recognition
M16	Frg	Figure recognition
M17	Wn	Object-Number recall
M18	Nf	Number-Figure recall
M19	Fw	Figure-Word recall

Mathematical-ability tests

N20	Ded	Deduction
N21	Puz	Numerical puzzles
N22	Rsn	Problem reasoning
N23	Ser	Series completion
N24	Ari	Woody-McCall mixed fundamentals, form I

previously learned ideas or things expressed verbally. The four speed tests all require "mental speed" in manipulating easy contents, the elements of which were quite familiar to all the children. The six memory tests fall into two categories: the first three are tests of memory that require only recognition of materials just learned in the test situation; the second three are also tests of memory of just learned materials, but they test memory by recall. Finally, the last five tests all seem to sample some form or other of mathematical ability.

Forming such composites in most problems has many practical short-comings and ambiguities. First, any classification that one might claim to be meaningful could be rejected easily as absurd or unreasonable by some argumentative authority in the field. Also, one might have some reservations about a particular grouping, because (as in the Holzinger-Swineford study) the different definers of the group may have so much diversity in other respects than the common property abstracted out in the group that the extraneous noncommon elements might completely overshadow the common. A third and more disastrous difficulty is the discovery that one is dealing with a collection of variables for which it is not possible to devise a set of rational categories or that categories can be found to accommodate part of the collection but not the rest of it.

Empirical clusters

Variables grouped together to form a composite must be shown objectively to be similar; they must also be shown to be different from the variables of other composites. These properties of "within-group similarity, between-group difference" distinguish empirical clusters and factors from traditional rational categories.

The completely objective feature that describes the similarity of the definers of a cluster and their difference from other clusters is the property of collinearity. This term means simply that the definers "fall on the same line," i.e., are collinear. Generally, collinearity is defined by the line graph of the correlation coefficients of two variables with all the variables in the study, their correlation profiles. Collinear variables have the same profile of correlations. The phenomenon of clusters of collinear variables is illustrated here in three very different studies. Clusters of collinear variables have two objective characteristics of similarity: they correlate positively with each other, and they follow the same pattern of correlations with other variables. They also are objectively different from other clusters of collinear variables because their common correlation profiles have a different shape from that of other clusters.

The degree of collinearity of the correlation profiles of any two variables can be measured objectively by a special index of collinearity (see Chap. 12), which measures the degree to which the correlations of two variables are consistently proportional across all the other variables of the study. The index P^2 is 1.00 when all their correlations with the other variables are the same; it is .00 when their correlations vary from each other in an unsystematic way; and the square root P is -1.00 if their correlation profiles are mirror images. Defining variables of empirical clusters reveal within-group similarity from the fact that the P^2 values between them approach 1.00, but they show between-group differences in collinearity because their P^2 values with the defining variables of other clusters are considerably less than unity.

Cluster analysis begins, essentially, with a comparison of variables using the index P^2 and defines clusters of variables by finding subsets of variables having values of P^2 within the subsets. On the other hand, factor analysis methods result in the definition of dimensions or factors without explicit reference to clusters or the index P^2. However, the mathematical-geometric model of the general variability in a set of variables, as calculated by a factor analysis, displays the collinearity of similar variables in a way equivalent to the procedures using P^2. After the variables are factored, they can be plotted in a spherical diagram (SPAN in BC TRY) or space. Variables with collinear correlation profiles will be collinear in this space; they are represented as points in the space lying on the same vector line drawn from the origin in the spherical diagram. We will turn to these relationships again when we take up factor analysis in detail.

What is exciting about locating collinear clusters objectively is the sense of discovery in doing so. The conclusion that collinear variables must be sampling the same domain of determinants of individual differences is a natural result of inspecting correlation profiles. In many problems, collinear clusters often turn out unsuspectingly also to be meaningful composites, so that one may easily speculate on what the substrate of causal components may be. In other cases, the causes may be obscure. For example, if the old saw about British villages were true that the variable "number of old maids" is highly inversely correlated with "number of field mice in the meadows that surround the villages," then these two variables might (after "optimal reflection") enter a collinear cluster, but it could take a Darwin to figure out the rational ecological connection. The imaginative cluster analyst might have solved the problem by including "number of cats," which, no doubt, would be collinear with number of old maids and number of mice.

Collinear clusters in the Holzinger-Swineford study

In Fig. 4.1 the high degree of collinearity of the correlation profiles of the defining variables of the first four rational groups of Holzinger-Swineford tests is easily seen (for simplicity, mathematic ability is left out here). The top chart presents the profiles of the four spatial tests. For example, the Loz line is simply a plotting of the line graph of the successive correlations of the Loz test with the first 19 variables of the study (see Table 4.1 for abbreviations of tests and their meanings). The correlations are read

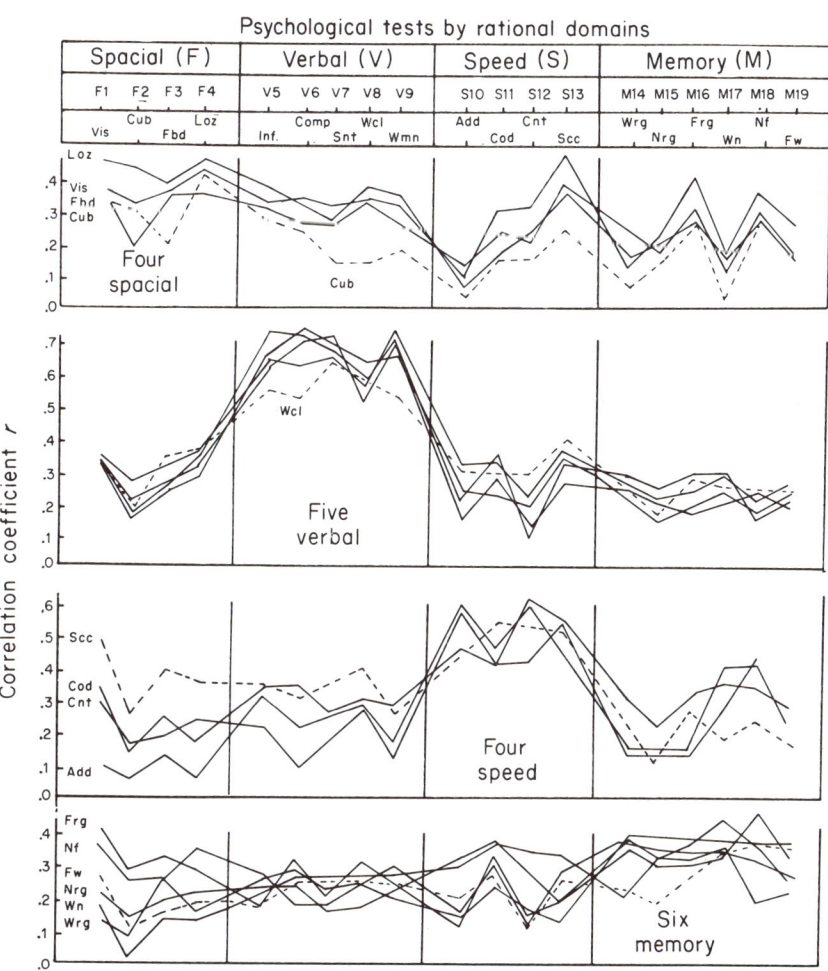

FIGURE 4.1
Collinearity of rational composites and empirical clusters in the Holzinger problem.

directly from the correlation matrix. At the top of the figure the tests are arrayed as successive points on a common abscissa, identified by their names, and grouped by the four rational categories. In general, all four space tests follow the same line; hence they are collinear. Even though one of them, Cub, departs from the others a bit, it preserves the same pattern. The second chart shows the profiles of the five verbal tests, all showing high collinearity (with a trivial departure of the Wcl tests) but a different line from that of the preceding four space tests. In the next chart are the four speed tests with their own distinctive collinearity (with some deviation of Scc). The final group of six memory tests is not sharply collinear. None, however, would be classed with the first three groups. They are, in fact, collinear in respect to these features: all generally reveal positive correlations with all tests, and each appears to correlate higher with tests of its own memory category than with tests in other rational groups.

Generally, then, when Holzinger and Swineford selected tests for their four rational groups they did in fact also select collinear clusters. Each group appears to sample a special domain of causes of individual differences, and the domains seem to be different from each other. The particular selection of tests was based on findings from an earlier project called the Spearman-Holzinger Unitary Trait Studies and from other earlier researches in which space, verbal, speed, and memory tests identical with, or similar to, those in this Holzinger-Swineford study were examined. The test selections were therefore expressly designed on the basis of prior knowledge to prove that the kind of collinearities seen in Fig. 4.1 would emerge in a new sampling of children.

The important scientific question is this: Without knowing the rational classes, can we objectively and empirically sort the 19 tests, without prior knowledge, into the four rational groups? The answer is that we can do so by objective cluster analysis methods using the objective index of collinearity P^2, discussed above. At this point we need to demonstrate that the four collinear clusters do in fact reveal high within-group P^2 values and somewhat lower between-group values. How this index is used with other procedures to discover the four empirical clusters that turn out also to be meaningful rational classes will be developed in the chapter on key-cluster factoring.

The average P^2 values between the sets of definers of the four rational groups are shown in Table 4.2. For the four spatial definers the congruence of their correlation profiles, so clear in Fig. 4.1, is seen in the table to have a high average P^2 of .94. The next set of five verbal tests has an even higher value of .97, certainly confirmed in their graph. The speed and memory groups have P^2 values just under .90. These within-group

TABLE 4.2 AVERAGE P^2 VALUES BETWEEN THE
SETS OF DEFINERS OF THE FOUR RATIONAL
CLUSTERS

	Spatial	Verbal	Speed	Memory
Spatial	.94	.77	.73	.77
Verbal	.77	.97	.66	.71
Speed	.73	.66	.87	.76
Memory	.77	.71	.76	.89

values are substantially larger than the between-group values listed in
cells off the diagonal. It should therefore be clear that if we set as a criterion
that the definers of a given cluster must mutually show P^2 values above
.80, then an empirical cluster selected by this criterion will include collinear
definers.

Collinear clusters in the social-area study

In order to illustrate the generality of this crucial concept of collinearity,
the empirical clusters discovered in two quite unrelated studies of group

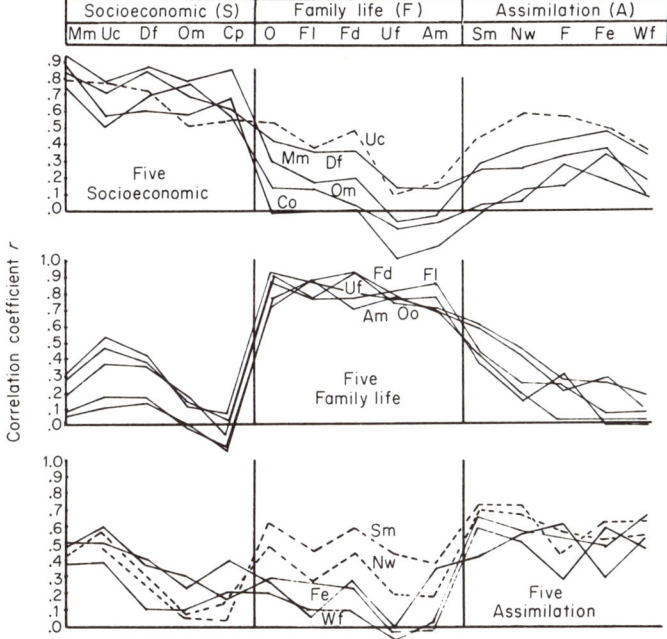

FIGURE 4.2
Collinearity of empirical clusters of demographic variables in the
social-area study.

differences are presented here. The first is a social-area study of the metropolitan San Francisco Bay Area (Tryon, 1955), in which the objects are more than 300 large neighborhoods (census tracts) observed in the base year of 1940. The Census Bureau groups census-tract characteristics by the three rational categories that were not intended to be the basis of composites, namely, population, occupation, and home variables. The neighborhoods were scored on 33 of these variables: 8 population, 13 occupation, and 12 home characteristics.

An empirical cluster analysis of the correlations among the 33 variables yielded the three salient collinear clusters whose definers are listed in Table 4.3 and whose correlation profiles are graphed in Fig. 4.2. When one looks at the top chart of the first group of five variables presented in Fig. 4.2 and reads the contents of these five definers from the table, it should be apparent that the general attribute measured by a composite score on them clearly refers to what is commonly recognized as the socioeconomic level of the neighborhoods.

The middle chart of the figure shows the correlation profiles of the second cluster, a strikingly collinear subset which, from Table 4.3, clearly

TABLE 4.3 DEFINING VARIABLES OF THE EMPIRICAL CLUSTERS OF THE SOCIAL–AREA STUDY (GROUP DIFFERENCES)

Socioeconomic level

Mm	Managerial-professional, males
Uc	Undercrowded
Df	Female domestics (living in)
Om	Own account males
Co	College education

Family life

Oo	Owner-occupied
Fl	Large families
Fd	Family-detached homes
Uf	Nonworking females (housewives)
Am	Older-aged males

Assimilation

Sm	Skilled males
Nw	Native-born whites
F	Females
Fe	Foreign from northwest Europe
Wf	White-collar females

refers to the familiar suburban-urban dimension among metropolitan neighborhoods. A high score on a composite of these definers identifies a suburban neighborhood, one characterized by a high level of family life.

The bottom chart is composed of two subclusters (one in broken lines) sufficiently collinear to collapse into one cluster. The rational conceptualization of this cluster is apparent if it is noted that a neighborhood with a *low* score on a composite of these five definers is characterized by unskilled workers, nonwhite and foreign, foreign from non-Protestant Europe and Asia, and blue-collar working women—clearly the unassimilated, segrated minorities. A *high* score on this cluster is thus identified as assimilation.

The collinear clusters here sample three domains of basic components that differentiate the neighborhoods. These three general domains were discovered by applying purely objective criteria that do not depend on theoretical constructs. Nevertheless, the results do reveal that collinear subsets of variables are meaningful dimensions and, incidentally, are those which social scientists have theoretically speculated about a priori for some years (Shevky and Williams, 1948).

Collinear clusters in the voting-attitude study

The second illustration comes from an investigation of the attitudes of small neighborhoods (precincts) as revealed by their voting on election issues in the city of San Francisco, 1954. In this study, a random sample of 200 such neighborhoods, drawn from more than 1,000 that composed the full supply, were the object of the analysis. The variables on which they were observed were 31 city and state propositions, including the vote for governor. After reading all the propositions and studying the preelection booklet giving arguments pro and con, it is still difficult to come up with any convincing set of rational categories that would be the basis of structuring a reduced set of composites representing general attitudinal attributes of these neighborhoods.

An objective cluster analysis of the intercorrelations between them readily revealed, however, three salient clusters. The definers of them are listed in Table 4.4, and their correlation profiles are shown in Fig. 4.3. In the top chart of Fig. 4.3 the four definers of the cluster described there are best understood from conceptualizing the neighborhood with *low* scores on them, i.e., with a low percent voting for the measures. Such a generally low-scoring neighborhood is against (freely interpreting the names of the variables in the figure) fringe benefits for city hospital employees, against aid to needy aged, for the Republican for governor,

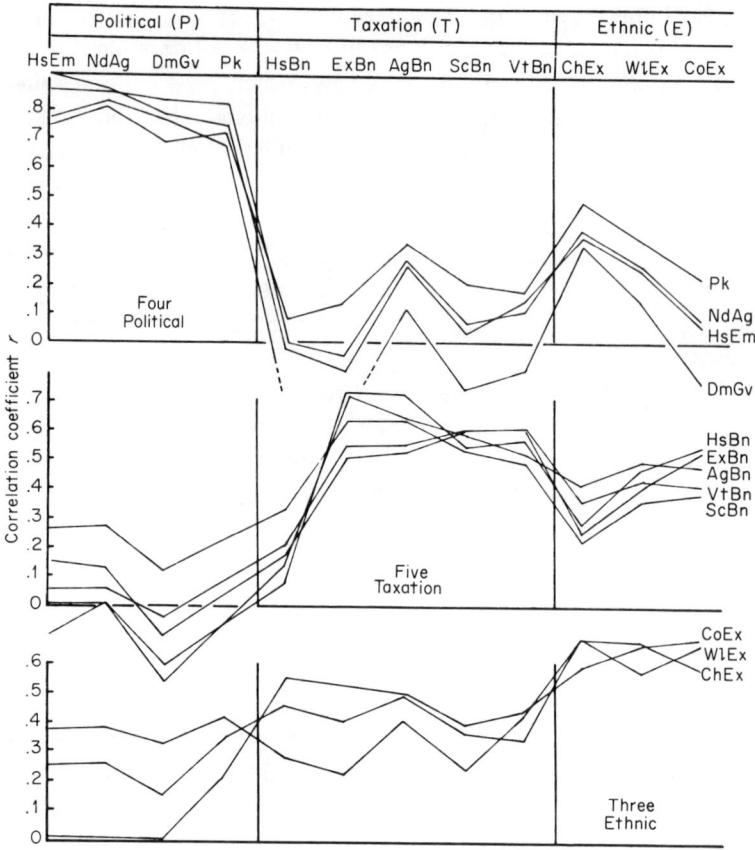

FIGURE 4.3
Collinearity of empirical clusters of election measures in the study of voting attitudes of precincts.

and against the state's going into the business of building parking facilities. The general attitude domain sampled by these four definers is quite clearly the dimension of liberal vs. conservative, Democratic vs. Republican, statism vs. rugged individualism. This cluster deserves the label "political."

The middle chart of profiles includes five bond issues for certain community enterprises. Since these enterprises were to be funded from property taxes, those neighborhoods with a higher vote in favor of them are neighborhoods of people readier to increase taxation for them. Those opposed are less enthusiastic and are composed more heavily of property owners. This is the property taxation cluster.

The bottom chart includes a cluster with three definers, high scores on which are earned by neighborhoods that favor exempting various institutions from certain taxes: churches, welfare institutions, and colleges

TABLE 4.4 DEFINING VARIABLES (ELECTION
MEASURES) OF THE EMPIRICAL CLUSTERS OF THE
VOTING–ATTITUDE STUDY OF PRECINCTS (GROUP
DIFFERENCES)

Political, P	
HsEm	Hospital employee benefits
NdAg	Needy aged aid
DmGv	Democrat for governor
Pk	Parking facilities by state

Taxation, T	
HsBn	Hospital bonds
ExBn	Exhibition hall bond
AgBn	Aged home bonds
ScBn	School bond
VtBn	Veterans' bonds

Ethnic, E	
ChEx	Church tax exempt
WlEx	Welfare institutions (religious, etc.) tax exempt
CoEx	College tax exempt

that are not already exempted, e.g., parochial schools. The strong inference here is that the attitude domain sampled by these three measures reflects a dimension that at one end favors ethnic minorities and at the other end is unsympathetic to them. This appears to be an ethnic cluster.

Rational composites and empirical clusters as domain samplings

In the Holzinger-Swineford study each of the rational categories of intellectual characteristics—spatial, verbal, speed, and memory—demark four rather large domains from which the selected tests are in each case only one finite sampling. For the spatial category, for example, the investigators could have utilized other spatial tests that could have been just as acceptable as definers of this domain as the four actually used. Also, in the verbal domain, in place of the five actually used, the investigators could have selected another set from the plethora of tests that psychologists have invented and called tests of verbal ability. Similarly for speed and memory abilities, there is a vast domain of possible tests in each rational group.

In each category not only are the tests seen to be samples from a rational domain, but from Fig. 4.1 virtually all of them are also seen to be, in addition, samplings from a collinear domain, more simply known as a cluster domain or, in orthodox factor analysis, a factor. With time, zeal, and a large budget one would be able to augment each of those categories with tests which, in addition to rational similarity, would have the same correlation pattern as those actually used in the study.

Validity of a cluster score as a measure of a domain

The point is emphasized that test samples are drawn from large domains, rational, or collinear, or both, in order to accent the fact that any composite score of a subject on a finite sample is *not* necessarily the exact score he would earn if it were based on a more extensive set of equally acceptable test samples drawn from a domain. A subject's observed, fallible cluster score inevitably suffers from limitations of domain sampling. Scores cannot be taken at face value: one must know how much rational or cluster-composite scores suffer from this type of sampling error. The degree of the limitation is quantified by the value of the correlation coefficient of the observed scores with domain scores that would be earned by the subjects on an indefinitely large battery of tests, all equally representative of the domain. Such a correlation coefficient is called the "domain validity coefficient" of the observed score. The expression "validity" carries its usual psychometric meaning, namely, the degree to which individual differences in fallible scores reflect individual differences in "true" scores—in this case hypothetical scores made by the subjects on an indefinitely large battery of tests drawn from the given domain. In orthodox factor analysis, the expression "accuracy of factor estimates" is also used in place of "domain validity of a cluster or factor score" (Harman, 1967, pp. 341ff) though the two expressions mean precisely the same thing.

How can such a validity coefficient be calculated, seeing that it is impossible to expose a subject to an indefinitely large battery of tests? Actually it cannot, but it can be estimated from available knowledge of the intercorrelations between observed definers of the cluster. The estimation formula is developed in Chap. 12. No assumptions are involved in the formula, only the definition of the score on a domain, or factor, as being composited from scores on many variables collinear with the existing lot. The relative contribution of each definer of a cluster to the validity coefficient of the composite score is indicated by the size of the definer's communality, an important index that is the main topic of the next chapter. To secure good estimates of the communalities of the variables in order to find their contributions to validity is one of the reasons why cluster and factor analysis is preoccupied by the communality problem.

The domain validities of observed composites, whether rational or cluster, are computed routinely in the Cluster Structure Analysis (CSA) program of the BC TRY System. In the Holzinger-Swineford study, the validity coefficients of the four rational categories, consisting of their full sets of definers, are listed in the first row of Table 4.5. The second row gives the validities of the most collinear set, i.e., after deleting the least collinear single variable from each category (these are the variables graphed in broken lines in Fig. 4.1).

TABLE 4.5 DOMAIN VALIDITIES AND RELIABILITY COEFFICIENTS OF COMPOSITE SCORES IN THE HOLZINGER–SWINEFORD STUDY

	Spatial	Verbal	Speed	Memory
Domain validities:				
Full sample	.83	.96	.91	.88
Most collinear set	.82	.96	.88	.88
Reliability coefficients (internal consistencies):				
Full sample	.69	.92	.83	.77
Most collinear set	.67	.92	.77	.77

Note that in Table 4.5 the domain validity of cluster scores based on the five verbal tests is the very high value of .96, signifying that individual differences among the subjects in their cluster scores would match individual differences among them almost exactly if they were to be exposed to the impossible hardship of taking an indefinitely large number of verbal tests collinear with the present lot. In short, it would not be necessary to expose subjects to such a hardship: the results on the present verbal battery of five tests are approximately equal to those which would be earned on a full domain.

In contrast, the composite scores of the subjects on the four spatial tests are a bit short on validity, their coefficient being only .83. It would be necessary to administer to the subjects additional spatial tests collinear with the present set of four if a really valid representation of individual differences was wanted.

One advantage of knowing the degree of collinearity of the different definers of a domain is that one can often weed out the least collinear without loss of validity of the total score. To illustrate, from the second line of coefficients in Table 4.5, it can be seen that by eliminating the least collinear test in each category the validities of the total score on the smaller composites are reduced only trivially.

Reliability coefficients of rational and cluster scores

A more familiar index of how much a composite score is subject to error because of limitations of test sampling is its reliability coefficient. The reliability coefficient of a cluster score, or of any composite, is defined as its correlation with a second composite consisting of definers "strictly comparable" to the existing first set. When the strictly comparable set is defined by variables collinear with the observed definers, the correlation is called an "internal-consistency" reliability coefficient. The internal consistencies of the Holzinger-Swineford clusters are given in Table 4.5. These values are simply the squares of the validity coefficients given above them (see Chap. 12). In psychological measurement one likes composites to have reliability coefficients well up in the .90s. By this standard, it will be seen that only the verbal cluster score has satisfactory internal consistency. However, the internal-consistency coefficient is a lower bound of the reliability coefficient of a composite. If a strictly comparable set of definers of a composite is not only collinear with the existing definers but also a set of "parallel forms" or repeated measures of the existing definers, then the estimated correlation between the observed composite and the parallel composite will necessarily be higher than the internal-consistency reliability. This higher coefficient is termed the "parallel-form reliability coefficient" or the "stratified reliability" of a composite (Tryon, 1957a, eq. 32). Here are some examples. Since Holzinger and Swineford report the parallel-form reliability coefficients of the individual definers, we can compute the parallel-form reliabilities of the first three cluster composites. The respective values are .86, .93, and .93, which are quite respectable values compared to their internal consistencies of .67, .92, and .77 given in the bottom row of Table 4.5.

Notice that we have now defined a third type of domain, the stratified domain, composed in this case of an indefinitely large number of parallel forms of the defining variables of the observed composite. The stratified domain is more restricted in scope than a collinear cluster domain, which in its turn is more restricted than a rational domain. It is important to keep in mind the domain from which an observed cluster score is a sampling. For example, since the internal-consistency reliability of the spatial cluster scores is only .69, the individual differences in these scores rather poorly represent differences in scores on a domain consisting of a large number of collinear spatial tests. On the other hand, the parallel reliability of .86 signifies that the observed cluster scores are quite reliable measures of a more restricted domain composed of a large number of tests that are parallel forms of the four definers, Vis, Cub, Fbd, and Loz. The only way to get a cluster score that is a good measure of a space cluster or factor not confined to measures of these restricted four space types is to add more collinear diversified tests to the observed cluster sample, enough to push the internal consistency of a new composite well up beyond .90.

Correlations of a cluster domain with the individual variables (factor coefficients)

It is helpful to estimate the degree of correlation of each variable of a study with the clusters (or factors) in the study. This estimate is known as a "factor coefficient" or "factor loading" (since they were first used in connection with the procedures of factoring, to be treated in a later chapter). For any given cluster domain the array of its factor coefficients with the full set of variables is a simple linear function of the mean correlation of its definers with the variables (see Chap. 12). We do not really have to look at the exact values of these factor coefficients because the mean correlation profiles graphically display monotonically the factor coefficients of, say, all the 19 variables on each of the four cluster domains of the Holzinger-Swineford study in Fig. 4.1. In short, looking once again at the mean rise and fall of the profile of correlations with the various tests will help build up a conception of each cluster domain. Thus, for the spatial domain, its factor correlations on the five verbal tests are only a shade lower than on its own four spatial definers. This fact leads to the possible conclusion that verbal components, and not "pure" spatial dispositions, may be utilized by the children in solving the problems in the four space tests. Said another way, the position can be taken that the spatial domain or factor appears in some way to involve verbal elements. Following this line of analysis for each of the domains will give considerable insight into the rational construction of the four general attributes measured in the Holzinger-Swineford study. Since the factor coefficients are more systematically considered in a later chapter on cluster structure analysis, we do not treat them in detail here.

Correlations between the cluster domains

More globally, the rational character of each cluster domain can be better understood if its correlations with other clusters can be estimated free from limitations of sampling. Such correlations are termed in orthodox factor analysis "correlations between factors." Their relative magnitude can be inferred from the general level of the correlation profiles between the definers of the different clusters. For example, in Fig. 4.1 the sets of definers of all four clusters generally have nonzero correlations at about the same positive level, the actual correlations between the domains all being of the general order of .50 or .60. This general positive level of correlation, termed g in orthodox factor analysis, means that the domains of space, verbal, speed, and memory appear mutually to embrace to some extent the same welter of causes of individual differences among the subjects. Contrast this finding with that in the social-area problem, where from Fig. 4.2 it is obvious that whereas the domains of socioeconomic level and family life appear to be expressions of independent causal matrices, the socioeconomic and assimilation domains do, on the other hand, seem to share some common elements or source of differences among neighborhoods.

Chapter 5

COMMUNALITIES OF THE VARIABLES

The communality of a variable is a number between .00 and 1.00 that measures the generalities of individual differences in the variable. The correlation profiles of the five verbal tests of the Holzinger study in Fig. 4.1 indicate that their communalities should be rather high because of the relatively high correlations between the variables and because of their generally positive correlation with all the other variables of the study. This pattern of correlations means that how any one of these tests ranks individuals is general; i.e., the ordering of individuals on the variable resembles the ordering on the other four verbal tests. The communalities of these verbal tests is of the order .70, except for the one test, word classification, which has a lower correlation profile than the others, on the order of .50.

The fact that these communalities are not 1.00 signifies that the five tests sample other kinds of variation besides the common verbal domain they jointly sample. The gap between .70 and 1.00, namely, .30, represents the "uniqueness" of each, i.e., variation that is not common to them. About three-tenths of the variation among individuals is not shared with any of the other variables but is a unique property of each; about five-tenths for the word classification test.

Definition of the communality of a variable

There are two alternative definitions of the communality h^2 of an observed variable. Both definitions are useful. In order to define its communality we hypothesize that the observed variable is only one of many possible definers of a collinear domain of variables—like the different definers of each of the four clusters of the Holzinger study. Specifically, the domain score D_i for the observed variable V_i is the hypothetical composite

$$D_i = V_i + V_i^{(1)} + V_i^{(2)} + \cdots + V_i^{(\infty)} \tag{5.1}$$

where superscripted V's are hypothetical values on many other variables collinear with the observed variable. Thus, a domain score is a hypothetical construct, but the figures of the last chapter demonstrate that domains of collinear variables are not only possible but common.

Communality as the common variance, or predictable variance, of a variable

The first definition of communality is the squared correlation of the variable and the domain of the variable

$$h_i^2 = r_{V_i D_i}^2 \tag{5.2}$$

The square of a correlation between any two variables is called the "index of determination" of one of the variables by the other; it is the proportion of variance of one of them predicted from the other. Therefore, by Eq. (5.2), the communality is that portion of the variance in the observed variable which can be predicted from the general domain. In the case of the verbal tests, the communality of .70 means therefore that about 70 percent of their observed variance is general or common variance, a variance shared with, and predictable from, a general attribute sampled by other variables. The variance of the verbal tests that is not common, i.e., residual, is about 30 percent of their observed variances. It is symbolized as u_i^2, which is

$$u_i^2 = 1 - h_i^2 \tag{5.3}$$

This so-called "unique variance" is to be explained by components that are not shared by any of the other variables of the study.

Communality as a correlation coefficient with a single collinear variable

The second definition of the communality of a variable is more down to earth. It is a special kind of correlation coefficient, namely, the correlation of the observed variable with another set of scores on a second hypothetical variable V' that is exactly collinear with V. Thus,

$$h_i^2 = r_{V_i V_i'} \tag{5.4}$$

where the correlations of the second variable V' are perfectly collinear with the observed array of correlations.

    Table 5.1 demonstrates with data from the Holzinger, social-area, and voting-attitude studies that variables commonly do have at least one other "reference" variable in the study with which they are highly collinear. The column headed Holzinger, for example, gives the relevant facts for each of the full set of 24 variables of the Holzinger problem. Opposite the ordinal number of each of these 24 variables is shown its highest P^2 value with any one of the other 23 variables. For example, the highest P^2 of variable 1 with another variable is .97. Note that for every one of the 24 variables the highest P^2 is .90 or more. The average, shown at the foot of the table, is .95. And similarly, for the full set of 33 variables in the social-area study: their highest collinearities average .93.

Where and how communalities are used in cluster and factor analysis

The need for an accurate determination of the communality of each variable in cluster or factor analysis can be demonstrated by a brief summary of the places in which it is used.

TABLE 5.1 IN THREE STUDIES, THE HIGHEST VALUE OF COLLINEARITY P^2 OF EACH VARIABLE WITH A REFERENCE VARIABLE

Vari-able	Study Holzinger	Social	Voting	Vari-able	Study Holzinger	Social	Voting
1	.97	.99	.94	18	.91	.97	.97
2	.96	.97	.95	19	.91	.97	.95
3	.94	.97	.95	20	.97	.97	.95
4	.97	.98	.93	21	.95	.97	.95
5	.98	.95	.99	22	.97	.95	.98
6	.99	.97	.94	23	.97	.96	.98
7	.98	.92	.92	24	.93	.95	.98
8	.97	.97	.93	25		.94	.98
9	.99	.92	.98	26		.99	.93
10	.90	.92	.91	27		.93	.96
11	.93	.89	.97	28		.94	.98
12	.90	.93	.93	29		.92	.99
13	.95	.97	.95	30		.91	.99
14	.94	.95	.98	31		.73	.96
15	.94	.96	.90	32		.68	
16	.94	.96	.94	33		.81	
17	.91	.92	.93	Mean	.95	.93	.95

Communality as an index of
the generality of a variable

Since communality describes the generality of individual differences in a variable, and since the main objective is to form composites of variables that measure general ways of ranking individual differences, it becomes important to discover from a variable's communality how much the variable can contribute to the formation of general clusters. In the first place a variable's usefulness as a definer in a cluster depends on the size of its communality, since its contribution to the size of the domain validity of the cluster and to the internal-consistency reliability of the cluster is a sensitive function of the size of its communality (see Chap. 12).

The magnitude of the communality of a variable is a function of its correlations with other variables of the study, and if these correlations are .00, the communality is .00; the variable will have factor coefficients of .00 with all clusters and factors and should therefore be discarded. In conducting a study that encompasses a very large number of variables (as in BIGNV analysis), the first step is therefore to estimate the communalities of all the variables and delete those which have trivial values. Discarding a variable means not that it does not have important unique variance but that it is irrelevant to the objective of cluster and factor analysis, which is to discover general attributes, not unique characteristics.

Communality as an indication of how much variance
is to be described by a cluster or factor analysis

Since our exclusive interest is in general variance, we are really concerned only with that portion of variance which is common variance. Thus, when the factor coefficient of a variable is low relative to its communality, we know that other clusters or factors are important correlates of the variable.

The contribution of each cluster or factor to the variance of a variable is discovered by the process of "augmenting" its factor coefficients, a matter we consider in Chap. 6 and one that requires knowing the value of its communality. Suffice it here to say that from the augmented factor coefficient we can assess the proportionate degree to which a cluster domain or factor determines the common variance. Indeed, it is standard practice to deal exclusively with augmented factor coefficients in some analyses where one ignores the unique variance of all the variables and deals only with this common factor variance.

Communality as an index of sufficiency (reproducibility) in factoring

In Chap. 6 we consider the matter of how many clusters or factors are needed, i.e., the sufficient number of composites to which one may wish to reduce all the variables. At this point, however, it should be evident that one would not want more clusters or factors than the number minimally sufficient to reproduce or account for all the common variances of the variables. The successive dimensions derived from factoring the correlation matrix "take out" decreasing proportions of the total variances of all the variables. Thus, before the factoring procedure is undertaken, one first estimates the communalities of all the variables and sets up the overall sum of the communalities across all variables, that is, Σh^2, as a criterion of when to stop factoring. To use this criterion, one computes after each dimension the cumulative proportion of Σh^2 accounted for up through that dimension. When a salient portion, say 95 percent or so, of the estimated communalities has been reproduced, or exhausted, the factoring is terminated.

Communality as a diagonal value in the correlation matrix

For each variable the correlation matrix gives the array of its correlations with the other $n - 1$ variables. The cell on the diagonal should contain the variable's "self-correlation," a term that does not mean the correlation of the scores with a duplicate series but its correlation with another variable that is somehow a replica of it. Such another variable is one exactly collinear with it, a variable that samples exactly the same general attributes that it does. That correlation is, by definition, $r_{V_i V_i'}$ or its communality by Eq. (5.4).

Estimating the communality of a variable

Like other parameters, such as population means, standard deviations, and correlation coefficients, direct calculation of the communality of a variable V_i is not possible. Since we do not have its domain score D_i on individuals, we cannot compute the communality by Eq. (5.2). Nor do we have a perfect second construct variable V_i' that is exactly collinear with it; therefore we cannot accurately compute the communality by the second equation, (5.4). The best we can do is to compute an approximation.

There has been a long disputatious history in factor analysis over the "problem" of how to estimate the "unknown communalities" (Harman, 1967, chap. 5; Thurstone, 1947, chap. 13). But since the concept is quite unambiguous on domain-sampling principles (Tryon, 1957b), Eqs. (5.2) and (5.4), the matter reduces simply to using methods of solving for the communality estimates that come closest to satisfying either Eq. (5.2) or (5.4).

No attempt is made here to present the detailed computing formulas used in estimating communalities. The derivative logic and formulas are presented in Chap. 12. The intention here is to show only how closely the estimates approximate good values.

There are 11 different methods of securing approximations to the communalities by using programs of the BC TRY System. They are summarized in Table 5.2, where the communalities of the 24 variables of the Holzinger study as approximated by each method are listed. The 11 methods are grouped into four main classes: those which compute estimates from (1) a single reference variable, from (2) subsets of reference variables, from (3) the full set of all the other $n - 1$ variables of the study, and (4) from the factoring procedures described in Chap. 6.

The most obvious conclusion to be drawn from an inspection of Table 5.2 is that whichever method is used, the value of the estimate of the communality of a variable is much the same. One may well ask: What has all the controversy been about? The main reason for it is that until the methods could be programmed on the computer, few students of the problem had the patience and budget to try out all the methods on a wide variety of different problems to see how closely they matched. To compute and carefully check the results in Table 5.2 would have taken weeks on a desk calculator: it now takes less than 6 min on the computer.

Now let us look at Table 5.2 with the aim of finally deciding which method gives the *best* approximations to the communalities. In order to make that decision we need to know the "true" communalities of the variables. With them at hand, we can find out which method gives the closest unbiased estimates of the true values. Actually, with real data there are no such values as the "true" communalities of the variables—except in artificial problems (see Tryon, 1957b). The situation is like asking: What is the "true" upper threshold of sensing auditory intensities? Even if the most elaborate, highly controlled instrumentation were used to measure a subject's upper limit of hearing on different experimental sessions, the results would vary. However, we might take the "true" limit to be the average of the separate limits. Similarly, we take as the "true" communality of each of the 24 tests its average value over the 10 independent estimates (the method called "MOD B" is not included because it is itself an average of two other estimates in the table).

TABLE 5.2 COMMUNALITIES BY DIFFERENT METHODS IN THE HOLZINGER STUDY

Tests		Single Reference Variable		Subset of Reference Variables				$n-1$ Reference Variables		Factoring			h^2 Average
		COLINR DVP60, COR2	HIGHR DVP50	APROXB DVP40	MOD B DVP41	PF(4) DVP60	PF(9) DVP60	SMR SMIS	QF DVP30	CLUST CC5	CENT CC5	PRIN FALS	AVCOM
F1	Vis	.45	.49	.39	.44	.47	.48	.49	.41	.42	.47	.50	.46
F2	Cub	.42	.42	.23	.33	.30	.28	.31	.35	.30	.24	.27	.31
F3	Fbd	.37	.39	.20	.30	.30	.36	.32	.37	.26	.29	.26	.31
F4	Loz	.45	.45	.49	.47	.48	.49	.45	.54	.52	.50	.48	.49
V5	Inf	.65	.72	.63	.67	.64	.67	.65	.75	.66	.63	.65	.67
V6	Cmp	.71	.72	.68	.70	.67	.65	.67	.54	.67	.67	.65	.66
V7	Snt	.69	.72	.69	.70	.66	.69	.68	.78	.70	.73	.71	.71
V8	Wcl	.57	.63	.50	.56	.59	.55	.54	.49	.49	.52	.52	.54
V9	Wmn	.71	.72	.75	.74	.74	.75	.69	.68	.76	.71	.75	.73
S10	Add	.59	.59	.67	.63	.59	.66	.55	.77	.71	.67	.69	.65
S11	Cod	.41	.53	.50	.51	.44	.50	.50	.60	.43	.42	.45	.48
S12	Cnt	.59	.59	.55	.57	.53	.66	.55	.68	.59	.64	.57	.60
S13	Scc	.43	.53	.45	.49	.49	.68	.54	.68	.61	.54	.60	.56
M14	Wrg	.39	.39	.43	.41	.39	.44	.36	.33	.40	.32	.35	.38
M15	Nrg	.39	.39	.33	.36	.31	.30	.28	.26	.36	.25	.29	.32
M16	Frg	.31	.42	.33	.37	.39	.38	.43	.35	.48	.45	.42	.40
M17	Wn	.35	.45	.46	.46	.36	.51	.42	.43	.43	.51	.50	.44
M18	Nf	.36	.45	.37	.41	.34	.38	.46	.42	.46	.43	.40	.41
M19	Fw	.37	.37	.24	.31	.28	.29	.31	.25	.19	.28	.24	.28
N20	Ded	.51	.51	.49	.50	.45	.48	.47	.53	.40	.41	.42	.47
N21	Puz	.43	.45	.39	.42	.41	.40	.48	.49	.44	.42	.43	.43
N22	Rsn	.51	.51	.42	.46	.45	.51	.46	.55	.39	.39	.42	.46
N23	Ser	.51	.51	.58	.55	.56	.56	.54	.50	.49	.49	.50	.52
N24	Ari	.41	.53	.50	.51	.45	.44	.53	.41	.48	.53	.49	.48

Method of Computing Communalities

The average communalities are listed in the last column of Table 5.2. The decision of which of the 11 methods gives the closest approximation to the true values is based on discovering the degree to which each column of estimates in the table matches the average values in the last column. It is not sufficient simply to find the product moment correlation coefficient between each column of estimates and the column of true values because if the relation were in fact curvilinear, the correlation would underestimate the relationship. Furthermore, for a given method it is possible for the correlation to be close to 1.00, and the absolute matching of estimated values with the true values could be seriously biased, e.g., by being constant underestimates or overestimates of the average values or by systematic compensatory underestimates at one level of average values and overestimates at another. We need a method of assessing not only the degree of correlation between estimates and the average values but also of the extent of constant or differential bias in the estimates. Such information is revealed (1) by computing and comparing the Pearson correlation coefficient r and the correlation ratio η in order to assess linearity and (2), to assess bias, by calculating the regression line of estimated values on average values. Furthermore, we would not want to put all our eggs in the Holzinger-study basket but should include in this appraisal the findings from the other two studies.

The results of testing each method to determine the degree of its matching of estimates to average values for the Holzinger, social-area, and voting-attitude studies are condensed in Table 5.3. Each value in Table 5.3 is the average of three values from separate analyses of the three groups (see the technical note below).

First of all, note that the 11 methods are bracketed under the four main classes, referring to the major kinds of computing formulas designed to estimate communalities, as in Table 5.2. For example, the first one includes two methods of estimating communality from a single reference variable, clearly referring to Eq. (5.4), namely, estimating h_i^2 from $r_{V_i V_i'}^2$, its correlation with a second variable collinear with it. The second class includes four methods of estimation from a subset of the most collinear reference variables. In this case, the intent is to designate a domain D_i of collinear variables from which a variable's communality can be estimated by $r_{V_i D_i}^2$, from Eq. (5.2). The third group, of two methods, also stems from Eq. (5.2), the intention being, by multiple correlation procedures, to devise a best-weighted composite of all $n - 1$ variables that will best predict a collinear domain from which $pari$ $passu$ the variable's communality can be estimated under Eq. (5.2). The fourth group, of three methods, is an interesting development from Eq. (5.4). For each variable a second hypo-

TABLE 5.3 ESTIMATION OF THE AVERAGE COMMUNALITIES OF THE VARIABLES BY
11 DIFFERENT METHODS IN THE HOLZINGER, SOCIAL–AREA, AND VOTING STUDIES

Method of Estimating the Communality h^2	Correlation between the Average and Estimated h^2 Values		Bias in Estimated h^2			$\Sigma\bar{d}$
			Mean Deviation d of Estimated h^2 from an Average h^2 Value of			
	r	η	.00	.50	1.00	
From a single reference variable:						
1. r with most collinear variable	.80	.87	.16	.03	− .07	.26
2. Highest r	.96	.97	.17	.06	− .05	.28
From a subset of the most collinear variables:						
3. Approximation B	.94	.96	− .15	− .07	.00	.22
4. Modified approximation B	.97	.98	.01	− .01	− .02	.04
5. Proportional fit on four reference variables	.97	.98	− .10	− .05	.00	.25
6. Proportional fit on nine reference variables	.97	.98	− .01	.01	.03	.05
From the $n - 1$ other variables of the study:						
7. Squared multiple R	.95	.96	.18	.08	− .02	.28
8. Quadratic formula	.92	.94	.02	.04	.05	.21
From factoring:						
9. Key cluster	.96	.97	− .13	− .04	.03	.20
10. Centroid	.97	.98	− .06	− .02	.01	.09
11. Principal axes	.98	.98	− .07	− .02	.02	.11

thetical collinear variable V' is constructed such that it has exactly the same factor coefficients on each of the factored dimensions as its mate, whence its communality is solved for directly by Eq. (5.4), that is, $r_{V_iV_{i'}}$.

For each method reported in Table 5.3 the first two columns give the average correlation between its estimates and the average values over the three problems. Except for method 1, the linear correlations are above .90, most being around .95 or higher. There is no curvilinear relation between estimates and average values; the values for the correlation ratio η are only trivially higher than their corresponding linear correlations. In sum, each method except the first produces ranked communalities of the variables closely corresponding to the rank orders of the average communalities.

The story is somewhat different with respect to bias, given in the last four columns of Table 5.3. Looking at the first column of biases, the case where the average communality is the value .00, we see that methods 1, 2, and 7 give communalities whose average deviation $\bar{d}$ from the average value of .00 is at least .16. These are methods that overestimate communalities. Methods that show negative deviations are those which underestimate the average communalities. In the second and third columns of biases, for average communalities of .50 and 1.00, over- and underestimation is generally less. Finally, the last column gives a numerical statement of overall bias, consisting of the sum of deviations at .00, .50, and 1.00, disregarding sign. Without going into detail, note that methods 1, 2, 3, 5, 7, etc., do poorly overall, but two of the methods, 4 and 6, are virtually without bias at any of the three levels of communality, low, middle, or high.

The method that wins the competition is the one that produces communality estimates that match best, in absolute magnitudes, the average communalities. There are, in fact, two of them: method 4, modified approximation B, or MOD B, and method 6, proportional fit on nine reference variables, or PF(9). Of these two, MOD B is superior in the practical sense of taking much less computer time to compute than PF, a matter of some moment in problems, say, involving 120 variables. Note that both methods produce estimates that correlate .97 with the average values but in addition are almost without bias, giving absolute values almost identical with the average. Method 1, highest r, is excellent for ranking variables by communality, but to secure good absolute values requires correcting for a severe overestimate in the lower ranges of communalities.

Technical note: method of computing relations between estimates and average communalities by RSCAT

The values of Table 5.3 are averages computed from the 11 correlation scattergrams of each group. These scattergrams give the relationship between the average values and the 11 estimates for each group. For example, in the Holzinger study the estimates and average values, as given in Table 5.2, were input as a data matrix to DAP2, the columns being treated as variables, rows as objects. The correlation-scattergram program, RSCAT, was then called, wherein the option was taken to compute all scattergrams between a "common variable," the average values, and the 11 other variables. Each scattergram gives both r and the η's, and the constants of the regression line of average values on the estimates, namely, the slope b and the intercept a at the value of .00. Thus, a is the estimate at a value of .00, $.50b + a$ at the value of .50, and $b + a$ at the value of 1.00. These values are those given in Table 5.3. Both the linear and curvilinear regression lines themselves are provided in RSCAT, thus also permitting a visual check on linearity.

Chapter 6

DISCOVERING SALIENT GENERAL DIMENSIONS BY
KEY-CLUSTER FACTORING

In this chapter we discuss means of discovering collinear clusters among a set of variables. The methods are purely objective, applicable by way of a computer procedure but capable of discovering the well-defined, meaningful clusters introduced in Chap. 4.

There are three primary objectives in key-cluster analysis: (1) to select from the variables of a study the mutually collinear variables defining each of the clusters; (2) to discover the minimal or salient number k of clusters sufficient to reproduce all the intercorrelations among variables and the communalities of the variables; and (3) when there are more clusters of variables than the minimally sufficient number, to provide information with which to select the most nearly independent (uncorrelated) clusters on which to score individuals in object cluster analysis, scores that will maximally differentiate the score profiles among the individuals.

Key-cluster factoring: a special case of the general principles of factoring (independent dimensional analysis)

The term "factor analysis" comes from the procedures of factoring, of which there are many variants. The different types of factoring are, however, all special cases of the general method of independent dimensional analysis (Tryon, 1959). Procedurally, in principle, the simple logic of the general method is as follows. Starting with the matrix of the correlations of the variables, with the communalities in the principal diagonal cells, the first dimension, or factor, is defined as a weighted composite of the variables. This composite is taken to be either a subset of them or the total set of them with a designated pattern of weights attached to the variables. Then, by a special form of adjustment to be described later, scores on this first factor composite are partialed out of each of the correlations in the matrix. The resulting matrix is called the "first factor residual matrix." A second dimension is then defined by a newly weighted composite of the variables, scores on which are partialed out of the first residual matrix, the result of which forms a second residual matrix. The procedure is continued until a final residual matrix is formed of entries that are of trivial magnitudes.

A matrix of trivial residuals is one in which the values are so small that they are not considered worth defining a new dimension on. In most problems there is not much difficulty in deciding on when to stop factoring by inspecting the residuals. Thus, in the Holzinger problem, after four dimensions are extracted, even the largest of the fourth factor residuals would not lead to a fifth dimension on which one would reasonably wish to score the subjects.

Another indication of how many dimensions to retain is simpler than scanning a large matrix of residuals: the proportion of the communalities accountable from scores on the dimensions provides a criterion of the sufficiency of the dimensions. The communalities represent the amount of common variance among the variables. As factoring proceeds, the portion of the variables' communalities accounted for by each dimension is computed. When a salient amount, say around 95 percent, is accounted for, it is usually not fruitful to continue factoring because any additional single dimensions would account for less than about 5 percent of the communalities of the variables. Such a dimension would have very narrow generality. In the key-cluster factoring program CC5 of the BC TRY System, called cumulative communality key-cluster analysis or, more simply, CC analysis, this statistic is kept track of. It has been programmed to quit factoring when 97.5 percent of the sum of all estimated communalities has been exhausted (although this criterion is under the control of the user of the program).

Key-cluster factoring is a special case of general factoring principles. Each dimension in key-cluster factoring is defined by a collinear subset of variables. Well-known other special cases (to be treated in a later chapter) are centroid and principal-axes factoring, which are alike in defining each dimension by the total set of all n variables with a specially selected pattern of weights attached to them. The methods differ from each other only with respect to how the weights are computed—centroid by simple sums, principal-axes by least-squares weighting. One of the oldest special cases is "two-factor" or "bifactor" analysis, in which the first dimension is defined by the total set of variables whereas the later residual dimensions are defined by subsets. Still another special case is that in which each dimension is defined by a single variable, a form variously known as "square root," "diagonal," or "pivot-variable" factoring. Confusion caused by this plethora of factoring methods can be avoided by remembering they are all merely special cases of the general principles stated above: they are simply variations on the same general theme.

The decision pattern of key-cluster factoring

The general method of multidimensional analysis can be thought of as consisting of eight "regions of decision." It is the pattern of special decisions across these eight regions that delineates a particular factoring method. We illustrate here the key-cluster pattern, making it concrete by showing the application of it to the Holzinger study. To demonstrate clearly how widely applicable key-cluster factoring is, we will show how it works out on the social-area and voting-attitude studies.

The Holzinger study

Reproduction of intercorrelations

The sufficiency of the factored dimensions is revealed, in one aspect, by how well they reproduce the correlation matrix. They are deemed sufficient as soon as a residual matrix contains trivial values. When the correlation of two variables is accounted for by a set of factor or cluster dimensions, i.e., reproducible in terms of those dimensions, the residual correlations after those dimensions are removed from the matrix will be zero. Just how residuals are specifically formulated will be considered later, but at this point we can assess how well the original correlations of the Holzinger problem are reproduced by noting the size of residuals after successive dimensions. These are shown in Table 6.1, Sec. B. The overall statistic that describes the average magnitude of the elements in a correlation matrix is the root mean square (RMS), secured by squaring each element in the

TABLE 6.1 SUFFICIENCY OF KEY–CLUSTER DIMENSIONS (FACTORS)
IN THE HOLZINGER STUDY OF ABILITIES

A. Reproduction of Communalities during First Factoring

	Verbal C_1	Speed C_2	Spatial C_3	Memory C_4
Proportion of sum of estimated communalities, each dimension	.49	.23	.16	.10
Cumulative proportion of communalities from dimension C_1 to C_4	.49	.72	.88	.98

B. Reproduction of Original Correlations after Fourth Iterated Solution

	Original	Residuals after			
		C_1	C_2	C_3	C_4
RMS	.33	.16	.10	.06	.04

matrix and finding the square root of the mean of all these squares. The RMS may be thought of as a slightly upward-biased estimate of the average of all the values, disregarding algebraic sign.

In Table 6.1, Sec. B, it is shown that for the 276 original correlations the RMS is .33. Looking now at the mean size of the elements in the matrix after extracting the first dimension, C_1, we see that the RMS of the first residuals is .16. The general combined sufficiency of four dimensions is revealed by the fact that after the fourth dimension, C_4, the fourth residuals have an RMS of only .04. This overall triviality does not necessarily mean that there might not be some residuals somewhat higher than .04. We consider this possibility later when we look at the distribution of residuals.

Reproduction of communalities

In Chap. 5 we emphasized that the main indicator of the generality of individual differences in each variable is its communality, namely, the portion of its variance that is predictable from the other variables of the study. Before factoring the matrix, we can make an estimate of the communalities of the variables and use the sum of these estimates as a target, an amount to be reproduced by a sufficient number of dimensions. As we take out one dimension at a time, we calculate the proportion of this total amount of variance that has been accounted for, up to and including each dimension, and stop factoring when no salient amount is left to factor.

The results are given in Table 6.1, Sec. A. The first dimension, C_1, defined as "verbal," takes out about half (49 percent) of the overall communalities, and the fourth, C_4, takes out only 10 percent. The next row of the table, giving the cumulative communalities accounted for up to each dimension, shows that by the fourth dimension 98 percent of the communalities has been accounted for and only 2 percent is left over for any additional dimensions. Such a residual communality seems hardly worth devoting one or more additional dimensions to.

The defining variables of each dimension

In key-cluster factoring we choose as definers of each of the dimensions the most collinear subset of variables that is also most nearly independent of the definers of other dimensions. Since the derivative basis of the objective procedures for locating such a cluster is given in Chap. 12, we sketch here only the simple logic of the procedures.

To locate the definers of the first dimension that are likely to be most nearly independent of succeeding clusters, we first look for a defining variable, called the "pivot variable" of the clusters, that meets the following conditions: (1) it must be relatively highly correlated with some other variables, in particular with additional definers-to-be of the cluster, and (2) at the same time it must tend to be uncorrelated with other variables, in particular, with definers-to-be of other clusters. Clearly such a pivot variable is one whose column in the correlation matrix has both high values in it and low values in it—in short, the variance of the absolute magnitudes or squares of its correlations is relatively large. From this reasoning, we therefore set up as an index of "pivotness" the variance of the squared correlations in each column of the correlation matrix. We arbitrarily accept that variable with the highest variance as the pivot variable around which to build the cluster that will define the first dimension.

Which additional variables should be included in the defining cluster along with the pivot variable? Obviously they should be those variables that are most collinear with it and with each other, as measured by their indexes of collinearity P^2. An objective criterion that has been developed and programmed into BC TRY sets three conditions for accepting a variable as a definer of the cluster: (1) it must show a relatively high average degree of collinearity with definers already accepted, a condition met only if the mean of its P^2 values with them is high, say not less than .81; (2) it must show substantial collinearity with each one of them, a condition met only if *all* its P^2 values are above a lower bound, say .40; and (3), the addition of it must preserve the "tightness" of the cluster, a condition met only if all its P^2 values with them lie within, say, twice the range of two P^2's of the pivot variable, namely, the value of the difference between its P^2 with the first definer added to the cluster and its P^2 with the last previous definer added. Note that there are three standard arbitrary bounds built into this programmed criterion of mutual collinearity. If a user of the BC TRY programs wants to set higher or lower bounds than the standard values, he can always opt to do so in using the programs.

At times the arbitrary index of pivotness may select a pivot variable for which one cannot find additional definers that meet the conditions of mutual collinearity. We have therefore built into our factoring program a trial-and-error search routine that looks for a pivot that can form a collinear cluster if one exists. This search procedure works as follows. If a given selected pivot fails to pick up additional definers, it is rejected. A new pivot is then selected, the variable with the second highest value of the pivotness index. If this variable should also fail, a third is tried out, and even a fourth. Four trials are taken to be enough, and if the fourth effort to find a pivot variable fails, no further search is made.

How the mutual-collinearity criterion works in the Holzinger problem

TABLE 6.2 DEFINING VARIABLES OF KEY–CLUSTER DIMENSIONS (FACTORS) IN THE HOLZINGER STUDY OF ABILITIES[a]

	Orthogonal Fc	h^2	$\bar{r}$		Orthogonal Fc	h^2	$\bar{r}$
C_1, *Verbal:*				C_3, *Spatial:*			
V9 Wmn	.86	.76	.72	F4 Loz	.64	.58	.46
V6 Cmp	.83	.69	.69	F1 Vis	.44	.38	.38
V7 Snt	.83	.69	.69	F2 Cub	.47	.30	.33
V5 Inf	.80	.65	.66	N21 Puz	.30	.46	.38
Other				Other			
V8 Wel	.68	.49	.37	N23 Ser	.39	.50	.42
				N22 Rsn	.30	.39	.35
				N20 Ded	.30	.42	.35
				F3 Fbd	.38	.28	.32
C_2, *Speed:*				C_4, *Memory:*			
S10 Add	.79	.76	.59	M14 Wrg	.56	.44	.40
S12 Cnt	.70	.56	.49	M15 Nrg	.46	.32	.34
N24 Ari	.46	.48	.46	M17 Wn	.47	.45	.38
S11 Cod	.50	.42	.43	M16 Frg	.43	.48	.38
Other				Other			
S13 Scc	.53	.54	.46	M18 Nf	.26	.43	.33
				M19 Fw	.19	.20	.25

[a] Abbreviations of test names are explained in Table 4.1.

is shown in Table 6.2. The definers of the first dimension, C_1, are the tightly collinear four verbal tests whose correlation profiles are displayed in Fig. 4.1. The fifth variable listed, V8, word classification, is not selected as a definer but is placed under the rubric "other," denoting that it is a runner-up as a definer in the sense that its correlation with the first factor (its factor coefficient, to be discussed below) is the highest it has with any of the four dimensions. For each of the definers and their runners-up in Table 6.2, we list three quantities of interest: (1) its factor correlation Fc, (2) its communality (after factoring) h^2, and (3) its mean original correlation with the selected definers $\bar{r}$. Each of the factor coefficients listed is the correlation of a variable with its dimension defined as independent of the other dimensions. (These so-called "orthogonal" coefficients are to be distinguished from oblique factor correlations, discussed in Chap. 7.)

Reflection of the defining variables of a dimension

It is not uncommon to discover that two variables selected as definers of a dimension are negatively correlated. In such cases (the Holzinger data do not provide an illustration), that variable also correlates negatively with other definers as well. Everything becomes straightened out if the offending definer is "reflected." Reflection is achieved by reversing the signs of its correlation coefficients both in its column and its row in the correlation matrix. Such an operation means changing the "direction" in which individuals have been scored on the variable: the individual that earned the highest score as originally scored is, after reflection, given the lowest, and vice versa. The result is that the name given the variable must be changed to its antonym or logical negative when interpreting its correlation coefficients with changed signs.

Sometimes reflection operations can get complicated when more than one definer needs to be reflected. Here is a real advantage of the computer as a high-speed clerk: it can execute the optimal reflection in seconds. This procedure is described in detail in Chap. 12. In general, optimal reflection means that all possible patterns of reflection of the defining variables (up to 10 definers) are tried out; for each pattern the sum of the correlations between the definers is computed. That particular reflection which yields a maximal sum of correlations between definers, as reflected, is finally retained. The result is that the definers are so reflected that one can be assured that the new directions in which the reflected variables are to be scored in a cluster score composite on individuals provide a cluster score that is rationally the most meaningful one possible.

Factor coefficients and partial communalities

The most important basic statistics computed by the factoring procedure are the factor coefficients (or factor correlations) of all the variables on all the dimensions. A factor coefficient of a variable is simply its estimated correlation with a hypothetical score on the dimension. For example, from Table 6.2, the factor coefficient .86 of V9, word meaning, on the first dimension, C_1, means that scores of individuals on this vocabulary test would correlate about .86 with a domain score on the first dimension, i.e., with a composite score on an indefinitely large number of tests perfectly collinear with the four variables that have been selected to define this first dimension. (For the logic of the factor coefficient, see Chap. 12.)

Table 6.3 gives the complete listing of the factor coefficients of all the 24 ability tests with the four dimensions. For example, V9, word meaning, correlates, as we saw above, .86 with the first dimension, of which it is a definer (see the Definers column in Table 6.3); with the remaining three dimensions its factor coefficients are trivial, namely, $-.07$, $-.01$, and .03.

Inspecting the values of the factor coefficients in the table gives one a notion of the weight of each of the dimensions in determining individual differences in each of the variables. A factor correlation is itself not the best index of the determination of individual differences by a dimension (or factor): the square of the factor correlation is the best estimate of the degree of variance predictable from the dimension. This squared factor coefficient is called the "partial communality" of the variable on the dimension. If, for example, one squares each of the four factor coefficients of word meaning, the resulting numbers give a more accurate notion of the variance in word meaning test scores determined by the successive dimensions than the unsquared coefficients would. Since the dimensions are independent, the factor coefficients are also multiple regression coefficients of the variable on the dimensions conceptualized as multiple predictors; the factoring procedures are merely a special case of multiple correlation and prediction.

The point here, however, is that the cumulated partial communalities of a variable on successive dimensions show the predictability (the squared multiple correlation) from the dimensions conceptualized as a battery of predictors. The cumulative partial communalities are given in Table 6.3 in the columns under the general heading of Communalities. Inspecting these values at any point in the factoring procedure shows how well the variance of each of the variables is predicted from the dimensions. The cumulative values on the last, or fourth dimension, C_4, are, in fact, the final estimates of the communalities of the variables from factoring, one type of com-

TABLE 6.3 FACTOR CORRELATIONS OF THE VARIABLES AND THE SUFFICIENCY OF THE FOUR DIMENSIONS TO ACCOUNT FOR COMMUNALITIES AND CORRELATIONS IN THE HOLZINGER STUDY OF ABILITIES

Tests		Definers	Orthogonal-factor Coefficients				Communalities					Correlations	
							Cumulative				Initial	Number of Residuals[a]	
			C_1	C_2	C_3	C_4	C_1	C_2	C_3	C_4		.0	.1
Spatial:													
F1	Vis	C_3	.39	.19	.44	.07	.15	.19	.38	.38	.43	23	1
F2	Cub	C_3	.26	.10	.47	−.04	.07	.08	.30	.30	.33	24	
F3	Fbd		.33	.10	.38	.12	.11	.12	.26	.28	.30	21	3
F4	Loz	C_3	.40	.07	.64	−.01	.16	.17	.58	.58	.47	23	1
Verbal:													
V5	Inf	C_1	.80	.08	.06	−.07	.64	.64	.64	.65	.67	24	
V6	Cmp	C_1	.83	−.01	.00	.08	.68	.63	.69	.69	.70	24	
V7	Snt	C_1	.83	.00	−.05	−.04	.69	.69	.69	.69	.70	24	
V8	Wcl		.68	.14	.08	.02	.47	.48	.49	.49	.56	24	
V9	Wmn	C_1	.86	−.07	−.01	.03	.75	.76	.76	.76	.74	24	
Speed:													
S10	Add	C_2	.29	.79	−.21	−.10	.08	.71	.75	.76	.63	24	2
S11	Cod	C_2	.37	.50	.06	.18	.14	.38	.39	.42	.51	22	1
S12	Cat	C_2	.19	.70	.12	−.10	.04	.33	.54	.56	.57	23	2
S13	Scc		.39	.53	.30	−.13	.15	.43	.52	.54	.49	22	
Memory:													
M14	Wrg	C_4	.31	.20	−.02	.56	.10	.14	.14	.44	.41	23	1
M15	Nrg	C_4	.25	.15	.17	.46	.06	.09	.11	.32	.36	24	
M16	Frg	C_4	.27	.20	.43	.43	.07	.11	.29	.48	.37	23	1
M17	Wn	C_4	.31	.35	−.11	.47	.10	.22	.23	.45	.46	23	1
M18	Nf		.25	.44	.33	.26	.06	.26	.37	.43	.41	23	1
M19	Fw		.29	.23	.15	.19	.08	.14	.16	.20	.31	23	1
Mathematical ability:													
N20	Ded	C_3	.52	.11	.30	.22	[a]	.29	.37	.42	.50	22	2
N21	Puz		.35	.46	.35	−.01	.27	.33	.46	.46	.42	24	
N22	Rsn		.52	.14	.30	.12	.12	.29	.38	.39	.46	23	1
N23	Ser		.53	.23	.39	.08	.27	.34	.49	.50	.55	23	1
N24	Ari	C_2	.51	.46	.03	.03	.28	.48	.48	.48	.51	22	2

[a] Tallied by first digit only.

munality estimate described in Chap. 5. This last column of values corresponds very closely to the initial estimates of communalities before factoring, listed in Table 6.3 in the column headed Communalities Initial. Indeed, as factoring proceeds, it is the overall sum of the cumulative partial communalities that is being compared with the sum of the initial estimates of the communalities. When the sums match up, factoring is terminated.

How much each dimension accounts for the communality of a variable can be discovered by dividing its partial communality on each dimension by its total communality. The resulting fraction is the proportion of common variance in a variable that is predictable from each dimension. For example, note that the partial communality of V9, word meaning, from the first dimension is .75 whereas its total communality from all four dimensions is only a little higher, .76. Clearly the fraction .75/.76 = .987 means that for this test the first dimension leaves little common variance for other dimensions to account for.

In practice, the factor coefficients are divided by the square root of the communality. These "corrected" coefficients are called the "augmented" or "normalized" factor coefficients. The augmented factor coefficients are, as we see below, the coordinates of each variable that determine its locus in the geometric configuration as plotted by the BC TRY component SPAN.

Residuals and reproduced correlations

Each of the dimensions after the first and the set of variables that defines it are determined from a matrix of residual correlations. For example, having extracted the amount of correlation between the variables that is accounted for by the first dimension, the verbal factor, we have a matrix of first factor residuals left over, in which the definers of the second dimension, the speed factor, are located. By way of illustration, take the case of the residual between the first two ability tests, F1 visual figure completion and F2 cube similarities. The first factor coefficients with the verbal dimension are .39 and .26, respectively, as shown in Table 6.3. The correlation between these two tests as produced by the verbal dimension is calculated by the general formula for the reproduced correlation, called the "cross-product" or "fundamental factor theorem" (see Chap. 12). The general formula for the reproduced correlation of variables V_j and V_k due to the ith dimension, with domain D_i, is

$$r^{(i)}_{V_j V_k} = r_{V_j D_i} r_{V_k D_i} \tag{6.1}$$

For example for V_1, V_2, and the first dimension in Holzinger study, Table 6.3,

$$r^{(i)}_{V_1 V_2} = (.39)(.26) = .10$$

The residual correlation between V_j and V_k, without the influence of D_i, is the original correlation minus the reproduced correlation

$$r_{V_j V_k \cdot D_i} = r_{V_j V_k} - r^{(i)}_{V_j V_k} \qquad (6.2)$$

Thus

$$r_{V_1 V_2 \cdot D_1} = .33 - .10 = .23$$

The matrix of such first residuals therefore represents the amounts of correlation among all 24 variables after we have extracted the amounts of variance among the variables due to the verbal factor. It follows that the speed dimension is independent of, i.e., uncorrelated with, the first dimension. Since the second and third residuals are similarly derived from the speed and spatial dimension correlations, the last two dimensions, spatial and memory, are also independent dimensions.

The sufficiency of the final set of dimensions to account for the original correlations can be appraised from the distribution of residuals after the final dimension. They must, of course, be quite trivial if the dimensions are to be considered sufficient. Because the dimensions are independent, the reproduced correlations for all dimensions are subtracted from the original correlation, for each pair of variables. For V_j, V_k, and the dimension domains $D_1, D_2, \ldots, D_K$

$$r_{V_j V_k \cdot D_1 D_2 \cdots D_K} = r_{V_j V_k} - \sum_{i=1}^{K} r^{(i)}_{V_j V_k} \qquad (6.3)$$

For V_1 and V_2 in the Holzinger data of Table 6.3

$r_{V_1 V_2 \cdot D_1 D_2 D_3 D_4}$

$$= .33 - (.39)(.26) - (.19)(.10) - (.44)(.47) - (.07)(-.04) = .00$$

Clearly, then D_1, D_2, D_3, and D_4 are sufficient to account for the correlation of V_1 and V_2.

Just how sufficient the four dimensions are in reproducing the raw correlations of all 24 variables can be seen in the last two columns of Table 6.3. For each variable we counted the number of final residuals with a first digit of .0 and the number with first digit .1 or greater. It is apparent from these two columns that quite generally all fourth-factor residuals are trivial.

Number of dimensions

A troublesome problem in the long history of factor analysis has been that of deciding how many dimensions to define. There is, of course, no completely general answer to this question. It depends on how salient one wants the dimensions to be. In key-cluster factoring by the BC TRY System the matter is left up to the investigator, though for programming purposes we put a stop on 15 dimensions. Experience has rather generally shown, however, that to factor into additional dimensions that account for less than, say, 5 percent of the estimated overall communalities is likely to result in dimensions whose defining variables either are difficult to conceptualize or are very narrow doublets that hardly seem worth dignifying as definers of a general dimension.

Number of iterations of factoring

Before factoring begins, estimates of the communalities are inserted in the diagonal cells of the correlation matrix. At the end of the first complete factoring procedure we have a revised set of communalities from the factoring itself. These new values are now inserted in the diagonal cells of the original correlation matrix, and a second factoring is made. This iteration process is repeated until a new set of communalities from factoring is the same as the last set, i.e., until the communalities converge.

The number of iterations undertaken is determined by the numerical precision to which one wishes to achieve convergence. Experience has generally shown that in most problems convergence of all communalities is better than a .05 criterion for all variables by about the fourth iteration. In the Holzinger problem the largest difference in any communality between its third and fourth iteration (printed on the diagonal in the final residual matrix) is .02; 14 of the 24 are less than .001.

Convergence through iteration is desirable not only to assure that stable factor coefficients are obtained: studies of artificial matrices where true values are known, because the matrix was produced by them, have indicated that only by iteration are the true values recovered (Tryon, 1957*b*, table 1, item 2 on the CC method).

The social-area study

For comparison with the Holzinger study, we present the general results of key-cluster factoring of the correlations among the 33 demographic characteristics of neighborhoods (census tracts) in the social-area study. Figure 4.1 shows sample variables from collinear clusters of three domains of demographic characteristics. How many dimensions does key-cluster factoring indicate would be sufficient to account for the generality among the 33 demographic attributes of neighborhoods?

The main data on this question are given in Table 6.4, which is parallel in form to Table 6.1 for the Holzinger analysis. Compared with the cognitive tests of children, the demographic characteristics of neighborhoods have more generality, as shown by the RMS original intercorrelation of about .44. After the third dimension the mean residual is only .06. Extracting a fourth does not substantially alter it. Sufficiency of the key-cluster dimensions to reproduce the initial estimates of communalities is shown in Sec. A of the table. After the third dimension is defined, 94 percent of the estimated communalities is exhausted.

The evidence indicates that there are clearly three salient demographic dimensions of neighborhoods. Should we accept the fourth one that contributes only 6 percent to the communalities of the variables and that reduces the residuals so insignificantly? Here is an example of a common dilemma in factoring: whether to accept a last factor that seems of borderline value. We leave the matter unresolved until the next chapter, where we examine this fourth dimension in the context of the whole cluster structure of the demographic features.

TABLE 6.4 SUFFICIENCY OF KEY–CLUSTER DIMENSIONS (FACTORS)
IN THE SOCIAL-AREA STUDY (DEMOGRAPHY)

A. Reproduction of Communalities during First Factoring

	Verbal C_1	Speed C_2	Spatial C_3	Memory C_4
Proportion of sum of estimated communalities, each dimension	.47	.32	.15	.06
Cumulative proportion of communalities from dimension C_1 to C_4	.47	.79	.94	1.00

B. Reproduction of Original Correlations after Fourth Iterated Solution

	Original	Residuals after			
		C_1	C_2	C_3	C_4
Root Mean Square	.44	.25	.12	.06	.05

The voting-attitude study

The comparable data on the sufficiency of key-cluster factors in the voting-attitude study of small neighborhoods (precincts) are given in Table 6.5. The RMS of the original correlations among the 31 election issues is .38. The residuals after the second dimension shrink to only .07, those after the third dimension to .05, signifying that in general the third dimension has removed little additional correlation from the matrix. The first and second dimensions account for only 84 percent of the communalities of all the variables, whereas the third raises the figure to 91 percent. The problem here is whether to retain the fourth dimension. It contributes only 4 percent to the communalities, and it does not reduce the residuals in any significant way. It looks here, therefore, as if there were only three salient attitudinal dimensions of small neighborhoods.

TABLE 6.5 SUFFICIENCY OF KEY–CLUSTER DIMENSIONS (FACTORS) IN THE VOTING–ATTITUDE STUDY

A. Reproduction of Communalities during First Factoring

	Verbal C_1	Speed C_2	Spatial C_3	Memory C_4
Proportion of sum of estimated communalities, each dimension	.44	.40	.07	.04
Cumulative proportion of communalities from dimension C_1 to C_4	.44	.84	.91	.95

B. Reproduction of Original Correlations after Fourth Iterated Solution

	Original	Residuals after			
		C_1	C_2	C_3	C_4
Root Mean Square	.38	.27	.07	.05	.04

Chapter 7

CLUSTER STRUCTURE ANALYSIS

Though the processes of key-cluster factoring indicate the salient number of cluster composites to be formed from the variables of a study and also what the definers of the clusters may be, they do not describe the structure of the relationships among the variables, a matter now to be considered.

The three main objectives of structure analysis are:

1 Inner structure of clusters: to provide information permitting improvement in the sets of definers of each cluster that have been selected by the factoring procedures.

2 Structural generality: to discover how important each cluster or factor is and how general is its kind of variation across all the variables of the study.

3 Structural relationships (organization) of clusters: to discover how the different clusters or factors are related to one other and to the variables of the study; knowledge of this total configuration aids in selecting the most nearly independent and meaningful clusters for O-analysis.

There are two ways to describe structure. The first is by various statistical quantities. The second is graphical and geometric. Both quantitative and graphical analyses are useful in understanding the cluster and factor structure.

The first step is to study the results of the initial key-cluster factoring, as described in Chap. 6. Then, if the definers of the factored dimensions require revision, one revises them and reruns key-cluster factoring but this time "presetting" the dimensional analysis on the revised set of definers. Both the original and the preset analysis include the Cluster Structure Analysis (called the CSA program in BC TRY), which gives the quantitative description of the inner structure of the clusters and their structural interrelationships and generalities. Also involved in both analyses is the BC TRY program component SPAN, which provides a graphical display of the configuration, namely, SPherical ANalysis.

Preset key-cluster factoring

Consider the results of the preset rerun on the Holzinger variables, in which the clusters are selected by the analyst from study of the empirical results. The first empirical key-cluster factoring described in Chap. 6 does not give the "best" definers of the four independent dimensions. The final metric description of structure is that given in the revised, preset run and *not* in the original blindly empirical factoring that usually does not give an optimal solution. We describe in a later section the specifics for application of general criteria used to improve the selection of definers for the preset solution. Here the reader should simply note the composition of the revised sets, listed in Table 7.1 in the column headed Definers (Revised). There it will be seen that F3, form board, has been added as a definer of the third dimension, C_3, and the two mathematical tests, N21, puzzles, and N24, arithmetic, have been deleted as definers, respectively, of dimensions C_3 and C_2. The results of factoring on the revised definers in this case differ little from those of the initial factoring. The revised orthogonal factor coefficients and communalities of Table 7.1 should be compared with the corresponding values from initial factoring, given in Table 6.3. The residuals given in the columns headed .0 and .1 are also about the same, a few less in the .1 category for the spatial and speed tests and a few more for the memory and mathematical tests.

Before leaving this matter we point out a great advantage of preset key-cluster factoring over the orthodox forms of factoring by the principal-axes and centroid solutions. Being able to preset the definers of key clusters makes it possible to define dimensions by any subset of variables. The analyst has complete command over the factoring process in key-cluster factoring.

TABLE 7.1 REVISED KEY-CLUSTER FACTORING OF 24 ABILITIES ON PRESET DIMENSION DEFINERS IN HOLZINGER STUDY

Tests	Definers (Revised)	Revised Orthogonal Factor Coefficients and Communalities					Augmented Factor Coefficients for SPAN				Reproduced Correlations Number of Residuals		Oblique Factor Coefficients			
		C_1	C_2	C_3	C_4	h^2	C_1	C_2	C_3	C_4	.0	.1	C_1	C_2	C_3	C_4
Space:																
F1 Vis	C_3	.39	.19	.50	.02	.44	.59	.29	.76	.03	24		.39	.33	.66	.39
F2 Cub	C_3	.26	.08	.42	−.05	.25	.52	.17	.83	−.10	24		.26	.18	.49	.21
F3 Fbd	C_3 New	.33	.15	.35	.09	.26	.65	.29	.68	−.17	22	2	.33	.27	.49	.36
F4 Loz	C_3	.40	.08	.63	−.05	.57	.54	.10	.83	−.07	24		.40	.23	.75	.33
Verbal:																
V5 Inf	C_1	.80	.09	.07	−.09	.66	.98	.11	.09	−.11	24		.80	.40	.54	.36
V6 Cmp	C_1	.83	−.02	.01	.09	.69	.99	.02	.01	.10	24		.83	.31	.48	.45
V7 Snt	C_1	.83	.00	−.06	−.03	.69	1.00	.00	−.08	−.04	24		.83	.32	.43	.35
V8 Wcl	C_1	.68	.15	.09	.01	.50	.97	.21	.13	−.01	24		.68	.41	.50	.40
V9 Wmn	C_1	.87	−.08	−.02	.04	.76	1.00	−.09	−.02	.04	24		.87	.27	.47	.40
Speed:																
S10 Add	C_2	.29	.75	−.21	.06	.70	.34	.90	.25	.07	24		.29	.81	.16	.31
S11 Cod	C_2	.37	.51	.06	.18	.43	.57	.77	.09	−.27	23	1	.37	.61	.37	.51
S12 Cnt	C_2	.19	.73	.15	−.12	.60	.25	.94	.19	−.16	24	1	.19	.75	.38	.29
S13 Scc		.39	.58	.36	−.18	.64	.48	.72	.44	−.23	23	1	.39	.68	.62	.33
Memory:																
M14 Wrg	C_4	.31	.17	.06	.55	.43	.47	.27	.09	.83	23	1	.31	.28	.26	.64
M15 Nrg	C_4	.25	.17	.17	.44	.32	.45	.30	.30	.79	24		.25	.25	.31	.56
M16 Frg	C_4	.27	.18	.44	.41	.47	.39	.25	.65	.60	24		.27	.27	.54	.62
M17 Wn	C_4	.31	.34	−.08	.48	.45	.46	.51	.12	.72	23	1	.31	.43	.19	.62
M18 Nf		.25	.40	.37	.25	.42	.38	.62	.57	.39	20	4	.25	.47	.52	.54
M19 Fw		.29	.18	.12	.23	.18	.68	.41	.28	.54	20	4	.29	.28	.30	.40
Mathematical:																
N20 Ded	C_3 Out	.52	.09	.23	.23	.39	.84	.14	.36	.38	21	3	.52	.28	.50	.51
N21 Puz		.35	.43	.37	−.02	.45	.53	.64	.56	−.03	22	2	.35	.53	.59	.39
N22 Rsn		.52	.12	.22	.14	.35	.88	.23	.37	.23	20	4	.52	.31	.50	.44
N23 Ser		.53	.20	.36	.08	.47	.78	.33	.53	.12	21	3	.53	.40	.64	.47
N24 Ari	C_2 Out	.51	.46	−.06	.06	.48	.74	.66	−.08	.09	22	2	.51	.63	.35	.44

Cluster structure analysis (CSA)

<u>Inner cluster structure</u>

The results of the revised analysis in the Holzinger study are given in Table 7.2 (copied from the CSA computer printout sheets). The first cluster, C_1, the verbal cluster, is defined by the same four variables (marked D in the column headed Definers) as in the first factoring. But the second cluster, C_2, speed, is defined now by only the three variables, because in the revision we have deleted N24, arithmetic, as a definer. The third cluster, C_3, space, is altered by replacing N21, puzzles, by F3, paper form board. But the fourth, C_4, memory, is defined as in the first factoring.

Oblique factor coefficients (from CSA) are of central interest in structure analysis. They are to be distinguished from the orthogonal factor coefficients computed in the CC factoring procedure. The distinction will be clear from Table 7.1, where the columns of revised orthogonal factor coefficients of all the variables and oblique factor coefficients are clearly labeled. For example, the verbal test V5, information, has as its four orthogonal coefficients the numbers .80, .09, .07, and −.09, whereas its corresponding four oblique factor coefficients are .80, .40, .54, and .36.

The cluster-defined dimensions in structure analysis are called oblique clusters because this term "oblique" has come to have the same meaning as the word "correlated" in factor-analytic jargon. They are oblique clusters because, being defined by composites of standard scores on their defining variables, they would normally be expected to yield nonzero correlations with other clusters. In contrast, the orthogonal dimensions derived by factoring, being defined by residuals from which variance from other cluster-defined dimensions has been removed, would necessarily correlate .00 with each other, hence be orthogonal. The terms oblique and orthogonal are really geometric terms, the meanings of which are quite clear when we come to the graphical geometric representation of cluster structure.

<u>Oblique unifactor structure</u>

As an aid in interpreting each oblique cluster we have programmed the cluster-structure component (CSA) to assign each variable of the study to that particular cluster with which its oblique factor correlation is highest. For example, in Table 7.2, under the four definers of C_1, verbal, marked D are listed three other tests, V8, N20, and N22, which have been put there because their highest oblique factor coefficients are with this cluster, a point that can be verified from the oblique factor coefficients in Table 7.1.

TABLE 7.2 INNER CLUSTER STRUCTURE OF THE FOUR BASIC AND EXPANDED OBLIQUE CLUSTERS IN THE HOLZINGER STUDY (OBLIQUE UNIFACTOR STRUCTURE)[a]

Cluster	Definers	Oblique F_c	h^2	$\bar{r}$	Expanded Reliability Single	Expanded Reliability Cumulative	Cluster Score on D's
C$_1$, verbal:							
V9 Wmn	D	.87	.76	.72			
V7 Snt	D	.83	.69	.69			
V6 Cmp	D	.83	.69	.69			
V5 Inf	D	.80	.66	.66			
V8 Wcl		.68	.50	.57	.90	.90(1)[b]	
N20 Ded		.52	.39	.43	.89	.90(2)	
N22 Rsn		.52	.35	.43	.89	.90(3)	
Domain validity							.95
Reliability							.90
C$_2$, speed:							
S10 Add	D	.81	.70	.58			
S12 Cnt	D	.75	.60	.54			
S13 Scc		.68	.64	.49	.84	.84(1)	
N24 Ari		.63	.48	.45	.82	.85(2)	
S11 Cod	D	.61	.43	.44			
Domain validity							.89
Reliability							.79
C$_3$, space:							
F4 Loz	D	.75	.57	.45			
F1 Vis	D	.66	.44	.40			
N23 Ser		.64	.47	.38	.75	.75(1)	
N21 Puz		.59	.45	.35	.75	.79(2)	
F3 Fbd	D	.49	.26	.30			
F2 Cub	D	.49	.25	.30			
Domain validity							.84
Reliability							.70
C$_4$, memory:							
M14 Wrg	D	.64	.43	.39			
M17 Wn	D	.62	.45	.38			
M16 Frg	D	.62	.47	.38			
M15 Nrg	D	.56	.32	.34			
M18 Nf		.54	.42	.33	.76	.76(1)	
Domain validity							.85
Reliability							.72

[a] Variable excluded because communalities are below .20: M19 figure-word recall.
[b] For explanation of figures in parentheses see text.

This kind of assignment is called an "oblique unifactor structure" because each variable is assigned to only one of the factors (Harman, 1967, p. 104). Such an allocation can be an aid in interpreting the oblique dimension, because a study of the constructed content of these nondefiners often helps in conceptualizing the rational meaning of the cluster. The variables in the unifactor listing of Table 7.2 have been ordered by the sizes of their oblique factor coefficients.

Expanding an oblique cluster for purposes of scoring individuals

There is another important use of the unifactor listing. In revising the definers of a cluster one may have included, at the time, what appeared to be the best definers of the given domain, but it may be discovered that a cluster or factor score on these definers yields a reliability coefficient that is too low for scoring individuals. Generally it is desirable to have the reliabilities of any composite score .90 or more if individuals are to be compared with each other on that score, as is done in object cluster analysis. Information on the nondefiners in a unifactorial listing may help in expanding the number of variables in a cluster in order to increase its reliability coefficient.

Below each cluster listing in Table 7.2 is information about the cluster score formed from a composite of the basic definers marked D. For example, the domain validity of a cluster score composed of the four definers of C_1 is .95, whereas its α reliability coefficient is the square of this value, .90. Expanding a cluster to include additional nondefining variables changes the reliability of the cluster score. The relevant information on the reliability of the clusters in the Holzinger study is given in Table 7.2 under Expanded Reliability. Under Single can be found the value of the reliability of the composite formed by adding each of the nondefining varibles singly to the basic composite of four definers. For example, if V8 is added to form a new five-variable composite for C_1, the cluster reliability is unchanged. If either N20 or N22 is added, the reliabilities of the two five-variable composites are .89. There is no reason here for adding any of the nondefiners to the basic cluster of four.

The reliability coefficient of the cluster composite when the additional nondefiners are added to it cumulatively and in the order of their single effects, as indicated by the ordinal numbers in parentheses, is given in the column headed Cumulative. Thus, when three nondefiners of the verbal cluster, V8, N20, and N22, are ordered by the sizes of the reliabilities resulting from adding them singly and then added in that order cumulatively to the basic four definers, the reliabilities of the expanded clusters remain unchanged at .90. The reliability coefficient of the composite score on the three definers of the speed cluster is only .79, but the table shows that it can be raised to .85 by adding the two nondefiners allocated to it, S13 and N24. With the space cluster, its basic reliability can be increased from .70 to near .80.

Table 7.2 includes the values of the communality h^2 of each variable, and also each variable's mean original correlation $\bar{r}$ with the basic definers of the cluster. Generally, a variable with low values on these two statistics does not contribute to the reliability or meaning of a cluster. This finding is so universal that the CSA component of BC TRY is programmed to exclude variables from unifactor allocation when their communalities are lower than .20, as indicated in the first footnote to Table 7.2 for the M19, figure-word recall, test.

Structural generality of an oblique cluster or factor

One way to assess the generality of a given cluster's variation is to note how much the intercorrelations among all the variables would be reduced if the cluster were a constant instead of a variable. This effect is precisely what is revealed in key-cluster factoring by the sizes of the first factor residuals after extracting the first dimension. To discover the generality of the different clusters we refactor the correlation matrix as many times as we have oblique clusters but on each refactoring merely define a first dimension, defining it by one of the sets of oblique clusters. Instead of computing first factor residuals in each case, however, we compute the reproduced correlations from the first factor coefficients, as described and illustrated in Chap. 6, in the section on residuals and reproduced correlations.

A simple index, then, of the generality of a given oblique cluster is to compare the reproduced correlations with the actual values of the original correlations. The index chosen is simply the ratio of the mean squares of the reproduced correlations of all variables divided by the mean squares of all their original correlations. If the reproduced correlations perfectly match the original correlations, the reproducibility ratio is 1.00. If no reproducibility occurs, the reproduced correlations are all .00 and the ratio is .00.

The results in the Holzinger study are given in Table 7.3, Sec. A. In the first row the verbal and space clusters have reproducibilities of the order .50, whereas those of the speed and memory clusters are nearer .30.

TABLE 7.3 STRUCTURAL GENERALITY OF THE FOUR OBLIQUE CLUSTERS
(FACTORS) AND THE RELATIONSHIPS AMONG THEM

A. Generality of the Oblique Cluster Dimensions

	C_1, Verbal	C_2, Speed	C_3, Space	C_4, Memory
Reproducibility of mean squares of correlations	.51	.32	.47	.34
Reproducibility of communalities	.50	.40	.48	.41

B. Correlations between Cluster Scores (Factor Estimates)[a]

C_1, *verbal*	(.90)	.33	.46	.38
C_2, *speed*	.33	(.79)	.31	.38
C_3, *space*	.46	.31	(.70)	.38
C_4, *memory*	.38	.38	.38	(.72)

**C. Estimated Correlations between Cluster Domains
(Common Factor Correlations)**[a]

C_1, *verbal*	(1.00)	.39	.58	.47
C_2, *speed*	.39	(1.00)	.42	.51
C_3, *space*	.58	.42	(1.00)	.53
C_4, *memory*	.47	.51	.53	(1.00)

[a] α reliability coefficients in parentheses.

Another index of generality is the degree to which the communalities
of all the variables are reproduced by each cluster treated as the first
independent dimension in key-cluster factoring. Recall that the amount
of variance of a single variable predictable from a given dimension is the
partial communality of the variable, namely, the square of the variables'
factor coefficient on the dimension. If the partial communality on a dimension equals the total communality, then that dimension fully accounts for
the common variance of the particular variable. Therefore, by treating
each cluster as a first dimension in factoring, another index of its generality can be formed by summing the partial communalities of all variables
on that dimension and dividing the sum by the sum of the total communalities of all variables. When the index value approaches 1.00, the given
cluster's variation is completely general; near zero, its generality is very
limited.

The communality reproducibility indexes of the four Holzinger clusters are given in Table 7.3. The findings on the generality of the clusters
by this index are about the same as by the reproducibility index of
correlations.

Structural interrelationships among the clusters

The definers of each of the oblique clusters are usually only a limited sampling of variables drawn from a collinear domain of variation. The cluster scores on each of these domains of variation are referred to in orthodox factor terminology as "factor estimates" or "measurements of factors" (see Harman, 1967, chap. 16). An understanding of the different kinds of cluster variation in a study is usually aided by computing the raw correlations between the cluster scores, as illustrated by Table 7.3, Sec. B. Solution for these correlations does not require actually scoring the children on the different clusters because the correlations can be computed directly from the matrix of correlations between all the variables (see Chap. 12).

Inspection of the values in Table 7.3 shows that they are all greater than zero, meaning that despite the fact that the different clusters appear to sample quite different types of cognitive abilities, they overlap some common substrata of causes of individual differences. It is ill-advised to put much stock in these correlations, because each cluster score is subject to error since it is based on a limited sample of variables. The net effect of this error of domain sampling, usually called the "error of measurement," is shown by the size of the α reliability coefficients of the cluster scores, given in parentheses down the diagonal of the raw correlation matrix in Table 7.3. The smaller (away from maximum of 1.00) these reliability coefficients are, the less one should depend upon the score intercorrelations as representing domain intercorrelations.

More important are the estimated correlations between domain scores, i.e., hypothetical scores on the children "freed of limitations of sampling each domain." These estimates, called the "correlations between factors," "common factor correlations," or, better, "interdomain correlations," are computed by the CSA program of BC TRY. They also have a geometric meaning, as we see in the next section when we plot the variables in a graphical model showing the configuration of all the relationships among the variables.

The values of the interdomain correlations are given in Table 7.3, Sec. C. They estimate what the correlations would be among the clusters if we had a great many tests in each of the clusters—additional samples of variables that were collinear with the observed set. They indicate what the relationships would be among the four different kinds of variations among the children if we were able to get a thorough measurement of the four kinds of variation in the verbal, speed, space, and memory clusters. The verbal cluster domain correlates highly with the space domain, and the space dimension seems to overlap on the other dimensions a bit more heavily than the verbal does. The speed domain seems to be a more specific kind of variation, but still it shares some common variance with the other three cognitive domains.

We obviously are dealing here with an oblique set of domains. This result is a common finding in the broad domain of cognitive abilities, one that has led some psychologists to assert the existence of a single common "underlying factor," often termed g (for general factor). It has provided such psychologists with the justification for lumping many kinds of tests like those of the Holzinger study into a single test battery, the total score on which is termed "general intelligence," scored in the form of the IQ. This simple idea is only one of many ways of interpreting such a matrix. A sounder one theoretically is to assume the operation of complex multiple environmental and genetic overlaps between the different cluster domains (Tryon, 1935).

Graphical representation of structure: SPAN

There is no substitute for a visual map as a means of describing the general structure of the relations among the variables. A pictorial display of the configuration shows at a glance the collinearity and structure of each cluster, its degree of generality with respect to all the variables, and the relationships among the clusters. The program of the BC TRY System that prints the configuration is the SPherical ANalysis component (SPAN). If the dimensionality of a matrix is three, SPAN presents the configuration of variables as a surface layout on a single sphere (a space of three dimensions), so that by capitalizing on clues for depth perception, one can "see" the configuration. If there are more than three dimensions, SPAN breaks up the configuration (which now lies in a hypersphere of more than three dimensions) so that only those parts of the configuration that can be visually represented on spheres are printed. Each such part is called a "dimension set" or "subspace." By cross-referencing the partial configuration on the different dimension sets one can usually come to "see" the whole configuration as it lies in the hypersphere of more than three dimensions.

Here we illustrate the graphical representation of the configuration of the variables of the Holzinger analysis. The orthogonal factor coefficients of the 24 tests on the four dimensions are given in Table 7.1. Since these dimensions are independent, they can be represented as four cartesian axes set at right angles to one other. Each of the 24 tests can therefore be plotted in this space by its four coefficients treated as coordinates. But this is not exactly the way SPAN proceeds. First, the factor coefficients of each test are augmented by dividing the four of them by the square root of the sum of their squares, i.e., by h, the square root of the communality. The resulting values are called the augmented factor coefficients of the test.

Table 7.1 gives the augmented factor coefficients of the 24 tests. For example, in the first test, F1, visual figure completions, the squared augmented values .59, .29, .76, and .03 sum to 1.00 (except for rounding errors). This relationship holds for all 24 tests, a fact that has an interesting consequence: when each test is plotted on four independent axes using its augmented values as coordinates, all 24 tests lie on the surface of a hypersphere in four dimensions. If the factoring procedure had given a three-dimensional solution, all 24 tests plotted by their three augmented values would have been spread out on the surface of a sphere. With a four-dimensional solution, those particular tests that have trivial squared augmented values, like the five verbal tests, would spread out on a sphere defined by the first three dimensions only, since their first three squared values add up to 1.00. In fact, some of these tests can be plotted on a "sphere in two dimensions," i.e., on the circumference of a circle defined by the first two dimensions; these are the tests whose last two squared augmented values are trivial, namely, the four verbal tests, and S11 and N24, whose first two squared coordinates approach 1.00. In short, one can usually organize the four-dimensional configuration into subspaces in each of which the full coordinates in four dimensions can be accounted for in three, two, or perhaps one dimension.

This manner of depicting a configuration of points that as a totality require representation in a nonvisual space of more than three dimensions, i.e., a hyperspace, can perhaps be developed better if we start from one dimension and work up to a hyperspace. In Fig. 7.1 the tests have been spread out on the first dimension according to the values of their augmented coordinates given in Table 7.1. The five verbal tests cluster at the right end of the axis near 1.00. Reliance on the configuration of all 24 tests represented on this single dimension would lead to some big mistakes about their relationships. Though the tests that lie near 1.00 will stay together as we add dimensions, others not near 1.00 but together in this

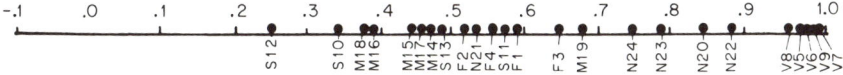

FIGURE 7.1
The configuration of 24 abilities (Holzinger study) plotted by their first augmented factor coefficients.

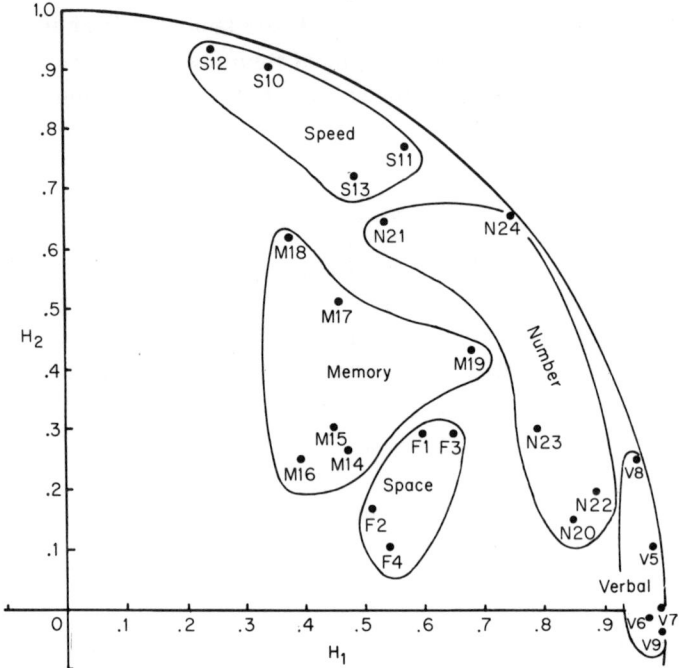

FIGURE 7.2
The configuration of 24 abilities (Holzinger study) plotted by their first and second augmented factor coefficients.

first dimension are likely to move apart. For example, the two speed tests S11 and S13 appear to cluster with the four space tests F1, F2, F3, and F4. Adding a second dimension radically changes this impression.

In Fig. 7.2 the configuration of the 24 ability tests is plotted by their first two augmented coordinates. The two tests of speed that previously clustered with the four tests of space have spread apart into different sectors of this two-dimensional space. The five verbal tests still remain clustered in this two-space. Figure 7.2 is a circle in a plane, the two agumented coordinate axes of which are set orthogonal to each other and the radius of which is 1.00, a unit circle. Variables that lie out near the end of the radius cannot be relocated by adding more dimensions. This is true for the four tests of speed that are near unit distance from the origin of the circle.

Figures 7.1 and 7.2 illustrate the real difference between a cluster and a factor. The variables in the verbal cluster, C_1, are all grouped near the value 1.00 on the factor dimension defined by the cluster. All other variables in the study are also located on the axis for the dimension, as represented in the figures. In Fig. 7.2, where a second factor dimension is defined by the speed cluster, C_2, the verbal variables are still grouped near the value 1.00 on the first axis while the other variables have scattered. The variables of C_2 are grouped together in the two-dimensional plot (they form a geometric cluster), but they do not fall so near the 1.00 terminus of the axis as C_1 variables do on the first axis terminus. The clusters in cluster analysis are selected to be as nearly independent among themselves as possible. If the clusters C_1 and C_2 in Fig. 7.2 were independent, they would appear grouped around the termini of their respective factor dimensions. However, the empirical relationship between C_1 and C_2 is not that of independence. The factor dimensions are strictly independent, and must be, because the factoring processes define each successive factor dimension in that way. The factor dimensions are orthogonal (at right angles) to each other while the clusters themselves are oblique (not at right angles) to each other with respect to the origin of the circle.

Figure 7.3 shows the variables plotted by their three agumented

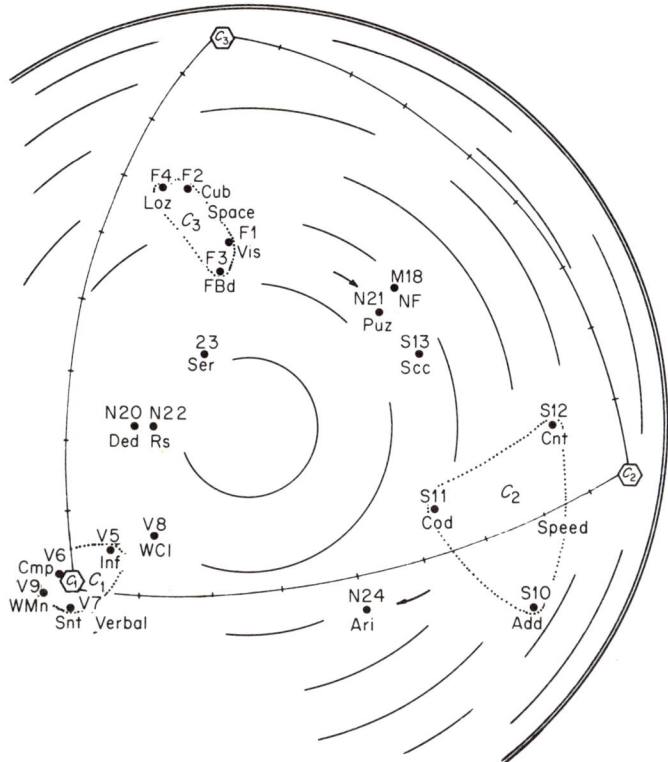

FIGURE 7.3
Configuration on the surface of a sphere: 19 abilities plotted by the three augmented coordinates H_1, H_2, H_3 accounting for their communalities.

coordinates. This time only those 19 tests that lie near the surface are included in the plot in this space of three dimensions. They all lie at or near the surface of a sphere, because any point the sum of squares of whose three factor coordinates equals 1.00 necessarily lies on the surface of a sphere. The sum of a variable's three squared augmented coordinate values equaling 1.00 also means in this figure that 100 percent of its communality is accounted for in this space.

In Fig. 7.3 circular contour lines help to visualize the third dimension. The configuration has been physically rotated (by the SPAN program) so that the swarm of test points is centered in the line of the eye. At the lower left the five verbal tests are clustered together at the end of their axis, labeled C_1, and set in a hexagonal box. At the lower right are the four speed tests. The orthogonal axis that they defined is shown nearby, boxed as C_2. The third dimension, C_3, is defined by the tests of space, so at the upper left are found the four F tests that define the third orthogonal axis clustered together, near the axis boxed as C_3.

The ends of the three axes are connected by arcs so that a surface right triangle is formed by them. The arc that connects C_1 with C_2 can be thought of as the place where a slice (or plane) through the sphere has been made; Fig. 7.2 is precisely this plane.

The important point to be made is that the configuration of these 19 tests in this subspace is going to stay put however more dimensions we add. It has already been seen that the five verbal tests remain clustered when a second dimension is added and both they and the speed tests stay put when a third is added. This means that all 19 will stay where they are (or change only trivially) if a fourth dimension is added. Therefore, a fourth dimension is unnecessary to describe their configuration; three dimensions are sufficient.

In Fig. 7.3 the definers of each of the three dimensions, as listed in Table 7.2, have been circled by dotted lines. The circled points represent the oblique clusters. The nondefiners unifactorially allocated to each oblique cluster in Table 7.2 are in fact near the oblique cluster to which they were allocated. In brief, the spherical configuration of Fig. 7.3 is merely a geometric model of the quantitative values given for the three clusters in Table 7.2.

There remains the fourth orthogonal dimension defined by four tests of memory. They do not lie in the space just discussed, but they do lie in another three-dimensional space defined by C_1, C_2, and C_4, shown in Fig. 7.4. Sixteen tests lie at or near the surface of this sphere. The verbal and speed clusters are carried over from the previous sphere because they lie in the C_1C_2 plane that is common to the two spheres. The orthogonal

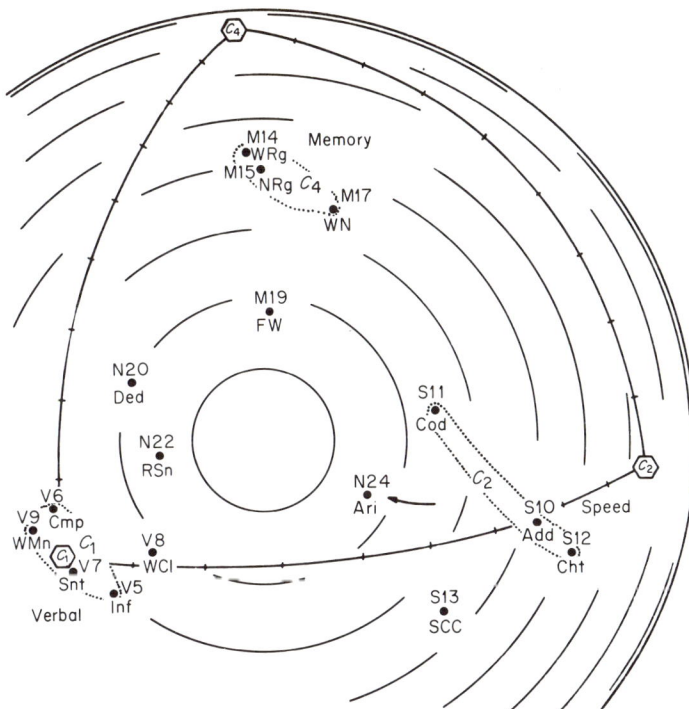

FIGURE 7.4
Configuration on the surface of a sphere: 16 abilities plotted by the three augmented coordinates H_1, H_2, H_4 accounting for their communalities.

dimension C_4 lies off 90° away from the first two axes (and also in a fourth dimension away from C_3 which of course is not represented in this figure).

The whole configuration of the 24 tests can be observed on these two spheres. Perhaps the reader can see it in his mind's eye in one perceptual grasp by butting together the common "verbal-speed" arc of the two figures but keeping the C_3 and C_4 apexes spread apart by 90°. The whole configuration lies within this frame. In any event, the whole configuration can be portioned into the two SPAN sets. The inner structure of each oblique cluster is pictorially evident in them, and the degree of relation between the oblique clusters is plainly displayed.

The distinction between factored orthogonal dimensions and the oblique dimensions becomes clarified in the configurations of Figs. 7.3 and 7.4. In Fig. 7.3, for example, the orthogonal dimension and its oblique counterpart defined by C_1 are identical, because the first factor dimension is represented by an axis that passes directly through the first oblique cluster, the verbal dimension. But the second orthogonal dimension is independent of the first (since it is derived from residuals); its definers are the tests of the oblique speed cluster, C_2. This factored orthogonal dimension is represented by an axis that passes as near to the second oblique cluster as it possibly can, subject to the condition that it must be 90° away from the first dimension. Looked at from the central origin of the sphere, the orthogonal factor dimensions derived from factoring are, as it were, pointers at oblique clusters, but they are prevented from pointing exactly at them because of the rigid requirement that they must be at right angles to each other. Thus, the third factor dimension points as accurately as it possibly can at the third oblique cluster, C_3, which constitutes its set of definers, subject to the condition that it must be 90° away from *both* the first and second factor axes.

The orthogonal factor coefficients of Table 7.1 are the correlations of the variables and hypothetical variables corresponding to the termini of the respective orthogonal factor dimensions. The oblique factor coefficients of Table 7.1 are the correlations of the variables and composite variables corresponding to a sum of the cluster variables but with communality of 1.00. The specific definitions of these coefficients by mathematical equations are given in Chap. 12.

Criteria for final selection of clusters

In many studies the configuration is not sharply structured into clusters of variables. In some parts of a study a clear cluster structure may be evident; in other parts the variables may form a graded series with no evident concentration that can form a cluster; or the whole configuration may form a graded series. Indeed, in object cluster analysis, where the points in the configuration are persons and not variables, the absence of any clear clustering may be the usual case. Nevertheless, whatever the structure of the configuration may be, the criteria of cluster selection that are programmed in the BC TRY System key-clustering procedure select pivot variables at the edge of the configuration and add definers to them by the mutual collinearity criterion.

When cluster structure is poor, the quantitative description given by the CSA component sometimes misleads one into assuming that a sharper structure is present than in fact does exist. Only by inspecting the whole configuration does the real structure become clear. The meaning and existence of the parts of a cluster structure are determined by the broad shape of the structure, and this broad shape can be observed properly only in the pictorial displays of SPAN and not by the quantitative metrics of CSA.

It is therefore of the greatest import that one assess how well the factored dimensions serve as pointers to clustered sectors of the configuration. If the factoring program does not pick the most salient cluster-defined dimensions, or if the definers of any particular dimension are not the best ones that could be selected, one should revise the factorial description. Revising it means refactoring the correlation matrix by presetting the number of dimensions or by revising the definers of one or more dimensions.

There are five criteria to keep in mind when assessing the results of the initial factoring procedure:

1   The degree of collinearity of the definers of each dimension
2   The degree of independence of the oblique cluster that defines each dimension
3   The meaning of each defining oblique cluster as a construct
4   The contribution of the definers to the reliability of a cluster score on the oblique cluster
5   The generality of each oblique cluster and of the variables that define it

Initial factoring in the Holzinger problem selected the verbal cluster as the first dimension, defined by the four verbal tests indicated in Table 7.1 by C_1 in the columns headed Definers (Revised). On the SPAN diagrams the definers are seen to be highly collinear (see also Fig. 4.1). The question of whether to add variable V8 to the four definers is resolved by applying the five revision criteria to it. Its collinearity is high with the basic four, and it fits in with the verbal meaning of the other four definers, but we find that it would be the least independent definer with respect to other clusters and that by expanding the cluster to include it the reliability coefficient of the cluster score would not be increased. Furthermore, its communality is much lower than the communalities of the four basic definers. Hence there appears to be no point in adding V8 to the verbal cluster. The pertinent information for these evaluations is all in Tables 7.1 and 7.2.

Initial factoring selects the speed cluster for the second dimension, defined by three speed tests plus the mathematical test N24, arithmetic. But, in both SPAN figures (Figs. 7.3 and 7.4) this mathematics test is not highly collinear with the other definers and fails a bit in independence since it lies in the direction of the verbal cluster. Furthermore, including arithmetic with "pure mental speed" seems inappropriate. The arrow drawn in the figures indicates that on these grounds this test was rejected as a definer of the second dimension. The reliability of the three-variable speed cluster (N24, arithmetic, left out) is .79. With S13, straight or curved capitals dis-

crimination (Scc), another test of speed, added to the cluster the reliability would be raised to .84. The SPAN figures reveal, however, that S13,
Scc, has only fair collinearity with the other definers. The communality
of S13, Scc, is as high as those of the three definers, but its independence
of other clusters is variable. In meaning, it fits in with the other definers
as a test of mental speed. The criteria, therefore, appear to encourage
adding this fourth test to the definers of the speed cluster.

A final word should be said on the criterion of generality. It commonly
occurs on the initial factoring that one or more of the last dimensions may
be defined by only two variables (a "doublet") or by definers that have only
very trivial residuals on the basis of which they are selected as a dimension.
Such dimensons have low generality, and in general should usually be
rejected in toto as dimensions, the reason being that the object of V-analysis
is to select salient dimensions, not trivia.

In revised factoring we excluded the five mathematical ability tests
from the four core special ability clusters because we wish (under the
broad design of the Holzinger analysis) in a later study of prediction from
clusters to discover the degree to which the mathematical tests represent
more general attributes predictable from the more special verbal, speed,
space, and memory abilities. From the nature of the overall cluster structure, such prediction looks promising because the mathematical tests
occupy a central position in the total configuration. This fact means that
they are correlated with the four core clusters. It also explains why, under
the revision criterion of seeking the most independent definers, they were
excluded from the core clusters. They, or some of them, can form a
dependent cluster, a composite score of which is highly predictable from
the four core clusters. We examine this matter of the predictability of
dependent clusters and variables from the minimally sufficient core clusters in Chap. 10.

Cluster structure of demographic characteristics—the social-areas study

At the beginning of the social-areas study of the San Francisco Bay Area
in the middle 1940s this question was asked: Of the large array of demographic characteristics of people living in urban neighborhoods, can we
empirically discover a minimal set of demographic clusters that sufficiently
describes differences between the neighborhoods? No one then knew the
answer—either about the Bay Area or anywhere else. The scientific quest
for an answer consists in observing many demographic characteristics of
the neighborhoods and then, following the empirical key-cluster factor-

ing of them, discovering their cluster structure by the methods just described. On each of the 243 neighborhoods of the Bay Area the 1940 census provides the 33 characteristics listed in Table 7.4 (for details see Tryon, 1955, app. A). The census categorizes these characteristics into three groups: population characteristics, occupations, and home features. No one would reasonably argue that three rational composite scores should be formed from these three heterogeneous categories. An empirical key-cluster factoring of the intercorrelations between the 33 variables, followed by cluster structure analysis, reveals clearly and objectively that three salient oblique clusters, named S, socioeconomic independence, F, family life, and A, assimilation, describe all the general variance among the neighborhoods.

The results of the analysis are depicted in the total configuration of the variables in the SPAN diagram of Fig. 7.5, which shows the revised

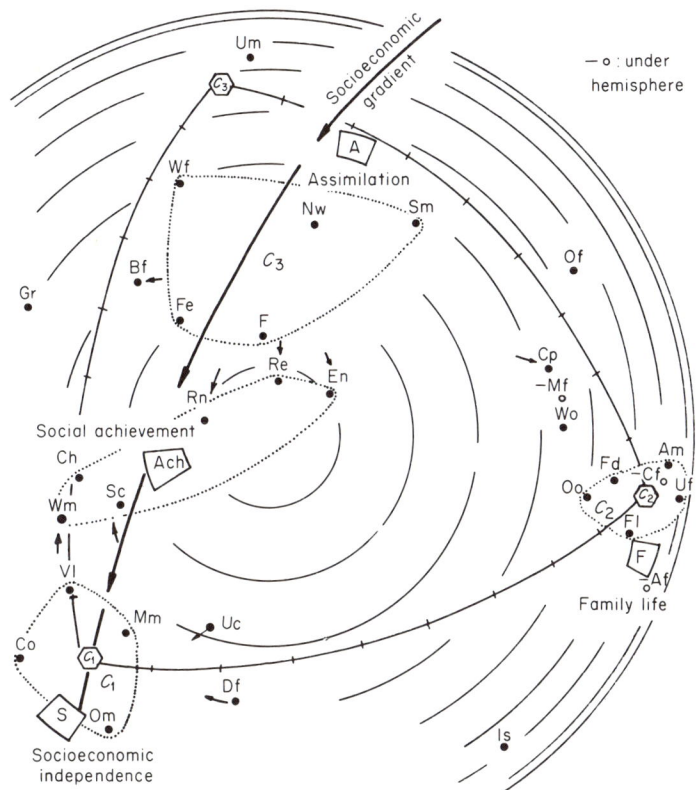

FIGURE 7.5
Spherical configuration showing the relationships among 33 demographic characteristics of Bay Area neighborhoods (1940 census tracts).

TABLE 7.4 THE SOCIAL–AREA VARIABLES

Original Rational Census Categories		Alphabetic Reference Listing	
Population characteristics:		Af	Older females
Af	Older females	Am	Younger males
Am	Younger males	Bf	Better fuel
Cf	Childless females	Cf	Childless females
Co	College education	Ch	Central heating
Fe	Foreign from northwest Europe	Co	College education
F	Females	Cp	"Nonfamily" couples
Nw	Native-born whites	Df	Female domestics (living in)
Sc	High school or better	Ef	Employed females
Occupations:		Em	Employed males
Df	Female domestics (living in)	F	Females
Ef	Employed females	Fd	Family-detached homes
Em	Employed males	Fe	Foreign from northwest Europe
Is	In school	Fl	Large families
Mf	Managerial-professional females	Gr	Good repair
Mm	Managerial-professional males	Is	In school
Of	Own-account females	Mf	Managerial-professional females
Om	Own-account males	Mm	Managerial-professional males
Sm	Skilled males	Nw	Native-born whites
Uf	Nonworking females	Of	Own-account females
	(housewives)	Om	Own-account males
Um	Nonworking males	Oo	Owner-occupied homes
Wf	White-collar females	Re	Refrigeration
Wm	White-collar males	Rn	High rent
Homes (Dwelling Units):		Sc	High school or better
Bf	Better fuel	Sm	Skilled males
Ch	Central heating	Uc	Undercrowded
Cp	"Nonfamily" couples	Uf	Nonworking females (housewives)
Fd	Family-detached homes	Um	Nonworking males
Fl	Large families	Vl	High home value
Gr	Good repair	Wf	White-collar females
Oo	Owner-occupied homes	Wm	White-collar males
Re	Refrigeration	Wo	Wood-fuel use
Rn	High rent		
Uc	Undercrowded		
Vl	High home value		
Wo	Wood-fuel use		

definers of the three oblique cluster dimensions S, F, and A circled by broken lines at or near the corners of the triangle. The meaning of each of these three core dimensions of urban social structure is apparent from the nature of the definers and of other variables with high oblique factor coefficients on them, as shown in Table 7.5. (The alphabetic listing of the variables in Table 7.4 will help in reading the figure.)

At the lower left is S, socioeconomic independence, the dimension of

TABLE 7.5 CLUSTER AND FACTOR STRUCTURE: THE SOCIAL–AREA STUDY
(OBLIQUE UNIFACTOR STRUCTURE IN CSA)

A. Inner Structure of the Clusters

Variables	Definers	Oblique F_c	h^2	$\bar{r}$	Reliability Single	Reliability Cumulative	Cluster Score on D's
C_1, socioeconomic independence, S							
2 Mm	D	1.02	1.05	.86			
3 Sc		.90	.97	.76	.94	.94(1)	
1 Wm		.90	.92	.76	.94	.96(2)	
27 Df		.81	.78	.68	.93	.96(3)	
26 Vl	D	.81	.68	.68			
29 Co	D	.79	.66	.67			
4 Ch		.79	.77	.66	.93	.97(4)	
28 Om	D	.73	.56	.61			
22 Uc		.70	.71	.59	.93	.97(5)	
21 Ef		.60	.49	.50	.91	.97(6)	
17 Mf		.59	.45	.50	.91	.96(7)	
Domain validity							.96
Reliability							.91
C_2, family life, F							
7 Fl	D	.93	.88	.83			
6 Oo	D	.92	.90	.82			
8 Fd	D	.88	.80	.78			
9 Uf	D	.87	.81	.77			
10 Am	D	.83	.72	.74			
14 Cp		.80	.82	.71	.96	.97(2)	
13 Wo		.70	.56	.62	.95	.97(4)	
16 Cf		−.69	.80	.62	.96	.97(3)	
15 Af		−.67	.89	.59	.96	.96(1)	
31 Is		.54	.50	.48	.95	.96(5)	
Domain validity							.98
Reliability							.96
C_3, Assimilation, A							
11 Sm	D	.87	.85	.64			
5 Bf		.81	.75	.59	.90	.93(3)	
12 Nw	D	.81	.66	.59			
18 Em		.80	.76	.58	.90	.92(2)	
20 Re		.80	.72	.58	.89	.94(4)	
19 Rn		.79	.79	.57	.90	.90(1)	
25 Wf	D	.77	.65	.56			
24 Fe	D	.62	.42	.45			
23 F	D	.58	.35	.42			
30 Gr		.42	.28	.31	.86	.94(5)	
32 Um		.39	.24	.29	.85	.93(6)	
Domain validity							.93
Reliability							.87

TABLE 7.5   CLUSTER AND FACTOR STRUCTURE: THE SOCIAL-AREA STUDY (OBLIQUE UNIFACTOR STRUCTURE IN CSA) (Continued)

B. Generality of Clusters

	C_1	C_2	C_3
Reproducibility of mean square of correlations	.41	.26	.45
Reproducibility of communalities	.43	.34	.45

C. Original and Interdomain Correlations between Clusters

	Original			Interdomain (cos θ)		
	C_1	C_2	C_3	C_1	C_2	C_3
C_1	(.91)	−.01	.35	1.00	−.01	.40
C_2	−.01	(.96)	.32	−.01	1.00	.35
C_3	.35	.32	(.87)	.40	.35	1.00

Notes: See Table 7.4 for explanation of abbreviated census categories. Variables excluded because communalities below .20: 33 Of.

wealth and social independence. Its meaning is evident in neighborhoods that would have high values of S: they include relatively more people working for self (Om), possessing more costly homes (Vl), having had an expensive college education (Co), and being managerial-professional males (Mm). The reliability coefficient of cluster scores on the four definers is .91, but it could be increased to .95 or higher by adding more variables, as shown in Table 7.5.

Virtually independent of S is F, family life, in the right middle of Fig. 7.5. Neighborhoods with high scores on F would orient around the family: in them are relatively more housewives (Uf), more children (Am), more owner-occupied (Oo) single-family homes (Fd), housing larger families (Fl). The concept of such neighborhoods as family-centered habitats is strengthened by two other characteristics, *not* containing childless females (Cf) or older females (Af), indicated by negative correlations of these attributes with F (in Table 7.5) and represented in SPAN as unfilled dots near F but on the underside of the hemisphere.

A, assimilation, in the upper sector of the configuration, refers to the degree to which a neighborhood includes persons belonging to the white Protestant middle-class majority class in America. The meaning of this dimension is clarified by noting the kinds of neighborhoods that would score low scores on A: they include relatively more persons who are foreign-born or nonwhite (Nw), foreign persons who do *not* come from Protestant Europe (Fe), more unskilled men among blue-collared workers (Sm), more blue-collared women among those at work (Wf), and an excess of males (F), as in Skid Row.

The demographic attributes that form the "socioeconomic gradient" range from those variables which roughly demark assimilation at one end to those which clearly denote socioeconomic independence at the other. In between is a group of variables that could serve as definers either of one or the other of the extreme clusters. A collinear "dependent cluster" is formed out of those in the middle, Ach, social achievement. On this "unnecessary" dependent dimension, neighborhoods with high cluster scores represent success in various forms of middle-class achievements: being employed (Em, Ef), men in white-collar work (Wm), in undercrowded homes (Uc), of high rent when rented (Rn), and equipped with modern appliances (Re, Ch).

The generality of the core three clusters, S, F, and A is statistically given in Table 7.5, Sec. B. The two dimensions S and A at the ends of the socioeconomic gradient are more general than F, since they encompass more variables. The correlations among the three clusters are given in Sec. C of the table, showing, what is also pictorially represented in SPAN, that family life is quite independent of socioeconomic independence: the zero correlation means that neighborhoods with high home life may be rich or poor, that one can live in solitary splendor or as an isolate in a hole in the wall. On the other hand, assimilation is mildly related to family life, there being some tendency for neighborhoods of the social majority, high E, to orient around family life, high F.

The steps by which the cluster structure of the social areas is unfolded are indicated by the data of Table 7.6. In the initial factoring, four orthogonal dimensions are found. The factor coefficients of all the variables are given in the table, as well as the distribution of their residuals. The sufficiency of the initial four dimensions is given at the bottom of the table. The fourth dimension accounts for only 6 percent of the estimated communalities. The definers of this fourth dimension are given in the column Definers (Revised). We concluded that they are a trivial meaningless doublet.

The factoring is therefore revised, following the criteria described earlier. For example, the variables Sc and Wm are deleted; they are marked by the notation "C_1 out" in Table 7.6 and by rejecting arrows in SPAN. Variables Co and Om are added to C_1. The purpose is both to increase the independence of this S cluster from assimilation and to put more stress on the meaning of social independence in the revised cluster. Similar principles governed the revision of the third cluster: by rejecting all the initial definers of this cluster and by replacing them with those circled as C_3 in

TABLE 7.6 EMPIRICAL VS. REVISED KEY-CLUSTER FACTORING: SOCIAL-AREA STUDY

Variables	Definers (Revised)	Empirical Factoring								Revised Factoring						
		Orthogonal Factor Coefficients and Communalities					Residuals			Orthogonal Factor Coefficients and Communalities				Residuals		
		C_1	C_2	C_3	C_4	h^2	.0	.1	.2	C_1	C_2	C_3	h^2	.0	.1	.2
Socioeconomic independence, S																
2 Mm	C_1 out	.97	.08	−.18	−.01	.98	33			1.02	.13	.07	1.05	33		
3 Sc	C_1 out	.94	.07	.12	.00	.89	33			.90	.12	.38	.97	33		
1 Wm	C_1 out	.95	−.09	−.10	−.05	.93	33			.90	−.04	.33	.92	32		
27 Df		.78	−.27	−.29	.13	.78	31	2		.81	.32	−.10	.78	31	1	
26 Vl		.84	−.05	.05	−.06	.71	33			.81	−.01	.13	.68	32	2	
29 Co	C_1 new	.79	−.18	.30	.14	.77	32	1		.80	−.15	−.04	.66	31	1	
4 Ch		.87	−.04	.24	−.05	.85	33			.79	.00	.39	.77	31	2	
28 Om	C_1 new	.68	.01	.08	.19	.56	31	2		.70	.03	−.16	.56	29	4	
22 Uc		.75	.31	.31	.33	.78	33			.70	.35	.30	.71	31	2	
21 Ef		.62	.14	−.31	.33	.60	32	1		.60	.17	.32	.49	24	8	1
17 Mf	C_1 new	.59	−.22	−.05	.02	.39	27	5		.59	−.19	.26	.45	22	9	2
Family life, F																
7 Fl	C_2	.07	.92	.04	−.13	.87	33			.04	.93	−.07	.88	31	2	
6 Oo	C_2	.24	.92	.07	.21	.95	33			.16	.92	−.12	.90	32	1	
8 Fd	C_2	.16	.89	−.01	.24	.87	33			.09	.88	−.10	.80	32	2	
9 Uf	C_2	−.17	.87	.11	.01	.79	31			−.20	.87	−.13	.81	31	2	
10 Am	C_2	−.12	.87	.00	−.33	.88	32	2		−.16	.83	.01	.72	30	2	
14 Cp	C_3 out	.29	.78	.45	.07	.91	31	1		.15	.80	.39	.82	26	7	
13 Wo		.22	.69	.20	.16	.58	32	1		.14	.70	.23	.56	31	2	
16 Cf		.49	−.71	.21	−.09	.79	33	1		.49	.69	.32	.80	31	2	
15 Af	C_4 out	.50	−.68	.35	.42	1.01			1	.41	.66	.53	.89	30	3	1
31 Is	C_4 out	.23	.53	−.29	−.40	.58	33			.30	.55	−.33	.50	25	7	
Assimilation, A																
11 Sm	C_3 new	.33	.52	.43	.18	.60	28	4		.15	.54	.73	.85	30	3	
5 Bf	C_4 out	.66	.02	.55	.06	.75	32	1		.51	.05	.70	.75	33		
12 Nw	C_3 new	.43	.31	.44	.54	.77	27	4	1	.25	.33	.70	.66	32	1	
18 Em	C_3 out	.66	.45	.50	−.14	.90	33			.53	.40	.49	.76	24	9	3
20 Re	C_3 out	.69	.34	.45	.15	.82	32	1		.55	.38	.53	.79	30	3	
19 Rn		.81	.22	.48	−.07	.93	33			.69	.26	.50	.65	26	7	
25 Wf	C_3 out	.47	−.04	.64	−.06	.64	30	3		.27	.06	.75	.65	28	5	
24 Fe	C_3 new	.52	−.10	.21	.19	.36	28	4		.42	.12	.48	.42	31	2	
23 F	C_3 new	.48	.21	.46	.00	.49	31	2		.35	.24	.42	.35	24	9	
30 Gr		.42	−.13	.44	−.06	.39	31	4		.34	−.11	.38	.28	27	7	
32 Um		.03	−.01	.51	−.03	.26	29			−.05	.01	.49	.24	25	5	1
Unclustered																
33 Of		.08	.28	.32	−.02	.19	29	3		−.04	.18	.23	.11	19	12	2
% communality reproducibility		.47	.32	.15	6					.43	.34	.23				
% of total residuals							96	4						88	11	1

the SPAN diagram of Fig. 7.5, its meaning was shifted from social achievement to assimilation. The fourth dimension, being trivial, was completely eliminated in the revision. The results of revision are shown in Table 7.6 under Revised Factoring. Of special interest is that the residuals, after the revised third dimension, are only slightly greater than the initial residuals after the fourth blind dimension.

To summarize the findings on the structure of demographic characteristics of these urban people: by empirical cluster methods we have discovered three meaningful salient cluster-defined dimensions of their social structure, S, socioeconomic independence, F, family life, and A, assimilation. The rational census categories, which group attributes by their population characteristics, occupations, and home features, provided no helpful clues about such dimensions; we find, indeed, that the definers of each of the empirically derived clusters cut across the rational categories. A new fourth cluster, social achievement, could also be used to describe the people, but its defining variables are largely statistically dependent on the three core clusters and therefore are not needed as a core dimension.

When we come later to the phase of locating social areas by O-analysis procedures, we must make a final decision about how to score each neighborhood. For scoring purposes we take the definers of the family life dimension as revised, since it is sharply differentiated in the configuration. Most of the other variables lie on a socioeconomic gradient that has no sharp cutoff into clusters. The assimilation end of the gradient is not as tight in its definers as we would like, but there seems to be no way to improve its scoring. At the independence end, we probably should eliminate VI, high home value, because it is scored in dollars, a metric that can change critically and rapidly over time due to the shifting recession-inflation tides of the economic well-being of the society. This would leave only three definers, too few for reliable scoring. We therefore add two other definers, Df, domestics (living in), and Uc, undercrowded. The cluster extended by these two attributes gives a highly reliable cluster score and also augments the independence of the central construct at the "high" end of the socioeconomic gradient.

The statistical description of the finally revised clusters (as computed in a rerun of CSA) is about the same as before for the three clusters S, F, and A. The new dependent cluster Ach, social achievement, composed of seven definers, forms a cluster score with a very high α reliability of .97. As is evident in the SPAN diagram, the Ach domain is highly correlated with S and A, its interdomain correlations with them being .80 and higher and its raw correlations not being much lower. The correlation of Ach with F, family life, is quite low, of the order .20.

Cluster structure of voting attitudes

The power of cluster structure analysis to reduce many variables to a small meaningful set of collinear clusters that accounts for the generality in them all is perhaps no better illustrated than in the voting-attitude study. The question asked is this: In a typical election, into how many attitude

TABLE 7.7   THE VOTING-ATTITUDE VARIABLES

Issues in Order of Presentation to Voters		Alphabetical Listing	
State issues:			
1 DmGv	Democrat for governor	28 AdOf	Administrative officer (city) appointment
2 VtBn	Veterans bonds	25 AgBn	Aged home bonds
3 ScBn	School bonds	4 AlCn	Alcohol control
4 AlCn	Alcohol control	19 AlnP	Alien property ownership ok'd
5 NdAg	Needy aged aid	26 BlJb	Blind city jobs ok'd
6 Vsl1	Vessel (carrier) tax-exempt	31 CbCr	Cable cars saved
7 LgPy	Legislators' pay increase	10 ChEx	Church tax-exempt
8 LnTt	Land title law	21 Chrt	Charters (county) loosened
9 Vsl2	Vessel (fishing) tax-exempt	15 CoEx	College (private) tax-exempt
10 ChEx	Church tax-exempt	1 DmGv	Democrat for governor
11 TrOf	Terms of office	12 DsVt	Disabled veterans tax-exempt
12 DsVt	Disabled veterans tax-exempt	22 ExBn	Exhibit hall bonds
13 ExFl	Ex-felon voting	13 ExFl	Ex-felon voting
14 VrCt	Vernon City charter	17 GvWt	Government water state control
15 CoEx	College (private) tax-exempt	24 HsBn	Hospital bonds
16 WIEx	Welfare institutions tax-exempt	29 HsEm	Hospital employee benefits
17 GvWt	Government water state control	7 LgPy	Legislators' pay increase
18 Pk	Parking facilities by state	8 LnTt	Land title law
19 AlnP	Alien property ownership ok'd	5 NdAg	Needy aged aid
20 NnLw	Nonlawyer judges ok'd	20 NnLw	Nonlawyer judges ok'd
21 Chrt	Charters (county) loosened	18 Pk	Parking facilities by state
City issues:		23 RcBn	Recreation center bonds
22 ExBn	Exhibit hall bonds	3 ScBn	School bonds
23 RcBn	Recreation center bonds	30 ShEm	Sheriff's employee benefits
24 HsBn	Hospital bonds	27 SpPy	Supervisors' pay
25 AgBn	Aged home bonds	11 TrOf	Terms of office
26 BlJb	Blind city jobs ok'd	14 VrCt	Vernon City charter
27 SpPy	Supervisors' pay	6 Vsl1	Vessel (carrier) tax-exempt
28 AdOf	Administrative officer (city) appointment	9 Vsl2	Vessel (fishing) tax-exempt
29 HsEm	Hospital employee benefits	2 VtBn	Veterans bonds
30 ShEm	Sheriff's employee benefits	16 WIEx	Welfare institutions tax-exempt
31 CbCr	Cable cars saved		

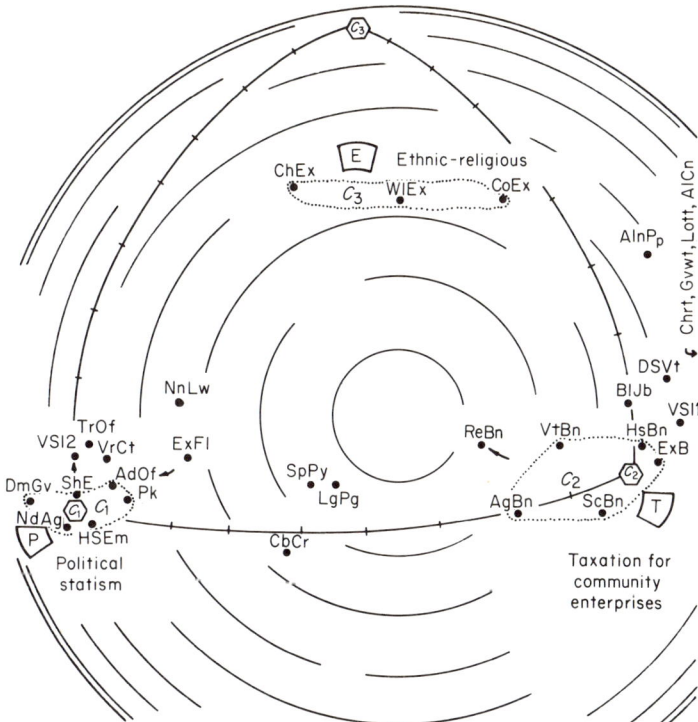

FIGURE 7.6
Spherical configuration showing the relationships among 31 election issues as voted on by 200 random San Francisco neighborhoods (precincts) in the election of 1954.

clusters do the wide array of issues on which the small neighborhoods (precincts) of a city vote meaningfully group? An empirical key-cluster factoring followed by cluster structure analysis gives a clear answer. The data consist of the 1954 voting in San Francisco of 200 random precincts drawn from more than 1,300 precincts. The issues studied were 30 state and city propositions plus the vote for governor. The code names and brief descriptions of the 31 issues are listed in Table 7.7 and for ready reference they are repeated alphabetically.

As one reads through the list it is difficult to think up any reduced set of meaningful rational categories into which to cast all 31 of these issues. However, the cluster structure of the issues becomes clearly revealed pictorially in the SPAN configuration of Fig. 7.6 and in the statistical unifactorial allocation of them given in Table 7.8. The tight collinear cluster marked P at the left on the SPAN diagram is rather sharply differentiated

TABLE 7.8 CLUSTER AND FACTOR STRUCTURE: THE VOTING–ATTITUDE STUDY (OBLIQUE UNIFACTOR STRUCTURE IN CSA)

A. Inner Structure of the Clusters

Variables	Definers	Oblique Fc	h^2	$\bar{r}$	Reliability Single	Reliability Cumulative	Cluster Score on D's
C_1, political statism, P							
5 NdAg	D	.96	.93	.86			
30 ShEm		.92	.85	.82	.96	.96(1)	
29 HsEm	D	.90	.80	.80			
1 DmGv	D	.86	.77	.77			
18 Pk	D	.83	.73	.74			
9 Vsl2		.82	.69	.73	.95	.96(3)	
13 ExFl		.76	.72	.68	.95	.96(2)	
28 AdOf		.63	.42	.56	.93	.96(5)	
14 VrCt		.61	.40	.55	.93	.95(6)	
27 SpPy		.56	.57	.49	.94	.96(4)	
11 TrOf		.53	.29	.47	.92	.95(10)	
20 NnLw		.51	.34	.46	.92	.95(9)	
31 CbCr		.51	.42	.45	.93	.95(8)	
4 AlCn		− .50	.44	.45	.93	.95(7)	
Domain validity							.97
Reliability							.94
C_2, taxation, T							
24 HsBn	D	.83	.71	.64			
25 AgBn	D	.82	.72	.63			
22 ExBn	D	.80	.67	.62			
23 RcBn		.77	.67	.59	.90	.90(1)	
3 ScBn	D	.73	.53	.56			
2 VtBn	D	.66	.44	.51			
21 Chrt		.64	.60	.50	.90	.92(3)	
26 BlJb		.60	.38	.46	.89	.93(6)	
8 LnTt		.59	.71	.45	.90	.91(2)	
17 GvWt		.58	.50	.44	.89	.93(4)	
19 AlnP		.56	.46	.43	.89	.93(5)	
12 DsVt		.46	.23	.35	.87	.93(7)	
6 Vsl1		.45	.22	.34	.87	.93(8)	
7 LgPy		.36	.22	.28	.87	.92(9)	
Domain validity							.94
Reliability							.88
C_3, ethnic-religious, E							
16 WlEx	D	.87	.77	.72			
10 ChEx	D	.81	.70	.67			
15 CoEx	D	.78	.66	.65			
Domain validity							.94
Reliability							.87

TABLE 7.8 CLUSTER AND FACTOR STRUCTURE: THE VOTING–ATTITUDE STUDY (OBLIQUE UNIFACTOR STRUCTURE IN CSA) (Continued)

B. Generality of Clusters

	C_1	C_2	C_3
Reproducibility of mean squares of correlations	.50	.42	.31
Reproducibility of communalities	.48	.44	.38

C. Original and Interdomain Correlations between Clusters

	Original			Interdomain (cos θ)		
	C_1	C_2	C_3	C_1	C_2	C_3
C_1	(.94)	.09	.29	1.00	.09	.32
C_2	.09	(.88)	.55	.09	1.00	.63
C_3	.29	.55	(.87)	.32	.63	1.00

from the almost completely independent cluster marked T on the right, whereas a third small cluster above labeled E is correlated with them both.

What do these three cluster-defined dimensions of social attitudes mean? The circled definers of the P cluster at the left consist of DmGv, Democrat for governor; HsEm, hospital employees benefits; Pk, parking lots operated as a state business; and NdAg, needy aged aid by state pensions, the old California "ham and eggs" perennial ballot issue that goes back to the Great Depression. No profound conceptualization is required to recognize this as the dimension of political statism vs. rugged individualism, one that is commonly believed to split Democrats from Republicans. Note that the circled second cluster T at the right consists entirely of bond issues that are supported by property taxes. There seems to be no doubt that this dimension means a readiness to suffer taxation for community enterprises. The correlated cluster E above consists of three issues that seek approval for relief from taxation of certain religious and ethnic institutions, hence designated as the ethnic-religious dimension.

The data on the inner structure of these three clusters, given in Table 7.8, show that all three have α reliabilities of the order .90. The political and taxation dimensions have the most generality, with political largest. The correlations among these clusters show what is also revealed in the SPAN configuration, namely, that the two main clusters are quite independent. (This finding comes as a surprise to some Democrats who expect heavily Democratic neighborhoods to tend to support taxation for community works. It still appears, however, that nobody, whatever his political hue, really loves the tax collector when voting in the sanctum of the election booth.) The fact that the ethnic dimension is more closely related to the taxation dimension than to the political dimension suggests that perhaps a tax money pinch has a stronger adverse effect on sympathy for minorities than a political conviction that one ought to support them.

How this empirical analysis, followed by revised factoring, produces the above findings is seen in the details of Table 7.9. At the bottom of the Empirical Factoring columns the third and fourth dimensions are shown to contribute only a small common variance as orthogonal dimensions. The definers of the fourth dimension do not lend themselves to ready interpretation. In the revised factoring the fourth dimension was dropped. The deletion and addition of definers to the first and second dimension, as shown in the column under Definers, was largely in the interest of increasing the independence of the two sets of definers or in increasing prima facie meaning. The three-dimensional solution under Revised Factoring gave about the same fit as the initial empirical four, and though the residuals are increased a little, they are quite trivial.

The general conclusion from this analysis need not be labored. Small neighborhoods vote mainly in terms of two general attitudes that differentiate them, P, political statism, and T, taxation. The attitude structure appears to be more sharply differentiated than the demographic. Almost all the 31 issues on which the people voted polarize around the two attitude clusters. On the other hand, many of the demographic characteristics form a graded series through the socioeconomic gradient. Perhaps this comparative result is to be expected. Attitudes are organized conceptions in the minds of people supported by strong emotions and are thus subject to simplifying stereotypy. Demographic characteristics, however, are not "mental" phenomena but the objective expression of a multitude of uncontrolled social and biological forces not so easily subjected to simplified structuring.

At a later point we score neighborhoods on the three voting-attitude clusters to find out how well variations between the neighborhoods in these attitudes are predictable from their three objective demographic scores on the socioeconomic independence, family life, and assimilation clusters. Since actual scores on individuals should be as prima facie meaningful as possible and should representatively sample a cluster domain, we change the definers of P, political, by deleting Pk, parking facilities by the state, and replace it by the broader issue, ExFl, one that favors giving ex-felons back their vote after paying their debt to society. For the T, taxation, cluster, one bond issue VtBn, veterans bonds, is deleted because the votes by the large block of veterans might be more a function of self-interest than of atti-

TABLE 7.9 EMPIRICAL VS. REVISED KEY-CLUSTER FACTORING: VOTING-ATTITUDE STUDY

Variables	Definers (Revised)	Empirical Factoring — Orthogonal Factor Coefficients and Communalities					Empirical — Residuals			Revised Factoring — Orthogonal Factor Coefficients and Communalities				Revised — Residuals		
		C_1	C_2	C_3	C_4	h^2	.0	.1	.2	C_1	C_2	C_3	h^2	.0	.1	.2
Political statism, P																
5 NdAg	C_1 out	.92	.02	−.08	.01	.86	31			.96	.02	−.04	.93	31		
30 ShEm	C_1 out	.93	.00	.02	.04	.87	31			.92	.02	.04	.85	31		
29 HsEm	C_1 new	.95	−.03	−.05	−.03	.90	31			.90	.00	−.02	.80	31		
1 DmGv	C_1 new	.85	−.19	−.01	.16	.79	31			.86	−.18	.01	.77	31		
18 Pk	C_1 out	.83	.19	.00	.09	.74	31			.84	.16	.05	.73	31		
9 Vsl2		.80	.01	−.11	−.01	.65	29	2		.82	.01	.13	.69	30	1	
13 ExFl		.73	.34	.18	.00	.68	29	2		.76	.34	.16	.72	31		
28 AdOf	C_4 out	.66	.14	.01	.51	.71	29	2		.63	.12	.05	.42	27	3	1
14 VrCt	C_4 out	.62	.10	−.11	.40	.57	31			.61	.10	.10	.40	28	2	1
27 SpPy		.57	.53	.04	.05	.61	29	2		.56	.50	.08	.57	30	1	
11 TrOf		.52	.06	.05	.05	.28	31			.53	.05	.09	.29	31		
20 NnLw		.50	.25	.19	.15	.38	28	3		.51	.21	.18	.34	26	5	
31 CbCr	C_4 out	.49	.41	−.09	.02	.42	31			.51	.40	−.05	.42	31		
4 AlCn		−.49	.45	.07	.35	.57	31			−.50	.43	.07	.44	25	6	
Taxation, T																
24 HsBn	C_2	.00	.85	.01	.11	.74	31			−.03	.84	.04	.71	31		
25 AgBn	C_2	.29	.79	−.04	−.05	.70	31			.28	.80	.03	.72	31		
22 ExBn	C_2 out	.08	.86	−.04	.01	.76	31			.08	.82	.00	.63	31		
23 RcBn	C_2 out	.29	.74	−.10	−.05	.65	31			.31	.74	.13	.67	31		
3 VtBn	C_2 new	.06	.68	−.03	.00	.47	30	1		.07	.72	.07	.53	30	1	
2 VtBn		−.15	.65	−.11	−.15	.47	29	2		−.13	.65	.06	.44	28	3	
21 Chrt		.27	.68	.31	−.03	.63	29	2		.25	.61	.25	.60	29	2	
26 BlJb		.04	.60	.16	.16	.39	26	4	1	.01	.61	.11	.38	27	3	1
8 LnTt		−.04	.64	.15	−.15	.71	30	1		−.11	.64	.11	.71	29	2	
17 GvWt	C_4 out	−.32	.61	.13	.40	.65	31			−.33	.58	.14	.50	27	3	1
19 AlnP	C_3 out	−.16	.61	.31	−.04	.49	30	1		−.15	.61	.32	.46	28	3	
12 DsVt		.04	.45	.07	.11	.22	28	1	2	.07	.46	−.09	.23	30	1	
6 Vsl1		.04	.44	.03	−.09	.20	29	1	1	−.08	.46	−.04	.22	30	1	
7 LgPy		−.39	.35	.07	.04	.27	23	7	1	−.32	.33	.05	.22	25	4	2
Ethnic-religious, E																
16 WlEx	C_3	.32	.53	.61	.00	.76	31			.28	.53	.64	.77	31		
10 ChEx	C_3	.43	.36	.60	.15	.70	31			.42	.35	.63	.70	31		
15 CoEx	C_3	.11	.61	.56	−.15	.71	31			.09	.60	.54	.66	31		
% communality reproducibility		46	41	8	5	96				49	43	8	95			
% of total residuals							96	3	1					95	4	1

tudes toward taxation. The E, ethnic, definers are left as they stand. These changes do not (as it happens) in any substantial way alter the statistical description of the structure (as computed by a rerun of CSA).

Domain sampling theory of structure

The term cluster analysis has one unfortunate connotation: it may imply that variables of a study *should* group themselves nicely into tight clusters; however, on domain sampling theory no sharp clustering is necessarily to be expected. The issue goes to the question of the basic causes of individual differences. In the domains of Holzinger abilities, for example, the causes of individual differences among the children are the thousands of genes lying in hosts of loci in many chromosomes; combinations of these determinants form the many varieties of genotypes that constitute the observed group of children. These genotypes are complexly influenced by innumerable varying environmental fields in which they develop and which they seek out. The 24 tests sample 24 different domains of these basic determinants, organized into five rationally designed special stimulating types of selected situations. As a consequence, these 24 selections produce the configuration in the SPAN diagrams of Figs. 7.3 and 7.4. But if different selections had been made, a different configuration might have been discovered. Ideally, if all varieties of abilities one could possibly think up had been used in the study (pity the children!), the configuration on the SPAN diagrams would have been a vast swarm of points grading across the surface of many spheres.

Similarly, the configuration of demographic characteristics of Bay Area neighborhoods seen in Fig. 7.5 is that of only one selection of many possible demographic features. With ingenuity (and a large budget) one might be able to find demographic characteristics that could fill any presently empty place in the configuration. Indeed, we find that in the socioeconomic gradient characteristics spread across the region like a Milky Way. The lacunae between this galaxy and the family life cluster may be due simply to failure to think of or presently be interested in demographic features that could fill these empty places. In the voting-attitude study, the differentiation into the general cluster groups seen in Fig. 7.6 does not necessarily mean that no social attitudes exist in people that could fill the empty places presently evident between the three cluster groups. Issues that would have elicited such attitudes happened to be absent from the 1954 ballot.

Generally then, one would not necessarily expect to discover sharply different clusters of attributes in any study, but if such a structure does exist, one would not necessarily impute to any of the clusters the property of it having a primary fundamental character. The clusterings that do appear result from complex selective forces at work at the time a study is undertaken, forces in the thinking of the investigator that cause him to measure some attributes and not others or social and biological forces currently at work generating special correlations between some attributes but not between others in the particular samples of subjects being studied (see Tryon 1935, 1940). On another occasion in time, or under the aegis of another investigator of a different theoretical bent, or in another quite different sampling of subjects, quite a different set of forces might generate a different configuration.

In contrast to this pluralistic theory of a vast number of biological and social domains or components being sampled selectively in any particular study there is the opposing monistic theory of one "underlying factor" with a few secondary special components (Spearman, 1927) or the neomonistic theory of a few general factors (Thurstone, 1935). The latter doctrine reached its most sophisticated form in the belief in a "simple structure" (Thurstone, 1947, chap. 14). There has been much confusion in efforts to understand this vague doctrine of *simplicité*. It should, however, be clear if its meaning is considered in terms of the configurations on SPAN diagrams (also see the relatively clear statement in Harman, 1967, pp. 112ff).

The simple structure theory is as follows. If variables were proper measures, they would be of "complexity 1," i.e., each would be a "pure" measure of one factor only and not of more than one. Such pure tests would therefore have a clear unifactorial structure and lie in tight clusters at corners of spherical triangles. The primary "factors" or axes at the corners might be orthogonal, whence the configuration would be termed an "orthogonal simple structure." But they are more likely to be correlated, in which case they would constitute an "oblique simple structure." With actually observed attributes of people, it seems impossible to find pure variables. One must therefore settle for the expectation that variables should be of complexity 2, i.e., lie in planes along the arcs connecting corners of the configuration, or be of greater complexity, i.e., lying in hyperplanes. However, variables should, if properly constructed, all be encompassed *within* the triangular figure, in which case their factor coefficients on the "primary" dimensions at the corners are all positive, a shape of things called a "positive manifold." Many efforts have been made to spell out the analytic conditions by which such a simple structure can be computationally identified (see Harman, 1967 pp. 290ff), but there does not appear yet to be any agreement on the matter. The belief in a parsimonious set of underlying factors that would generate a simple structure appears to be unsupported by any of the basic biological or social sciences.

A few special points are now made here on aspects of structure analysis which were not treated above, because they would have encumbered the flow of thought.

The clustered correlation matrix

From knowledge of the oblique structure one can arrange the original correlations among the variables so that the correlation matrix itself reveals the cluster structure. Such a clustered correlation matrix presents a meaningfully organized set of data which would otherwise be a chaotic mass of numbers.

In the clustered correlation matrix the variables along the borders are arranged by clusters. The order of the clusters themselves can usually be meaningfully arrayed in accordance with their structural interrelationships (as revealed in CSA and SPAN). Often the cluster order, however, is arbitrary. The actual process of clustering the matrix is performed in the BC TRY System by one quick pass of a special program called REDE (which means REorder, DElete, or reflect rows and columns of a matrix).

To demonstrate how meaningful the clustered matrix is, Table 7.10 shows the clustered matrices of the three studies with which we have been dealing. In the abilities study, Sec. A, the ordering of the variables comes from our final decision on the definers of the V, S, F, and M clusters and on the configuration of their relationships (as revealed in the SPAN configurations of Figs. 7.3 and 7.4). As a result, the submatrices along the principal diagonal show relatively elevated correlations among the definers of clusters. The definers of each cluster are highly collinear, as described and illustrated in Chap. 4 (see Figs. 4.1 to 4.3). The relations between clusters are also clearly revealed in the cross-correlation submatrices sectioned off in the off-diagonal submatrices.

We have included all the variables of the three studies in their clustered matrices in Secs. A to C of Table 7.10 in order to have in one place the full set of data on which the various analyses given in this book are based. We do not discuss these three matrices in detail but summarize below the salient advantages of the clustered matrix, which can be confirmed by a study of the three matrices in Table 7.10.

TABLE 7.10 CLUSTERED CORRELATION MATRICES

A. The Holzinger Study of Abilities

	Verbal					Speed				Space				Memory					Unclustered					
	V9 Wmn	V7 Snt	V6 Cmp	V5 Inf	V8 Wcl	S10 Add	S12 Cnt	S11 Cod	S13 Scc	F4 Loz	F1 Vis	F2 Cub	F3 Pfb	M14 Wrc	M17 Wn	M16 Frc	M15 Nrc	M18 Nf	N20 Ded	N22 Rsn	N23 Ser	N21 Puz	N24 Ari	M19 Fw
Verbal																								
V9 Wmn		.69	.71	.72	.54	.18	.12	.25	.27	.35	.32	.19	.26	.25	.29	.24	.21	.21	.45	.49	.50	.27	.42	.28
V7 Snt			.72	.65	.63	.20	.20	.25	.36	.24	.31	.16	.27	.23	.25	.20	.16	.18	.46	.40	.40	.31	.44	.24
V6 Cmp				.62	.52	.10	.10	.36	.34	.18	.31	.23	.26	.32	.36	.31	.24	.35	.43	.39	.43	.26	.43	.29
V5 Inf					.57	.07	.17	.18	.34	.37	.49	.25	.39	.19	.19	.22	.14	.24	.43	.44	.43	.32	.42	.20
V8 Wcl						.24	.27	.30	.41	.38	.33	.16	.33	.24	.21	.30	.17	.26	.43	.36	.49	.36	.39	.26
Speed																								
S10 Add							.59	.47	.42	.16	.13	.08	.25	.16	.30	.24	.21	.20	.16	.25	.20	.40	.53	.20
S12 Cnt								.42	.53	.14	.17	.01	.13	.13	.28	.20	.16	.35	.15	.20	.25	.36	.41	.11
S11 Cod									.53	.29	.42	.29	.31	.32	.28	.29	.25	.33	.20	.32	.35	.40	.41	.29
S13 Scc										.37	.33	.19	.19	.19	.21	.18	.19	.33	.25	.29	.38	.43	.36	.25
Space																								
F4 Loz											.45	.42	.37	.37	.14	.29	.21	.35	.33	.38	.34	.41	.23	.18
F1 Vis												.33	.37	.38	.17	.42	.22	.36	.36	.41	.47	.37	.27	.27
F2 Cub													.19	.39	.01	.29	.13	.26	.30	.23	.34	.29	.21	.11
F3 Pfb														.20	.13	.31	.19	.26	.21	.16	.29	.33	.12	.15
Memory																								
M14 Wrc															.37	.31	.39	.20	.31	.25	.25	.18	.29	.22
M17 Wn																.34	.35	.45	.28	.30	.30	.19	.34	.33
M16 Frc																	.31	.35	.39	.29	.35	.35	.27	.25
M15 Nrc																		.32	.26	.25	.25	.23	.16	.18
M18 Nf																			.31	.32	.27	.36	.41	.36
Unclustered																								
N20 Ded																				.48	.51	.40	.37	.17
N22 Rsn																					.51	.37	.38	.35
N23 Ser																						.45	.43	.30
N21 Puz																							.45	.33
N24 Ari																								.37
M19 Fw																								

TABLE 7.10 CLUSTERED CORRELATION MATRICES (Continued)

B. The Social-area Study (Demography)

Column groups — **Socioeconomic Independence:** 2 Mm, 27 Df, 29 Co, 26 Vl, 28 Om, 22 Uc · **Social Achievement:** 19 Rn, 3 Sc, 4 Ch, 1 Wm, 18 Em, 20 Re · **Family Life:** 7 Fl, 6 Oo, 8 Fd, 9 Uf, 10 Am

	2 Mm	27 Df	29 Co	26 Vl	28 Om	22 Uc	19 Rn	3 Sc	4 Ch	1 Wm	18 Em	20 Re	7 Fl	6 Oo	8 Fd	9 Uf	10 Am
Socioeconomic Independence																	
2 Mm																	
27 Df	.81																
29 Co	.83	.59															
26 Vl	.81	.66	.63														
28 Om	.73	.67	.55	.61													
22 Uc	.75	.69	.55	.58	.48												
Social Achievement																	
19 Rn	.72	.56	.45	.66	.48	.68											
3 Sc	.92	.73	.78	.75	.56	.79	.80										
4 Ch	.78	.56	.59	.75	.52	.64	.85	.82									
1 Wm	.89	.67	.69	.82	.62	.67	.80	.89	.86								
18 Em	.59	.48	.31	.52	.36	.67	.87	.70	.68	.61							
20 Re	.63	.52	.35	.51	.35	.73	.82	.74	.76	.67	.81						
Family Life																	
7 Fl	.14	.34	−.16	.04	.13	.35	.32	.10	.01	.00	.50	.35					
6 Oo	.28	.40	−.02	.17	.11	.50	.43	.28	.25	.14	.60	.52	.85				
8 Fd	.23	.35	.00	−.05	.01	.45	.29	.25	.14	.05	.47	.49	.76	.91			
9 Uf	−.09	.12	−.30	−.17	−.12	.07	.00	−.14	−.23	−.23	.20	.12	.80	.74	.76		
10 Am	−.04	.15	−.24	−.19	−.06	.15	.08	−.03	−.21	−.20	.35	.18	.85	.69	.68	.75	
Assimilation																	
11 Sm	.27	.24	.02	.20	.01	.44	.52	.42	.35	.34	.57	.57	.43	.60	.58	.42	.36
12 Nw	.38	.25	.12	.29	.05	.57	.54	.52	.47	.41	.57	.66	.25	.45	.42	.17	.15
25 Wf	.33	.08	.17	.34	.08	.36	.70	.53	.60	.55	.67	.60	.06	.19	.09	−.08	−.01
24 Fe	.46	.37	.37	.37	.20	.47	.50	.59	.47	.51	.43	.48	.04	.27	.27	−.03	.04
23 F	.41	.33	.15	.36	.27	.56	.61	.56	.45	.45	.67	.60	.26	.27	.21	−.04	.32
Unclustered																	
21 Ef	.57	.40	.44	.54	.46	.49	.69	.61	.60	.56	.72	.55	.23	.30	.19	−.05	.05
30 Gr	.32	.20	.26	.38	.18	.32	.56	.42	.50	.43	.47	.34	−.01	.08	−.11	−.26	−.21
5 Bf	.55	.42	.38	.49	.30	.59	.79	.73	.73	.67	.68	.75	.06	.21	.16	−.15	−.08
14 Cp	.26	.35	−.09	.21	.11	.50	.63	.35	.34	.25	.77	.68	.81	.81	.69	.60	.65
13 Wo	.23	.31	.00	.12	.09	.46	.37	.30	.18	.15	.51	.51	.64	.68	.66	.57	.55
−16 −Cf	−.41	−.13	−.47	−.43	−.32	−.26	−.35	−.46	−.51	−.53	−.12	−.21	.61	.49	.52	.79	.68
−15 −Af	−.36	−.14	−.39	−.44	−.18	−.31	−.39	−.48	−.56	−.55	−.16	−.30	.63	.42	.45	.75	.73
−17 −Mf	−.59	−.34	−.65	−.46	−.30	−.42	−.34	−.62	−.43	−.51	−.25	−.27	.23	.07	.05	.26	.26
31 Is	.32	.34	.25	.14	.30	.16	.19	.24	.04	.16	.33	.13	.55	.42	.39	.41	.63

B. The Social-area Study (Demography)

	Assimilation								Unclustered					
	11 Sm	12 Nw	25 Wf	24 Fe	23 F	21 Ef	30 Gr	5 Bf	14 Cp	13 Wo	-16 -Cf	-15 -Af	-17 -Mf	31 Is
Socioeconomic Independence														
2 Mm	.27	.38	.33	.46	.41	.57	.32	.55	.26	.23	-.41	-.36	-.59	.32
27 Df	.24	.25	.08	.37	.33	.40	.20	.42	.35	.31	-.13	-.14	-.34	.34
29 Co	.02	.12	.17	.37	.15	.44	.26	.38	-.09	.00	-.47	-.39	-.65	.25
26 Vl	.20	.29	.34	.37	.36	.54	.38	.49	.21	.12	-.43	-.44	-.46	.14
28 Om	.01	.05	.08	.20	.27	.46	.18	.30	.11	.09	-.32	-.18	-.30	.30
22 Uc	.44	.57	.36	.47	.56	.49	.32	.59	.50	.46	-.26	-.31	-.42	.16
Social Achievement														
19 Rn	.52	.54	.70	.50	.61	.69	.55	.79	.63	.37	-.35	-.39	-.34	.19
3 Sc	.42	.52	.53	.59	.56	.61	.42	.73	.35	.30	-.46	-.48	-.62	.24
4 Ch	.35	.47	.60	.47	.45	.56	.50	.73	.34	.18	-.51	-.56	-.43	.04
1 Wm	.34	.41	.55	.51	.45	.56	.43	.67	.25	.15	-.53	-.55	-.51	.16
18 Em	.57	.57	.67	.43	.67	.72	.47	.68	.77	.51	-.12	-.16	-.25	.33
20 Re	.57	.66	.60	.48	.60	.55	.34	.75	.68	.51	-.21	-.30	-.27	.13
Family Life														
7 Fl	.43	.25	.06	.04	.26	.23	-.01	.06	.81	.64	-.61	-.63	-.23	.55
6 Oo	.60	.45	.19	.27	.27	.30	-.08	.21	.81	.68	-.49	-.45	-.07	.42
8 Fd	.58	.42	.09	.27	.21	.19	-.11	.16	.69	.66	-.52	-.45	-.05	.39
9 Uf	.42	.17	-.08	-.03	-.04	.05	-.26	-.15	.60	.57	-.79	-.75	-.26	.41
10 Am	.36	.15	-.01	-.04	.32	-.05	-.21	-.08	.65	.55	-.68	-.73	-.26	.63
Assimilation														
11 Sm						.40	.18	.53	.66	.53	-.14	-.02	-.20	.15
12 Nw	.71					.34	.30	.60	.57	.50	-.13	-.31	-.26	-.16
25 Wf	.63	.54				.55	.49	.66	.44	.21	-.27	-.42	-.31	-.02
24 Fe	.58	.50	.48			.40	.26	.58	.23	.21	-.23	-.35	-.41	.04
23 F	.42	.54	.51	.28		.37	.31	.58	.54	.37	-.29	-.28	-.23	.11
Unclustered														
21 Ef	.40	.34	.55	.40	.37		.42	.50	.39	.23	-.23	-.20	-.28	.28
30 Gr	.18	.30	.49	.26	.31	.42		.52	.23	.04	-.41	-.47	-.17	-.14
5 Bf	.53	.60	.66	.58	.58	.50	.52		.45	.35	-.43	-.53	-.43	-.02
14 Cp	.66	.57	.44	.23	.54	.39	.23	.45		.68	.31	.22	.07	.32
13 Wo	.53	.50	.21	.21	.37	.23	.04	.35	.68		.29	.27	-.01	.24
-16 -Cf	-.14	-.13	-.27	-.23	-.29	-.23	-.41	-.43	.31	.29		.84	.45	.35
-15 -Af	-.02	-.31	-.42	-.35	-.28	-.20	-.47	-.53	.22	.27	.84		.41	.52
-17 -Mf	-.20	-.26	-.31	-.41	-.23	-.28	-.17	-.43	.07	-.01	.45	.41		.00
31 Is	.15	-.16	-.02	.04	.11	.28	-.14	-.02	.32	.24	.35	.52	.00	

TABLE 7.10 CLUSTERED CORRELATION MATRICES (Continued)

C. The Voting-attitude Study

	Political						Taxation						Ethnic		
	5 NdAg	30 ShEm	29 HsEm	1 DmGv	18 Pk	13 ExFl	24 HsBn	25 AgBn	22 ExBn	23 RcBn	3 ScBn	2 VtBn	16 WlEx	10 ChEx	15 CoEx
Political															
5 NdAg		.85	.87	.83	.81	.71	-.01	.28	-.04	.29	.06	.12	.26	.38	.06
30 ShEm	.85		.91	.78	.74	.66	.01	.26	-.08	.26	.04	.16	.31	.39	.12
29 HsEm	.87	.91		.77	.74	.62	.00	.27	-.11	.22	-.11	.14	.25	.37	.06
1 DmGv	.83	.78	.77		.69	.63	-.21	.11	-.26	.14	-.14	-.05	.15	.33	-.04
18 Pk	.81	.74	.74	.69		.74	.09	.33	.13	.46	.20	.18	.35	.41	.22
13 ExFl	.71	.66	.62	.63	.74		.25	.46	.23	.50	.30	.34	.51	.51	.37
Taxation															
24 HsBn	-.01	.01	.00	-.21	.09	.25		.74	.74	.57	.55	.47	.46	.29	.54
25 AgBn	.28	.26	.27	.11	.33	.46	.74		.62	.62	.56	.52	.49	.40	.47
22 ExBn	-.04	-.08	-.11	-.26	.13	.23	.74	.62		.67	.56	.50	.40	.23	.51
23 RcBn	.29	.26	.22	.14	.46	.50	.57	.62	.67		.56	.53	.56	.49	.54
3 ScBn	.06	.04	-.11	-.14	.20	.30	.55	.56	.56	.56		.60	.37	.23	.39
2 VtBn	.12	.16	.14	-.05	.18	.34	.47	.52	.50	.53	.60		.42	.34	.41
Ethnic															
16 WlEx	.26	.31	.25	.15	.35	.51	.46	.49	.40	.56	.37	.42		.71	.69
10 ChEx	.38	.39	.37	.33	.41	.51	.29	.40	.23	.49	.24	.34	.71		.59
15 CoEx	.06	.12	.06	-.04	.22	.37	.54	.47	.51	.54	.39	.41	.69	.59	
Unclustered															
9 Vsl2	.76	.73	.74	.71	.71	.64	-.01	.23	-.07	.27	.07	.13	.34	.42	.13
28 AdOf	.61	.64	.58	.47	.60	.51	.17	.25	.08	.31	.10	.11	.30	.25	.22
14 VrCt	.60	.59	.55	.37	.57	.56	.14	.21	.05	.26	.08	.12	.27	.31	.24
27 SpPy	.54	.55	.51	.37	.57	.57	.42	.51	.41	.62	.42	.37	.49	.46	.38
11 TrOf	.48	.42	.45	.44	.50	.44	.03	.21	.01	.24	.05	.10	.26	.26	.14
20 NnLw	.47	.41	.40	.41	.54	.56	.19	.29	.21	.40	.16	.16	.40	.38	.26
31 ChCr	.49	.44	.45	.34	.53	.48	.34	.49	.34	.40	.29	.25	.35	.33	.24
-4 AlCu	.46	.44	.47	.57	.29	.22	-.47	-.23	-.42	-.15	.26	-.09	-.15	.06	-.28
21 Chrt	-.27	-.24	-.25	-.39	-.13	.06	.60	.44	.61	.45	.42	.40	.40	.24	.58
26 BlJb	-.02	.05	.05	-.07	-.01	.19	.53	.47	.51	.47	.39	.42	.36	.29	.43
8 LnTt	-.50	-.47	-.50	-.59	-.31	.14	.58	.36	.57	.33	.41	.34	.25	.05	.42
17 GvWt	-.31	-.28	-.33	-.46	-.09	.00	.51	.38	.51	.33	.47	.34	.33	.10	.47
19 AlnP	-.18	-.12	-.18	-.20	-.01	.21	.51	.43	.54	.40	.40	.26	.46	.31	.55
12 DsVt	-.04	-.07	-.01	-.14	-.07	.12	.40	.36	.43	.30	.25	.31	.28	.25	.26
6 Vsl1	-.07	-.03	-.04	-.20	-.02	.02	.41	.32	.36	.28	.32	.31	.25	.16	.27
7 LgPy	.32	.31	.31	.16	.35	.29	.32	.36	.32	.36	.21	.17	.30	.31	.23

C. The Voting-attitude Study

Unclustered

	7 LgPy	6 Vsl1	12 DsVt	19 AlnP	17 GvWt	8 LnTt	26 BlJb	21 Chrt	−4 AlCu	31 ChCr	20 NnLw	11 TrOf	27 SpPy	14 VrCt	28 AdOf	9 Vsl2
Political																
5 NdAg	.32	−.07	−.04	−.18	.31	.50	−.02	.27	.46	.49	.47	.48	.54	.60	.61	.76
30 ShEm	.31	−.03	−.07	−.12	.28	.47	.05	.24	.44	.44	.41	.42	.55	.59	.64	.73
29 HsEm	.31	−.04	−.01	−.18	.33	.50	.05	.25	.47	.45	.40	.45	.51	.55	.58	.74
1 DmGv	.16	−.20	−.14	−.20	.46	.59	−.07	.39	.57	.34	.41	.44	.37	.47	.47	.71
18 Pk	.35	−.02	−.07	−.01	.09	.31	.01	.13	.29	.53	.54	.50	.57	.57	.60	.71
13 ExFl	.29	.02	.12	.21	.00	.14	.19	.06	.22	.48	.56	.44	.57	.56	.51	.64
Taxation																
24 HsBn	.32	.41	.40	.54	.51	.58	.53	.60	−.47	.34	.19	.03	.42	.14	.17	−.01
25 AgBn	.36	.32	.36	.43	.38	.36	.47	.44	−.23	.49	.29	.21	.51	.21	.25	.23
22 ExBn	.32	.36	.43	.54	.51	.57	.51	.61	−.42	.34	.21	.11	.41	.05	.08	−.07
23 RcBn	.36	.28	.30	.40	.33	.33	.47	.45	−.15	.40	.40	.24	.62	.26	.31	.27
3 ScBn	.21	.32	.25	.40	.47	.41	.39	.42	−.26	.29	.16	.05	.42	.08	.10	.07
2 VtBn	.17	.31	.31	.26	.34	.34	.42	.40	−.09	.25	.16	.10	.37	.12	.11	.13
Ethnic																
16 WlEx	.30	.25	.28	.46	.33	.25	.36	.40	−.15	.35	.40	.26	.49	.27	.30	.34
35 ChEx	.31	.16	.26	.31	.10	.05	.29	.24	−.06	.33	.38	.26	.46	.31	.25	.42
15 CoEx	.23	.27	.26	.55	.47	.42	.43	.58	−.28	.24	.26	.14	.38	.24	.22	.13
Unclustered																
9 Vsl2	.44	−.01	−.05	−.08	−.24	−.34	−.06	−.19	−.40	.39	.53	.50	.47	.51	.54	
28 AdOf	.33	.12	−.01	.00	−.07	.16	−.12	−.03	.07	.36	.43	.39	.51	.63		.54
14 VrCt	.23	.22	−.04	−.06	−.05	.18	−.05	−.03	.13	.29	.38	.31	.37		.63	.51
27 SpPy	.56	.03	.10	.28	.09	.03	.29	.22	.08	.46	.35	.25		.37	.51	.47
11 TrOf	.21	.03	.09	−.06	−.08	−.18	−.10	−.09	−.08	.10	.23		.25	.31	.39	.50
20 NnLw	.43	.17	.12	.23	.00	.00	.05	.17	.02	.36		.23	.35	.38	.43	.53
31 ChCr	.29	.27	.19	.12	.10	.03	.18	.12	.00		.36	.10	.46	.29	.36	.39
−4 AlCu	.06	−.18	−.27	−.39	−.58	−.59	−.16	−.44		.00	.02	−.08	.08	.13	.07	−.40
21 Chrt	.20	.30	.37	.69	.54	.69	.53		−.44	.12	.17	−.09	.22	−.03	−.03	−.19
26 BlJb	.09	.25	.49	.34	.23	.36		.53	−.16	.18	.05	−.10	.29	−.05	−.12	−.06
8 LnTt	.18	.41	.29	.58	.66		.36	.69	−.59	.03	.00	−.18	.03	.18	.16	−.34
17 GvWt	.10	.34	.25	.44		.66	.23	.54	−.58	.10	.00	−.08	.09	−.05	−.07	−.24
19 AlnP	.24	.26	.27		.44	.58	.34	.69	−.39	.12	.23	−.06	.28	−.06	.00	−.08
12 DsVt	.08	.18		.27	.25	.29	.25	.37	−.27	.19	.12	.09	.10	−.04	−.01	−.05
6 Vsl1	.35		.18	.26	.34	.41	.25	.30	−.18	.27	.17	.03	.03	.22	.12	−.01
7 LgPy		.35	.08	.24	.10	.18	.09	.20	.06	.29	.43	.21	.56	.23	.33	.44

123

The clustered correlation matrix has the following advantages:

1 The final results appear in an intelligibly organized correlation matrix, not difficult even for a neophyte to understand. The format is one in which the correlations or selections from them can be published effectively or otherwise presented.

2 The matrix is itself divorced from the derivative factoring process and from the complications of geometric and statistical oblique structure analysis, procedures that may be difficult for some people to understand.

3 The clustered matrix includes observed correlations and thus represents the reality of correlated variations among the variables, whereas the configurations from SPAN refer to theoretical augmented common factor correlations. For example, in the abilities study, though the correlation matrix tells the same relative story as SPAN, it shows plainly that there is in fact considerable specificity in the observed variations in the abilities; except for the verbal cluster, even the correlations among the definers are rather low.

4 The meaning and inner cluster structure of each of the clusters emerges readily from a study of the submatrix of defining variables near the diagonal and from their relations to other cluster definers as given in the cross-correlation submatrices away from the diagonal.

5 If one is interested in an illustration of the derivation of the various formulas used in oblique cluster structure analysis, the clustered table with communalities in the diagonals organizes the coefficients in precisely the form used in these formulas, all of which are variants of the "correlation of sums" formula (Guilford, 1950, pp. 586ff). For example, the domain validity and α reliability of a cluster score come only from its submatrix of definers on the diagonal. The common factor (interdomain) correlation between two clusters requires, in addition to their two diagonal submatrices, only their off-diagonal cross-correlation submatrix; the real correlation between the cluster score involves exactly the same data but with unities in the diagonals. The oblique factor coefficients of the n variables on a given cluster domain involve only the columns of correlations of its definers, as grouped together in the matrix. The index of generality (reproducibility) of a cluster requires these same columns and, in addition, either the total of the squares of correlations over the whole table or the sum of communalities across all n variables. Chapter 12 presents the formulas necessary to calculate each of these statistics.

Physical model of the spherical configuration

Investigators who have difficulty seeing the SPAN configuration in three dimensions usually find the matter solved when the configuration is transferred to a real globe or balloon. The printout of any SPAN set gives the coordinates by which each point on the globe can be quickly located by triangulation.

Relation between two different indexes of collinearity, P^2 and $\cos \theta$

We use two different indexes of collinearity in separate places in objectively locating clusters having an inner structure of high collinearity in the full cycle analysis. Selecting the most collinear subset of variables by the mutual collinearity criterion during key-cluster factoring of the first dimension involves the use of P^2 as the index of collinearity in the correlation matrix. However, in gauging the collinearity of variables in structure analysis, we use the nearness of variables as points on the sphere. Nearness is represented by the arcs between variables, and the arcs in their turn are determined by the central angles between them, more specifically by the cosines of the central angles. These cosines are the common factor correlations between the variables.

The problem is this: Do these two indexes tell the same story? We find out simply enough by comparing them in each of the three illustrative studies, as follows. In the Holzinger study, for example, for each variable we list its P^2 values with all the other 23 variables of the study, an option in DVP41. This series represents the collinearity of a variable's pattern of correlations with the 23 patterns of the correlations of the other variables. Beside this listing we write the corresponding values of the cos θ of the variable with the other 23 as computed in the CSA component: these are the cos θ values in the geometric model of SPAN (computed in CSA by opting the correlation matrix to be reproduced from the unaugmented factor coefficients, from which the SPAN figures are produced, and then calling for the augmented correlation matrix, the cos θ values, to be calculated on the reproduced matrix).

With the two separate sets of indexes listed (key-punched) side by side, we input them as two variables into RSCAT in order to determine the relation between them. The results are shown in Fig. 7.7 for the Holzinger study (top) and for the social-area and voting-attitude studies (bottom). In all three studies there is an obvious very high relation between the two indexes, the correlations between them being respectively .92, .90, and .94. Since P^2 (written PSQ in RSCAT and in Fig. 7.7) is computed from the original correlations whereas cos θ comes from the reproduced correlations, which deviate from the original values by the residual values, we cannot expect the scattergram correlations to reach unity unless all the residuals are exactly zero. The values found are probably smaller than the values possible, considering that the residual errors in the cos θ values are not exactly zero.

The conclusion is that when one looks at a SPAN configuration, a representation of cos θ values, one is looking at collinearity relations among the variables which, relatively speaking, are almost precisely those in the correlation matrix, as indexed by the P^2 values. They even match in absolute magnitudes for high values: note that when either is above .90 so is the other in most cases. In many studies the definers of oblique dimensions

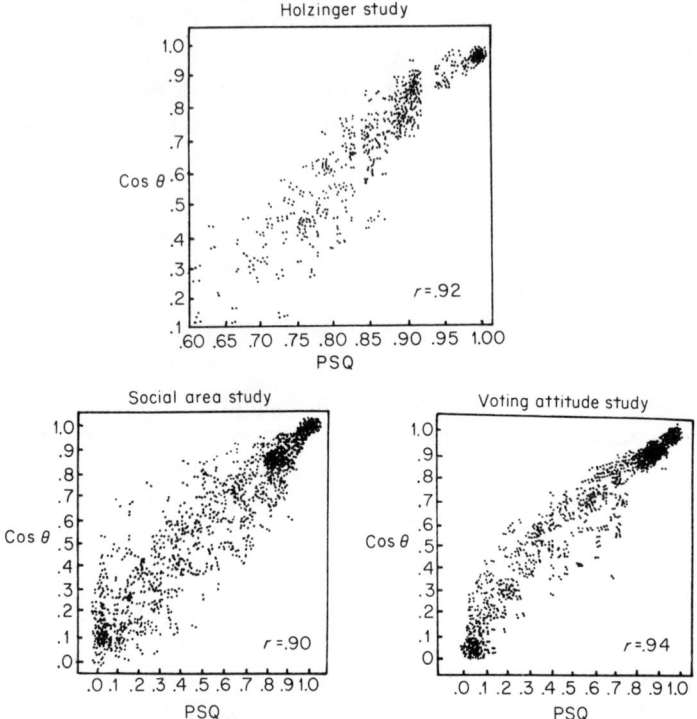

FIGURE 7.7
In three studies the relationship between P^2 and cos θ as separate indexes of the degree of collinearity of the variables with each other.

usually have P^2 values with each other of the order at least .90. In Table 4.2, for example, see the diagonal average P^2 values of the definers of clusters in the Holzinger study.

Structure analysis by alternative methods

This section is concerned with the comparison of some of the main methods of factor analysis in describing the structure ("organization" or "configuration") of the relations among variables of a problem.

Structure from key-cluster factoring

As an anchor analysis, we use the configuration of the 24 Holzinger abilities described by key-cluster factoring and displayed on the SPAN diagram, shown in Fig. 7.8, upper left sphere. The configuration there is virtually identical with that given earlier in Fig. 7.3, deviating from it only by virtue of minor revisions of the definers of the speed and memory clusters. Despite the revision, the configuration shown on the surface of the sphere has remained relatively invariant.

Circled in broken lines are the defining variables of the dimensions.

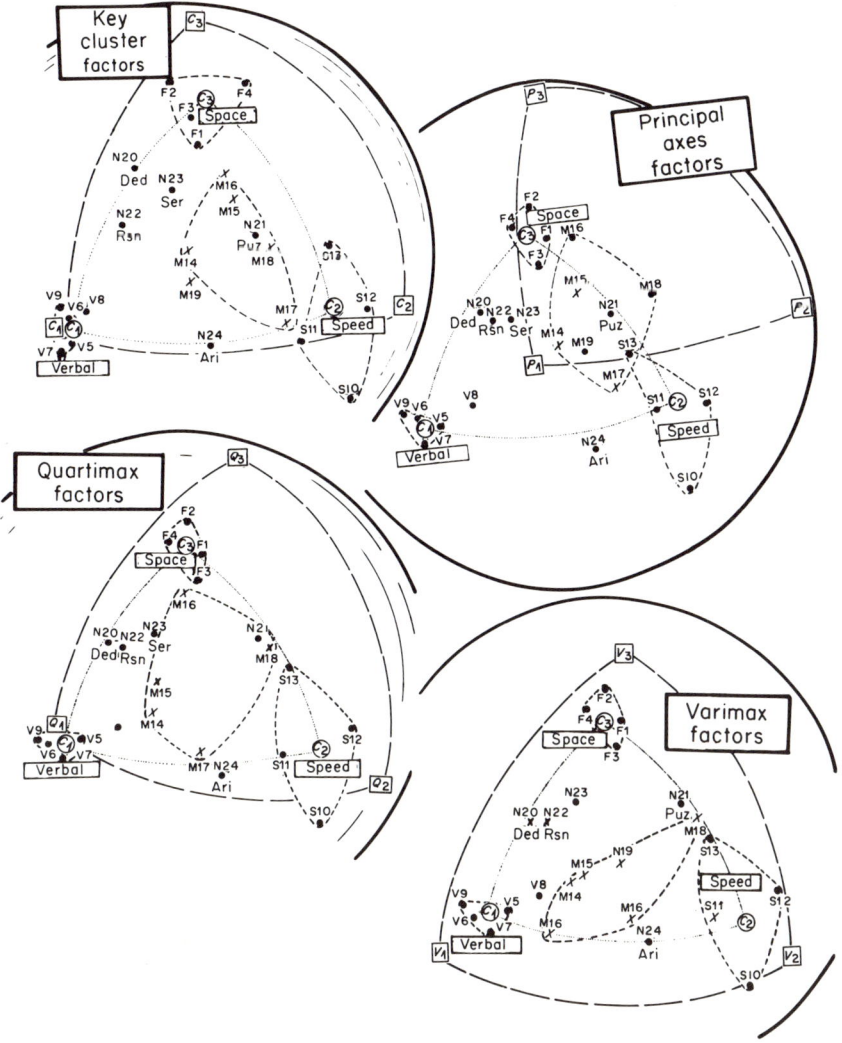

FIGURE 7.8
Invariant spherical configuration by four methods of factor analysis of the 24 abilities in the Holzinger problem.

The orthogonal dimensions themselves are, of course, 90° apart, as shown by the symbols C_1, C_2, and C_3 in boxes at the three vertexes of the large spherical triangle drawn in long broken lines. Recall that four dimensions are required in the Holzinger problem. We avoid presenting a second sphere, however, that includes the fourth dimension (as in Fig. 7.4) by showing those variables which lie in the fourth dimension also on the surface of this single sphere, designating their loci as X's, meaning that they project into a fourth dimension, C_4, and are therefore not shown in this three-dimensional figure. The six tests that lie in the fourth dimension are memory tests (M), shown in the center of the figure.

Orthogonal factors vs. oblique factors

In order to compare structure analyses performed by different methods, we must define or recall several important terms common to all such analyses. Orthogonal factors are the independent dimensions or axes derived from residuals by the factorization processes. These points are termini of the three orthogonal axes that pass through the origin of the sphere (see Fig. 7.2). Oblique factors are correlated dimensions or axes whose termini are not orthogonal but are, for example, at the centroids of the definers (circled by short broken lines in the figure) represented in the figure by the circled points. They are oblique because the scores of children on these three clusters would correlate positively. The values of the correlations between them are the "interdomain correlations" computed by the CSA component of the BC TRY System. Note that the three oblique factors are connected by dotted lines, forming an oblique spherical triangle that describes the oblique factor structure in this problem.

The larger, outer spherical triangle connecting the termini of the orthogonal axes forms the orthogonal factor structure. The orthogonal structure is not a very meaningful one because the second and third factors, C_2 and C_3, are outside the configuration, and are therefore not directly defined by actual tests, whereas the oblique factor structure is meaningful because all three oblique factors are well defined by actual tests, namely, by the definers of the three oblique clusters.

The important concept to stress here is the distinction between factors and configuration (or structure). The configuration is the general organization of the variables as represented by the structural swarm of points; factors are any loci in the configuration that one may wish to consider critically important and through which one may wish to pass axes. The selection of factors does not, of course, alter the configuration in any way.

Rotation of factors

It is important to stress that the orthogonal factors in this sphere do not pass through tests whereas the oblique factors are in the middle of highly collinear clusters. These two oblique factors may be thought of as "rotations" of the orthogonal factors to the oblique positions. Thus, the C_2 cluster centroid may be thought of as an oblique rotation of the orthogonal axis C_2, a process symbolized by the arrow in the figure. The actual rotation of key-cluster factors is unnecessary because discovering and locating oblique factors is "directly" achieved during factoring by identifying the oblique factors as the collinear subsets of variables selected by the mutual collinearity criterion of key-cluster factoring. But dimensions derived by other forms of factor analysis, such as principal-axes factoring, ordinarily *must* be rotated if they are to be meaningful, as we shall see below.

Structure from principal-axes factoring

Just as a key-cluster dimension (or factor) is defined by a subset of variables selected from all the n variables, a principal-axis dimension (or factor) is defined on the total set of all n variables. The score of a child on the first key-cluster factor, for example, is his composite verbal score on the subset of only four verbal tests, V5, V6, V7, V9, whereas his score on the first principal factor is his composite score on the total set of scores on all 24 tests of the study. The first principal axis is graphically represented in the sphere at the upper right in Fig. 7.8 as the point P_1 in the middle of the configuration. The sphere plotted there is a direct tracing of the SPAN sphere defined by the first three dimensions of the principal-axes solution computed by program FALS (Factor Analysis Least Squares) of the BC TRY System.

    A score on a principal-axis factor can be thought of as a specially weighted sum of scores on all 24 variables. The first three such principal axes are represented in the Holzinger problem by the points P_1, P_2, P_3 at the vertexes of the large spherical triangle plotted in short broken lines in the upper right sphere of Fig. 7.8. The point P_1 lies at that place in the configuration such that the sums of squared deviations of all 24 points from it are a minimum. The least-squares weights attached to each of the 24 standard scores are such as to place P_1 at the dead center of the total configuration (including the fourth dimension). A point located any other place would have a larger sum of its 24 squared deviations. One computational procedure that locates the position of P_1 and computes the factor coefficients of all 24 tests on P_1 starts its solution by setting the locus of

P_1 by "trial values," then "waggles" the trial point, as it were, in such a systematic iterative fashion that at each successive position its sum of squares of 24 deviations is less than on the previous trials. The iterative process stops whenever the point P_1 settles down on a given locus (within an arbitrary convergence criterion).

The second principal axis P_2 is at right angles to the first, as shown in the figure. Trial values of P_2 can be thought of as the termini of spokes of a wheel the axle of which goes through P_1 and the origin of the sphere. The finally selected locus of P_2 is at the end of that spoke from which the sums of squared deviations of all 24 test points is a minimum. In locating the third principal axis, a point is selected from all possible points that are 90° from *both* P_1 and P_2; there is only one such point for which the sums of squares of the 24 deviations is a minimum. That point is P_3. And so the process continues until factoring is terminated.

Though this solution for principal factors may sound elegant mathematically (Harman, 1967), the meaning of the principal factors, such as P_1, P_2, and P_3, is usually obscure. Thus, in this Holzinger configuration, though a total-set score on the first principal axis P_1 can be meaningfully interpreted as "general intelligence," the meanings of P_2 and P_3 are quite ambiguous. The reason is that there are no actual tests with high projections (factor coefficients) upon them, so that we do not know what such factors can mean rationally. For this reason factor analysts rarely estimate such meaningless scores of subjects on principal factors after the first. They "rotate" them first, as in the quartimax and varimax solutions discussed below.

The configuration from principal axes

Though the principal axes may be rationally meaningless as dimensions, the configuration of the 24 tests plotted from the factor coefficients on the principal-axis factors beautifully portrays a structure that is virtually identical with that given in the key-cluster sphere. The oblique factor structure on the principal factor sphere, denoted by the smaller, inner spherical triangle plotted in dotted lines in Fig. 7.8, is in essence the same as that directly given without rotation in the key-cluster solution shown in the sphere at upper left. In short, the oblique cluster structure can be derived from the principal-axes solution, provided, of course, one has a computer program like SPAN that can depict the configuration from the orthogonal principal factor coefficients.

In one sense, the configuration plotted from principal-axis factors is slightly superior to that given by the key-cluster solution. More variables are likely to be represented on the principal-axis sphere defined by P_1, P_2, P_3 than on the first key-cluster sphere defined by C_1, C_2, C_3. For example, note that in the principal axis sphere there are only three memory tests "lost" into the fourth dimension, i.e., marked X, whereas five are lost into the fourth dimension of the key-cluster sphere. But superiority in this particular is trivial in most problems compared with the more substantial advantages of key-cluster factoring, whose factors point in the general direction of oblique clusters, whereas the principal-axis factors point at no meaningful places in the configuration. Another disadvantage of the principal-axis solution is that some factors are "negative" dimensions (on which *all* variables may have negative factor coefficients) that must be reflected in order to provide a configuration that is meaningful.

Structure from rotated principal-axis factors:
the quartimax and varimax orthogonal factors

One way to develop meaning of the principal-axis factors is to "rotate" them to places in the configuration that may give them meaning. Two such methods of rotation, computed by the BC TRY component GYRO and plotted graphically on SPAN spheres, are the quartimax rotation (Neuhaus and Wrigley, 1954), depicted in Fig. 7.8, lower left, and the varimax rotation (Kaiser, 1958), given in the lower right sphere. Inspection of these two configurations indicates that they are the *same* configuration; indeed, they are exactly the configuration given on the principal axes in the upper right sphere in which the axes P_1, P_2, P_3 have merely been swung to the new positions designated as Q_1, Q_2, Q_3, respectively, in the quartimax sphere, and as V_1, V_2, V_3, respectively, in the varimax sphere. Nothing important is different in all four spheres of Fig. 7.8. The same oblique factor structure is depicted in them all, being denoted by the dotted oblique spherical triangle bounded by the verbal, speed, and space clusters, with the cluster of memory tests represented as projected into a fourth dimension. The noticeable but unimportant difference is that the orthogonal dimensions have been put at different places in or around the configuration in the four solutions.

The orthogonal factors Q_1, Q_2, Q_3 of the quartimax solution are fairly close to those in the key-cluster solution. From the origin of the sphere, the first factors in both, namely, Q_1 and C_1, point as vectors at the oblique verbal cluster. The second, Q_2 and C_2, point at a place as close as possible to the oblique speed cluster while still being orthogonal to the first dimensions; similarly with the other two of the four dimensions. Generally, the key-cluster factors are better pointers than the quartimax, for the simple reason that in the key-cluster factoring procedure the cluster-search routines are expressly designed to be the best pointers, whereas the quartimax solution is a general analytic procedure not expressly designed to locate clusters but one which nevertheless tends to do so.

The varimax factors, V_1, V_2, V_3, are worse placed and hence less meaningfully placed than the quartimax in relation to the oblique cluster structure bounded by the verbal, speed, and space clusters. None of the varimax vectors directly points at oblique clusters. The quartimax factors start off excellently as pointers but do the job decreasingly well as one moves from the first to the second, and thence to the third, and so on. In short, quartimax, like key-cluster factors, orders the factors by importance or weight in such a fashion that the earlier, more heavily weighted quartimax factors, being rotations of the more heavily weighted principal axes, are more meaningfully interpreted. But varimax rotation, on the other hand, tends to emphasize the meaningfulness of the least important principal axes as much as they do the more heavily weighted. The consequence is that varimax represents the total configuration on its sphere less well than quartimax, shown by the fact that no less than nine (one-third) of the 24 tests are lost into the fourth varimax dimension, whereas only six are lost in the quartimax (and key-cluster) sphere, and only three in the unrotated principal-axis sphere. This disadvantage of varimax is somewhat matched, however, by the advantage that the configuration is better presented aesthetically on varimax dimensions than on the other three. The symmetry of the oblique structure (the inner, smaller dotted triangle) to the outer orthogonal broken-line spherical triangle suggests that the rotated varimax factors would converge on the oblique and meaningful cluster structure if the orthogonal varimax factors were moved inward at the vertexes of the triangle.

Conclusion

The general conclusions on structure derived by different forms of factoring and rotation is that the orthogonal dimensions from any particular factoring method, whether key-cluster, principal-axes, quartimax, varimax, or any other (there are many, e.g., Thurstone centroid, diagonal, canonical, bifactor, averoid, alpha—all available in the BC TRY System) are, as such, not uniquely critical in describing the oblique, meaningful structure of the variables. These different forms of factor structure describe the same configuration, as is obvious in the four spherical representations of Fig. 7.8. Where orthogonal axes are located in the configuration is unimportant.

But some method of factoring must be used in order to set up the scaffolding, as it were, on which to hang the invariant configuration. The key-cluster solution is generally preferable because its factors are designed to point at or near natural clusters among the variables, and, just as important, the program routinely leads to the program CSA which describes directly, and with metric precision, the nature of the oblique factor structure as depicted graphically by the dotted triangles in Fig. 7.8. But, if an analyst has a predeliction, say, for the principal-axis varimax rotation, he can use it with accuracy to describe the configuration. From the resulting SPAN diagram he would make a decision on what should be the defining variables of oblique clusters. He would then use these definers to execute a preset key-cluster solution in order to secure a more precise description of the oblique structure.

Bifactor structure and other forms of factor structure

In the light of the conclusion that for a given number of dimensions that yield trivial residuals, the configuration is the same whatever the type of factoring employed may be, it is no longer necessary or desirable to quarrel over which factors are "best" in some fundamental sense. Factors are *any* places in the configuration that one may want to run an axis through, and different types of factoring put axes through different places. It is therefore seriously misleading to call any such arbitrarily placed dimension an "underlying" factor, as if its position referred to some deep-seated biological, social, or psychological "cause," or "source" trait. To be sure, such a dimension *may* refer to such an entity, but there is absolutely no basis inherent in the procedures of factoring to justify such an inference. Evidence for causation must come from other sources than mere procedures of factoring correlations.

During the first quarter of this century, psychology was wracked by the question of whether the factoring procedure "proved" the existence in cognitive traits of the "g factor," often referred to as "general intelligence." The main advocate of this belief, Spearman, always dealt with oblique configurations of abilities like those of the Holzinger problem. In modern terms, what was being said was that one can put an axis through the grand centroids of such configurations, like the first principal axis P_1 and call it g if one wishes. Such a positioning of an axis does not prove that it represents an "underlying," "source" entity.

The traditional factorial solution of this sort was called, in its original awkward formulation by Spearman, the "two-factor theory," but it was finally generalized with some finesse by Holzinger as the "bifactor theory" (see Harman, 1967 chaps. 7 and 8). Today, it is sometimes interesting or convenient to perform a bifactor solution, using, say, the BC TRY System. If, for example, one wants to remove from the correlations among the 24 Holzinger abilities a total-set, or total-score, dimension and study the structure of the correlations among them left over after the removal, one would call the principal-axis solution (FALS), preset it to one dimension, compute residuals (by FAST), and then perform a full-cycle key-cluster analysis of the residual matrix. This solution is a bifactor analysis in modern dress.

With the MMPI items, a common theory is that one reason why the items tend to be generally positively correlated is a desire of many respondents to give favorable responses to them and to reject the obviously unfavorable alternative responses. One can think of the first principal axis as representing such a "favorability" continuum, remove it from the intercorrelations among the items by taking first-factor residuals, and then study the matrix of residuals for the clusterings of items "freed" of the favorability principal axis.

These and many other factoring designs can lead to very interesting solutions. Another specially useful one is to define dimensions by *single* meaningful variables. This form of factoring is called "diagonal" factoring or "pivot-variable" analysis (an option in CC factoring in the BC TRY System). All these variants should be thought of not as devices that inexorably reveal underlying truths but as various designs to extract variance of different, defined sorts from the correlation matrix, leading to different ways of deriving meaningful scores on the individuals who compose the group being studied.

Chapter 8
OBJECT CLUSTER ANALYSIS

From the time of Hippocrates, and doubtless even before, there has been an urge to describe the salient characteristics of a person by identifying him with a "type." Though forming typologies may sometimes be decried, it is an important scientific pursuit. The medical profession could not exist without typologies, the medical term being "syndromes." In the general field of biology, a typology is called a "taxonomy"; plants and animals could not be conceptualized were they not cast into typological classes called families, genera, species, and so on. These groups of organisms are not fanciful constructs, they are genetically differentiated. Indeed, in biology the term "type" is recognized as vitally important. An organism's configuration of genetic factors is its "geno*type*," the chromosome photograph of which is its "karyo*type*." The basic scientific problem is not really a theoretical dispute over whether typologies "exist." Rather, the problem involves the more difficult and important question of developing an objective method of forming a typology and of assigning individuals to their proper groups within it.

There are a number of reasons why typologies are desirable: (1) Since a particular type includes many individuals, a considerable amount of information derived from experimental or field observations on the individuals accumulates to the type. Any new individual who fits the type can be better understood than if no such cumulative information were available. (2) A given type possesses special characteristics that differentiate it from other types; hence, the strengths and weaknesses of its members can be conceptualized in terms of the distinctive high and low elevations of their profile of characteristics. (3) Among members of a given profile type the absolute level of an individual's score on any particular characteristic can be properly assessed in the light of high and low elevations on *other* attributes of his profile. For example, if two men have the same high score on Strong's Scale of Lawyers' Interests, this fact carries an implication for their relative success in law that becomes quite different if one has a high score on verbal ability and the other has a very low score. (4) Since a given type describes a collection of individuals who have the same standing on multiple attributes, other behavior characteristics of these individuals are better predicted than would be the case if prediction were based only upon standing in one general attribute.

The scientific problem involved in classifying many individuals into groups having distinctive profiles on many attributes was forbidding before the advent of the computer. Computer programs now can do the job in a matter of minutes. In the field of psychology such programs were originally called "inverse analysis," but now the more appropriate term is "person cluster analysis" or simply "O-analysis." In biology, where the classification of species and varieties is important, such programs are called "numerical taxonomy" (Sokal and Sneath, 1963). In electrical engineering they are called programs of "pattern recognition."

The term pattern recognition provides a clue to how a typology can be formed. An individual's profile is his pattern of scores on a multiple set of attributes. Instead of depicting a profile as a line, however, we can express it more efficiently as a point in a space of scores. Chapter 2 gives a simple example, in Fig. 2.1, where 301 individual children are expressed as points that give their positions on two cluster scores that measure verbal and mathematical abilities. Each of the 301 individuals could have been described by his profile on these two scores, but the whole lot of 301 profiles would be difficult to conceptualize by one visual scanning, whereas when they are expressed as a swarm of points in this space of two dimensions, their detailed similarities and differences can be sized up visually. The number of dimension scores can be increased as much as we wish, in which case the 301 points would spread out in the hyperspace of the many dimensions. Though we cannot visually observe the cloud of points in a space of more than three dimensions, we can write computer programs by which the computer can "visualize" them in the higher-order space. Conceptualized in this fashion, the problem of forming a typology of a group of individuals on scores on many attributes is a simple one in principle: one writes a computer program to represent the individuals as a swarm of points in a hyperspace of scores and to locate within the swarm regions of high density; there are as many types as there are regions of high density in the space of scores. In practice, the problem usually is a bit difficult because individuals expressed as a swarm of points often do not show clear-cut regions of high density.

Figure 8.1 illustrates a computer-program-discovered typology in the form of 15 profile types in the Holzinger problem. The 15 different classes of profiles in the figure were actually discovered by the BC TRY program OTYPE, which located the 15 types as 15 areas of density in a score space

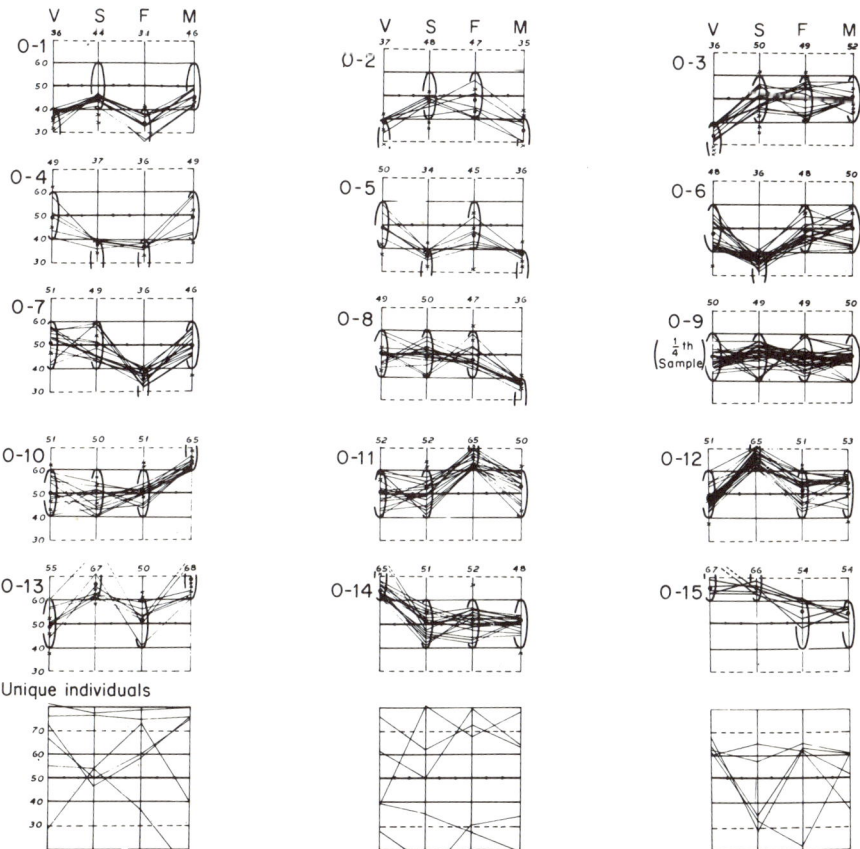

FIGURE 8.1
Profiles of 301 children in 15 core O-types in the Holzinger problem.

of four dimensions, these being the four cluster scores on V (verbal), S (speed), F (form or space), and M (memory). Each of the four dimension scores was expressed in standard score form with a mean of 50 and a standard deviation of 10, as shown in the form of vertical axes in Fig. 8.1. The "unique individuals" at the bottom of Fig. 8.1 had profiles too deviant to fit any of the 15 types shown above them.

This illustration points up some of the difficulties in O-analysis. The "real" configuration of the profiles is the swarm of points of the individuals in hyperspace. One can, of course, actually see such a swarm of points and pick up areas of density in it if there are only two dimensions. When there are three dimensions, one would observe the swarm within a sphere or cube, as in a room. In four or more dimensions there is no easy way to "see" the areas of density. Obviously, some profile types are more like each other than they are like others. In any actual problem it may be possible or necessary to combine some types into higher order clusters. To do so, it is necessary to observe the relationship among the different profile types and combine those which are most similar. In the Holzinger problem

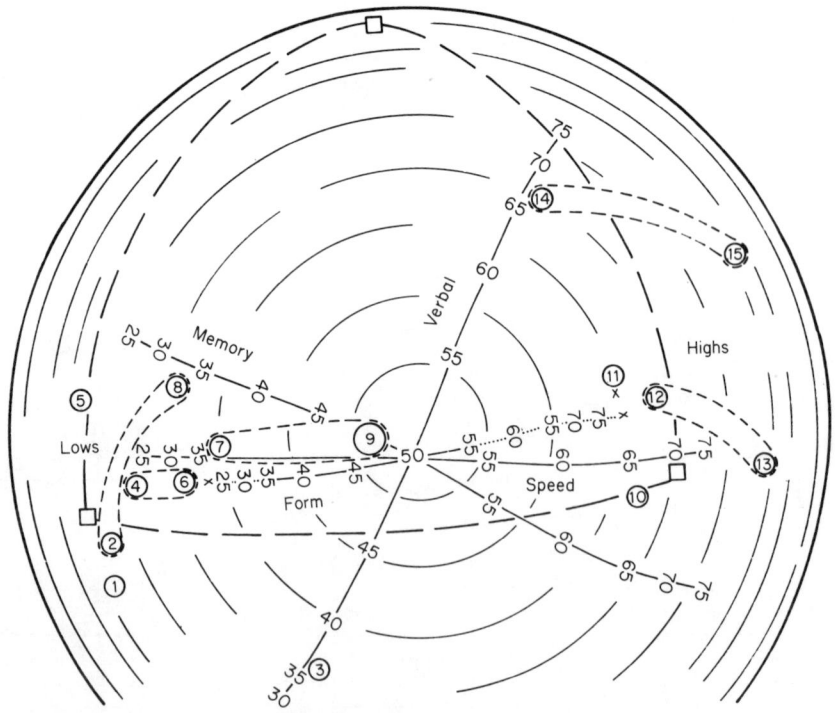

FIGURE 8.2
Spherical structure of the 15 core O-types of the Holzinger problem.

the relations among the 15 types are revealed in Fig. 8.2, where the 15 different types are described by the 15 centroids, shown in the spherical configuration. Each of the 15 different centroids is expressed by a point on the surface of a sphere, the familiar SPAN spheres described earlier in the key-cluster solution of variables in Chap. 7. The configuration of these centroids in relation to the four score axes also drawn on the spherical surface shows that types 1 and 2 are rather similar, being close together as points. These two types are both "lows" in most of the four dimensions, i.e., below the mean. By cross-referencing similarities and differences in Figs. 8.1 and 8.2 it can be discovered that these two methods are merely alternative ways of describing the total configuration. Profiles that have opposite shapes, i.e., are mirror images, lie at opposite sides of the sphere, e.g., types 1 and 15.

The advantage of expressing the configuration of types in the spherical representation of Fig. 8.2 is that it can be visualized, whereas it cannot easily be seen in the profiles of Fig. 8.1. Though there are actually four dimensions, we represent the overall structure in the two-dimensional plane of a piece of paper. In many problems typologies in many dimensions can be represented on only a few subsets of spheres like the one in Fig. 8.2.

Still another way to describe the relationships among the profiles of the individuals or of their centroids is in the form of a "hierarchical chart" showing the hierarchical structure of the O-types. We turn to this matter a little later, but in the meantime Fig. 8.6 shows the hierarchical chart of the 15 core O-types of the Holzinger problem.

To sum up the methodological problem, there are four different ways to describe the typology of individuals: (1) types can be described as centers of density in score space (Fig. 2.1), (2) they can be described as profile charts (Fig. 8.1), (3) they can be described by configurations on the surface of a hypersphere (Fig. 8.2), and (4) the typological structure can be described summarily by a hierarchial chart (Fig. 8.6). All four ways have their particular advantages and disadvantages.

Methods in object cluster analysis

Deciding on the score space

Since object clusters are formed on the basis of profiles of scores of individuals on a given set of variables or dimensions, the first major step is to decide upon the dimensions of the score space. They can be the n raw scores of all the variables of a study. This form of clustering, sometimes called Q-analysis, is usually, however, an undesirable undertaking. Using single variables as dimensions denies one the power and information derived from a V-analysis of all the n variables. Except in those studies where each specific variable may have an important meaning in its own right and should not be included in a composite derived from a V-analysis, the dimensions of a typology should be the cluster or factor scores derived from a prior V-analysis. A central reason for doing so is to secure the gain in generality that such dimensions necessarily possess over individual variables. A greater degree of generality in typological prediction can be expected where a typology is based on cluster-composited dimensions.

There are different forms of V-analysis, namely, key-cluster, principal-axes, and sundry rotations of these. When oblique dimensions are chosen as the dimensions of the score space, the form of factor analysis that was used in the V-analysis makes little difference, for, as we showed in Fig. 7.8, the oblique structure of the variables is invariant with respect to the type of factoring procedure employed.

Once a decision is made on the dimensions, the individuals are scored on the different dimensions. In the BC TRY System, this component is FACS (Factor And Cluster Scores), which is capable of computing dimension scores on individuals in any form one wishes. These can be oblique scores on the dimensions derived by the prior V-analysis or orthogonal scores on such rotated principal axes as varimax or quartimax. Whatever the type of dimension scores desired, it is usually advisable to convert the scores on each dimension to standard score form, with a mean of 50 and standard deviation of 10. This conversion means that all dimensions have equal weight in determining the typology. If, however, it is desired to give greater weight to some dimensions than to others, a scoring component such as FACS permits one to weight the dimensions differently. One may wish, for example, to weight the dimensions that have the greatest generality more heavily than those which are more specific. The more general dimensions therefore stretch out the individuals more in score space than the less general dimensions.

The problem of missing data

Many studies are crippled by the problem of missing data, i.e., by some individuals having no scores on some variables. This missing-data problem may not be serious in V-analysis because, in preparing the correlation matrix for factoring, the pairwise correlations between variables can be computed only for those individuals which have complete scores (as in the BC TRY correlation program COR3). But in computing cluster or factor scores on each dimension, if there is a missing score on any single variable for a given individual, that individual may be excluded from the entire typological analysis, because absence of a score on any variable prevents his profile or his locus in score space from being computed. There is one exception to this principle: when cluster scores in contrast to factor scores are being used as measures of the dimensions. Recall that a cluster score on any dimension is a composite of the standard scores on a collinear sub-set of defining variables [Eq. (2.2)]. These definers of a dimension sample the same general domain of variation, are positively correlated, and there-fore can replace each other if any one of them is missing. Thus, in the FACS program of the BC TRY System, if the standard score on any definer of a given dimension is missing, it is replaced routinely with the mean standard score on the other definers whose scores are present. Since it is unusual for scores on all defining variables of a given dimension to be missing, when dimensions are defined by cluster scores, it is rare for any individual to be missing from the typological analysis. When a score is missing on any variable that does not measure a cluster-defined dimen-sion, it is of course of no consequence.

The problem is serious when the V-analysis is performed by any of the orthodox methods of factor analysis such as the principal-axis solution any any orthogonal or oblique rotation of its dimensions, for factor esti-mates of these dimensions are based on a least-squares weighting of scores on all n variables. Hence if any of the n scores is missing for a given individual, his factor scores cannot in principle be computed.

Typology from the overall configuration
in spherical analysis

The best way to form a typology is to do so from a study of the entire configuration of individuals, in which one can see regions of density in the score space if there are any. When there are more than two dimensions, it is not possible to see this configuration in the score space proper. One must therefore represent the configuration in some other fashion. The device for doing so has already been presented in V-analysis by the general procedures of projecting many observations as points on the surface of a hypersphere of k dimensions. In the Holzinger problem, for example, the configuration is presented in Fig. 8.2 but only as points representing the centers of density, or centroids, of individuals.

Since this form of typological analysis is of crucial importance, the detailed logic of the procedure is fully explicated here on a simple fictitious problem. Six clusters of 100 individuals are defined by their scores on two dimensions, Z_1 and Z_2, as depicted in Fig. 8.3. Thus, cluster A consists of

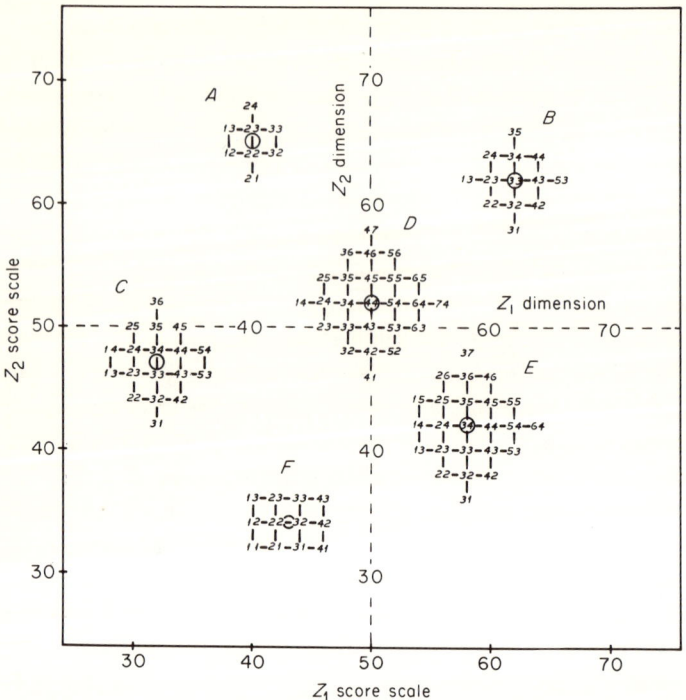

FIGURE 8.3
Fictitious example showing 100 individual objects clustered together to form six O-types based on their scores on the two dimensions Z_1 and Z_2.

the eight individuals in the upper left sector of the figure whose numbers are 12, 13, 21, and so on. We can represent the eight individuals in cluster **A** by the identification symbols A12, A13, and so on, to the eighth individual, A33. Similarly, those of B would have numbers B13, B22, and so on. The locus of each of the 100 individuals in all six clusters is determined by its coordinates on the two dimensions. These two-dimensional scores are listed for the first two individuals in each of the six clusters in the top two rows of scores given in Table 8.1. For example, the Z_1 and Z_2 coordinates of A12 are shown there to be 38 and 64, respectively.

The exact arrangements of the individuals in this space are represented by the spatial distances between them in the two-dimensional score space, i.e., by their euclidean distances, represented by D. Thus the value of D for the first two individuals, A12 and A13, of cluster A is simply

$$D = \sqrt{(38 - 38)^2 + (66 - 64)^2} = 2$$

which is small, as it should be, since these points virtually sit on top of each other in this space.

The BC TRY System includes a program called EUCO, which can compute the entire matrix of distances among all the 100 objects depicted in Fig. 8.3. This basic matrix constitutes the data on which the typology of the 100 individuals is formed. The distances among the objects in a given cluster will approach zero, but the interspace differences between objects in different clusters will not be zero. The problem therefore is to devise a method of locating the centers of density by working with the distance matrix, to find the six submatrices in which the distances approach zero.

We already have the procedures for doing so in full-cycle key-cluster analysis explicated in Chap. 7 on V-analysis. We can utilize this method provided the distance matrix is converted to a correlation matrix. The process of conversion is illustrated in Table 8.1. First, for each dimension we select a given set of hypothetical objects that span the dimension. Thus, for dimension Z_1, we choose reference-marker objects, called OMARKs, at equal distances (steps of one standard deviation) on the Z_1 dimension. These are the six individuals whose Z_1 scores are 20, 30, 40, 60, 70, 80 (and whose Z_2 scores are perforce 50 at the Z_2 mean) given in the six rows under "On dimension Z_1" in Table 8.1. An analogous set of six OMARKs is selected which span the Z_2 dimension. The final OMARK is at the grand origin whose coordinates are (50,50). Now, instead of working the distances between all objects pairwise, we can merely work the distances of each object with the 13 reference OMARKs that span the score space. Any two individuals that have the same pattern of distances with all the other 99 objects will have the same pattern with the 13 reference objects. This can be verified by noting that for the first two objects in each of the six clusters their pattern of distances with the 13 OMARKs is virtually identical, signifying that they are together in this score space.

The final aspect of conversion is to compute the correlations between the columns of 13 distances of the 100 points in pairs. These correlations are given for the first two objects in each cluster in Sec. B of Table 8.1. Correlations between the two objects within each cluster are virtually 1.00, and as the clusters separate in space the correlations go down to .00 and become negative when the coordinates tend to go to mirror images.

The 100 by 100 distance matrix is thus transformed to a full correlation matrix. We may then perform a full-cycle key-cluster analysis on this matrix, with diagonal values set to 1.00. In this full sequence, the final component called is the spherical analysis, SPAN. The actual result is shown in Fig. 8.4, where the configuration of objects displayed on the surface of a sphere is in almost precisely the form they took in the score space of Fig. 8.3.

The point of all this is that, in general, the configuration of objects

TABLE 8.1 CONVERSION OF EUCLIDEAN DISTANCES TO CORRELATION COEFFICIENTS IN EUCO ANALYSIS

	First Two Individuals in Each Object Cluster											
	A12	A13	B13	B22	C13	C14	D14	D23	E13	E14	F11	F12
Dimension scores:												
Z_1	38	38	58	60	28	28	44	46	54	54	40	40
Z_2	64	66	62	60	46	48	52	50	40	42	32	34
A. Euclidean Distance from Dimension Scale Reference Points												
On dimension Z_1 (6 reference points):												
20	22	24	39	41	8	8	24	26	35	34	26	25
30	16	17	30	31	4	2	14	16	26	25	20	18
40	14	16	21	22	12	12	4	6	17	16	18	16
60	26	27	12	10	32	32	16	14	11	10	26	25
70	34	35	16	14	42	42	26	24	18	17	34	34
80	44	44	25	22	52	52	36	34	27	27	43	43
On dimension Z_2 (6 reference points):												
20	45	47	42	41	34	35	32	30	20	22	15	17
30	36	37	32	31	27	28	22	20	10	12	10	10
40	26	28	23	22	22	23	13	10	4	4	12	11
60	12	13	8	10	26	25	10	10	20	18	29	27
70	13	12	11	14	32	31	18	20	30	28	39	37
80	20	18	19	22	40	38	28	30	40	38	49	47

General origin
(1 reference
point):
50

B. Correlation Coefficients Between Columns of Euclidean Distances (r_{DD})

	18	20	14	14	22	22	6	4	10	8	20	18
A12	1.00	.99	.52	.38	.57	.62	.76	.63	−.15	−.06	−.07	−.00
A13	.99	1.00	.55	.40	.51	.56	.71	.58	−.20	−.11	−.15	−.08
B13	.52	.55	1.00	.98	−.26	−.22	.46	.47	.15	.26	−.40	−.36
B22	.38	.40	.98	1.00	−.34	−.31	.41	.45	.26	.37	−.36	−.32
C13	.57	.51	−.26	−.34	1.00	.99	.65	.57	.11	.11	.59	.64
C14	.62	.56	−.22	−.31	.99	1.00	.66	.57	.06	.07	.54	.59
D14	.76	.71	.46	.41	.65	.66	1.00	.97	.47	.54	.44	.51
D23	.63	.58	.47	.45	.57	.57	.97	1.00	.63	.69	.52	.58
E13	−.15	−.20	.15	.26	.11	.06	.47	.63	1.00	.99	.72	.73
E14	−.06	−.11	.26	.37	.11	.07	.54	.69	.99	1.00	.66	.67
F11	−.07	−.15	−.40	−.36	.59	.54	.44	.52	.72	.66	1.00	.99
F12	−.00	−.08	−.36	−.32	.64	.59	.51	.58	.73	.67	.99	1.00

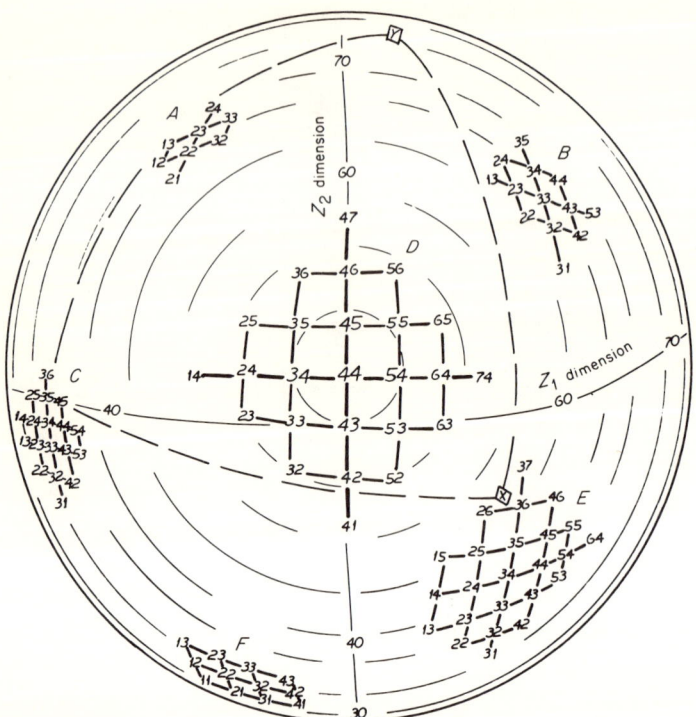

FIGURE 8.4
The six object clusters of 100 individuals projected onto the surface of a sphere by EUCO analysis.

within score space is spread out on the surface of spheres, where it can be inspected for cluster structure. If there is a clear cluster structure showing definite areas of density, i.e., O-types, they will leap to the eye as they actually do in Fig. 8.4, where the six object clusters are immediately evident.

Thus, EUCO analysis is a form of pattern recognition, in which the pattern of objects in k-dimensional space is displayed for recognition on a spherical surface. Since it clearly shows the configuration of the objects in their interrelationships, one can also discover the hierarchical structure among the objects. In Fig. 8.4, for example, clusters D and E are most highly related, E is slightly negatively related to A, and so on. The exact hierarchical relations are given metrically in the CSA component, where, just as in V-analysis, the interdomain, i.e., intertype, correlations are given, and where in the oblique factor coefficients the relation of each object to each domain, i.e., to each type, is given.

There is one special difficulty with forming a typology from the SPAN configuration given by EUCO analysis. If there is no clear cluster structure, the analyst may have difficulty in making up his mind about the object clusters that constitute a proper typology. But if no structure does appear, this is a fact that he must live with. In this case, he will quite arbitrarily break up the configuration into sectors whose centers of density form patterns of scores that are meaningful to him. For analysts who have difficulty making up their minds, we need methods (given below) that will always provide a finite set of object clusters when cluster structure is not clear, methods which discover "natural" areas of density in those regions of the configuration where such densities do exist.

Another difficulty is the expense of EUCO analysis if there are many individuals in a study, a likely possibility. To compute a correlation matrix between the columns of distance values with six reference points on each of k dimensions, say, for 1,000 individuals can be a forbidding undertaking, even with modern computers. Though shortcuts can be employed, such as forming a composite SPAN on random samples drawn from the full supply and increasing the number of samples to secure a converged configuration that describes the cluster structure of the full supply (Tryon, 1966), EUCO analysis is really not a practical solution if N is large. It is, however, practical for representing the spherical configuration of centroids only, as found by other methods described below, or for studying the configuration of the cluster structure among a more limited number of individuals that may lie in a sector of the configuration in hyperspace (Chu, 1966).

Typology by iterative condensation on centroids

A practical solution relatively unaffected by the number of individuals in a study, and one not requiring large-scale distance matrices or correlation coefficients, is an iterative procedure starting with trial O-types and ultimately converging on O-types at centers of density. If a clear cluster structure does not exist, the procedure nevertheless provides an arbitrary set of O-types. This procedure is essentially a program of pattern recognition that discovers a pattern of clusters if such a pattern exists in the cloud of individual points in the score space. This new method is illustrated fully in real data in later parts of this and following chapters.

We may simply illustrate the logic of this iterative method of forming a typology on the fictitious problem of 100 individuals organized in six distinctive clusters, as depicted in Fig. 8.3. Programmed in the component OTYPE of the BC TRY System, the method begins with iteration 0, called I_0, that selects an arbitrary set of trial "core O-types" from the swarm of points. In the two-dimensional space of the fictitious problem it computes the coordinates of the centroids of each core O-type and then reassigns each one of the 100 points to the centroid of that particular core O-type with which it has its smallest euclidean distance. After all 100 points are reassigned, each O-type is thus changed; hence in the second pass, I_1, the coordinates of the new centroid of each O-type are computed, and all 100 points are again reassigned to the new centroids with which they are closest. New centroids are thus formed on which, in I_2, all 100 points are again reassigned. The process continues until the final iteration, at which all 100 points remain unchanged in their reassignment to O-types.

During successive iterations the trial O-types wander about a bit, moving closer to centers of density. During iteration, if two trial O-types approach closely enough (according to an arbitrary criterion), they merge together to form a single O-type. At any reassignment of objects, if any one point is too far away from any centroid (according to an arbitrary criterion) it is cast out as "unique," but it can come back and join a cluster if on any later iteration it again falls within the criterion.

The results of this process of iterative condensation on centroids are shown in Fig. 8.5 for eight separate OTYPE runs on the fictitious problem. Besides the different criteria mentioned above, one of the most critical parameters is the one that provides the initial starting trial core O-types in iteration I_0. This is the set of "partitions," or "cutoff points" on each dimension that establishes the initial arbitrary "sectors" in the score space within which the initial core O-types are located. For example, in Fig. 8.5, top left, each dimension is dichotomized to form four sectors, hence called "Sect-4." Starting from the four core O-types resulting from I_1, labeled Trial 1 in the figure, the trial O-types formed by I_1 are the four shown enclosed in broken lines. At the end of Trial 8 the process converges on the final four O-types shown in unbroken lines: this Sect-4 run therefore fails to converge on the "true" clusters. In the next run, however, shown in the next lower diagram labeled Sect-9, where core O-types are initially formed by trichotomizing each dimension, I_1 forms seven O-types, but on I_4 the process converges exactly on the true cluster structure. The rest of the runs (including Sect-100, not shown) converge on "truth" except Sect-49 and Sect-64, each of which, on convergence, picks up a seventh spurious O-type from the two edges of true O-types D and E.

A variety of other parameters affect the convergence process of the centroid condensation method, but we cannot discuss the details of them here. In general, however, when there is a clear cluster structure, the OTYPE component of BC TRY finds it. The best procedure is to make several runs under different options, and if the same set of converged O-types does not appear, to input the coordinates of the different O-types (output on cards after each OTYPE run) as a final batched run on OTYPE.

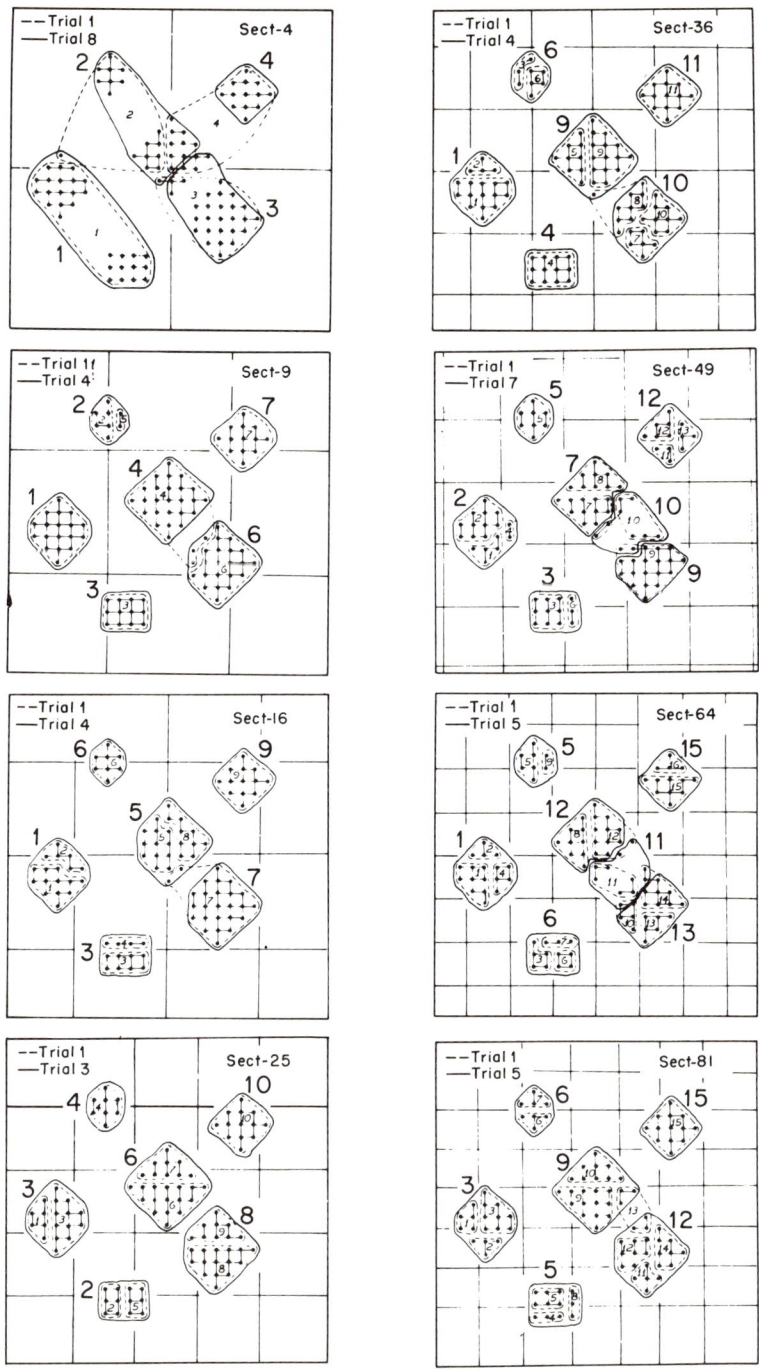

FIGURE 8.5

In the example of six clusters of 100 fictitious individuals, converged O-type "discovered" by the centroid condensation iterative method, for eight different starting trial O-types (the problem of pattern recognition).

The converged result is the final set of O-types accepted. To show the power of the batched run, when we batch the output O-type cards from the erroneous runs of Sect-4, Sect-49, and Sect-64 with the other correct outcomes, the single batched run on OTYPE converges exactly on the true cluster structure. As a final check on typological structure, the analyst will usually run a EUCO analysis of the data in order to compare the results of the OTYPE run with those given by the overall configuration in SPAN.

Typology by other procedures

It has been pointed out above that another method of representing the cluster structure of objects is to form a hierarchical chart of the resemblances between them as measured by the interobject distances. In the OTYPE program the hierarchical order of the O-types is also routinely output as a guide to understanding the intercluster structure of the centroids. An illustration is given in Fig. 8.6, which gives the hierarchy of the initial

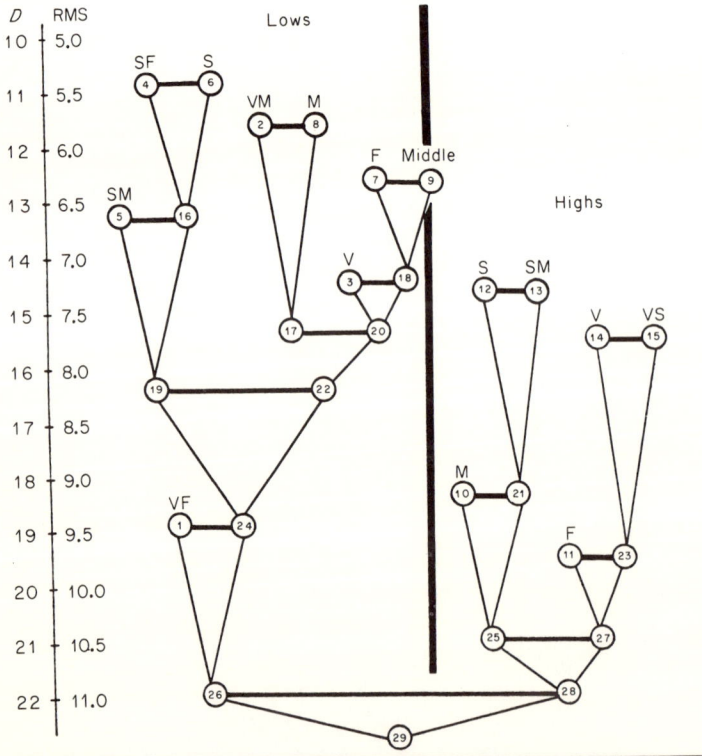

FIGURE 8.6
Hierarchical structure of the 15 core O-types in the Holzinger problem.

15 core O-types. The meaning and use of the knowledge conveyed by the hierarchical graph are discussed later in the chapter.

Mention should also be made of the "single-bond" method of object clustering of Sneath (Sokal and Sneath, 1963, p. 180). By this method, the individual points in score space are hierarchically, i.e., successively, condensed on the basis of smallest interobject distances. The principle is the same as that illustrated in Fig. 8.6, except that individual points and whole clusters are merged on the basis of the smallest distances between individuals but not between centroids. In the fictitious problem of Fig. 8.3, the single-bond principle first forms a hierarchical chart showing that, on the basis of smallest distances, the six true clusters are separately formed at the same level of interobject distances.

The single-bond method gives a good hierarchical chart between single members of the total configuration. It has the defect of uniting clusters into a higher-order cluster only on the basis of a single bond of objects on edges of clusters. The result could be that if a true cluster includes a long string of points, progressive condensation could follow the string and miss the fact that another large cluster might exist nearby.

The method of average linkage of Sokal and Michener (Sokal and Sneath, 1963, p. 182) joins individual points to clusters and clusters to each other on the basis of the average distances rather than single bonds. This method is therefore virtually identical with that reassignment aspect of the centroid condensation method in which each cluster is represented by its centroid as defined by the average of the coordinates of all the points that constitute it. Since the average distance between a given individual and all the individuals that compose a cluster is the same as the distance between the individual and the centroid of all the members of the cluster, the condensing operation by the two methods should be the same.

An example: intelligence, the Holzinger problem

In this section we present an illustration of an object cluster analysis, a typological study, of the 301 grade school children in two Chicago schools who took 24 pencil-and-paper tests sampling a variety of intellectual abilities. The basic data are those already used in previous chapters, the Holzinger study. The original scores of all children in both schools are reported in the original reference. The tests of intellective capability of the study are listed in Table 4.1. The results of an empirical key-cluster analysis on the variables of the study are reported in Chap. 6, with the primary results of the analysis being reported in Tables 6.1 to 6.3. The revised cluster analysis results and the results of the cluster structure analysis of the revised clusters and their scores are reported in Chap. 7. The primary results of these analyses are given in Tables 7.1 to 7.3 and Figs. 7.3 and 7.4.

For the purposes of this analysis we selected four clusters of variables: verbal, V5, V6, V7, and V9; speed, S10, S11, S12, and S13; space, F1, F2, F3, and F4; and memory, M14, M15, M16, M17, and M18. The group profiles presented in Fig. 8.1 are profiles of scores on these clusters for the 15 groups of children that were selected by the O-analysis procedures described here. In the interest of brevity and simplicity we use the symbols V, S, F, and M to stand for the verbal, speed, space (form), and memory clusters and the corresponding cluster scores.

Persons in cluster score space

The first step in O-analysis is to compute scores for each subject on the four dimensions defined by the clusters V, S, F, and M. A variety of techniques for preparing these scores have been proposed (see Harman, 1967). In practice we have found a very high degree of correlation among the various methods, and the cluster composite method is the method we choose. The cluster composite method gives scores that correspond to the cluster composites of cluster structure analysis. The score on a cluster dimension is defined as a simple sum of the standardized scores of the variables in the cluster, hence a simple additive composite. For example, the score of a given subject on the verbal cluster, V, is the sum of the standard scores for that subject on V5, V6, V7, and V9. This composite score ordinarily is transformed into standard score form, e.g., with mean of 50 and standard deviation of 10. Other transformations are used for a variety of purposes, but for our illustration we use means of 50 and standard deviations of 10.

A given individual's profile on the four abilities V, S, F, and M is defined by his four standard scores on the four clusters. It is these data which are depicted in Fig. 8.1.

The O-analysis of the subjects in terms of the four abilities of the Holzinger study is based on the distances between the subjects, taken pairwise, as defined by cluster score dimensions and the subjects' scores. For example Table 8.2 shows hypothetical standard scores for two individuals A and B and the computation of the distance between them D. The similarity between the profiles of A and B is represented by the one value, the distance D. D is simply the square root of the sum of the squared deviations on the four scores as shown in the computations at the bottom

TABLE 8.2 CLUSTER SCORES AND DISTANCE CALCULATION

	V	S	F	M
Cluster scores of A	36	44	34	46
Cluster scores of B	37	48	47	35
Score difference d, A's *score minus* B's *score, each cluster*	−1	−4	−13	11
Squared difference d²	1	16	169	121

$$D = \sqrt{\Sigma d^2} = \sqrt{307} = 17.5 \qquad RMS = \frac{D}{\sqrt{K}} = 8.8$$

of the table. For these two individuals $D = 17.5$. Any computer program that can compute composite scores, standardize them, and calculate the matrix of distances between all pairs of individuals will do the basic work for the condensation method of O-analysis. In the BC TRY System special programs are employed, FACS, EUCO, and OTYPE.

The euclidean distance D given by the equation in Table 8.2 is not a metric of similarity that is consistent across studies, since its magnitude depends on how many dimensions there are in a study. A useful index that is not a function of the number of dimensions is the square root of the mean square difference in cluster scores. We call this index RMS. The index is a simple function of the distance

$$RMS = \sqrt{\frac{1}{K} \Sigma d^2} = \frac{D}{\sqrt{K}}$$

where K is the number of dimensions.

Core O-types, O-clusters, and unique persons from arbitrary sectors in the score space

A preliminary stage in clustering individuals is to cast them temporarily into core O-types based on arbitrary sectioning of the cluster score space. In general, the number of sectors is determined by how many broad categories one chooses on each of the dimensions, how many dimensions there are, and how many subjects are available. In the Holzinger problem, three broad categories were used on each of the four dimensions, the result being that their joint cluster score space is sectioned into $3^4 = 81$ sectors.

For studies with limited numbers of subjects it is desirable to keep the number of sectors to a small number. The number of sectors is a simple function of the number of categories on each of the dimensions and the number of dimensions. If there are K dimensions and C categories on each of the dimensions, the number of sectors S is equal to

$$S = C^K$$

The broad categories may, of course, cut the scores at any place. With three categories, convenient cutoff points are at 1 standard deviation above and below the mean. Thus for mean of 50 and standard deviation of 10 three categories give low scores below 40, middle scores of 40 and above to 60, and high scores of above 60. The BC TRY System provides a component OTYPE to do all this work. Standard decisions based on these principles are built into the program, but an analyst can opt for any other cutoff points he wishes. If there are more than six dimensions in a problem, it is usually desirable to use not more than the most salient six; e.g., in the MMPI problem with seven dimensions, only the most salient four are used. But, if the analyst wishes to retain more than six dimensions, he is advised to perform separate O-analyses on selected sets of dimensions, each set including the most highly correlated or rationally related dimensions, not exceeding six.

The next step is to sort the individuals into sectors in cluster score space. The results are shown in Table 8.3 where, for comparative purposes, data from the analysis of the MMPI are also given. Under "Score Patterns," the first sector is LLLL, meaning a score pattern all low (L), i.e., below 40 on the four dimensions. The column marked Incidence and Core Types in the Holzinger problem shows that there are only two children with such a generally low profile of scores. The next score pattern, LLLM, is one with a profile of scores that are low on V, S, and F, but middle (M) with scores from 40 to 60 on the last dimension M. Two children have this profile. All told there are 81 patterns, the last of which is HHHH, all highs (H) across the four scores, i.e., scores above 60 on all four dimensions. There were three such children with this pattern.

A computer program like OTYPE of the BC TRY System is desirable for this work, but actually it can be quickly achieved on an old-fashioned card sorter. One sorts all 301 cards on the first two digits of the last dimension M, ranking them from low to high. This deck is then ranked on the first two digits of F from low to high, then on S and finally on V. If, before this sorting is made, one puts on the top of the deck 81 "dummy" cards, each with standard scores corresponding with one of the 81 score patterns shown in the list on Table 8.3, these dummies take a position in the sorted deck which marks the beginning of each of the 81 sectors so that on a listing of the sorted deck all the cases in the 81 sectors are nicely marked off.

TABLE 8.3 COMPARISON OF THE FREQUENCY OF PERSONS IN 81 DIFFERENT SCORE PATTERNS IN THE HOLZINGER AND MMPI PROBLEMS, AND THE SELECTION OF CORE TYPES FROM THEM

Left half:

Holzinger: V S F M / MMPI: I B S T				Holzinger Frequency	Holzinger Core No.	MMPI Frequency	MMPI Core No.
L	L	L	L	2		10	(1)
L	L	L	M	2		0	
L	L	L	H	0		0	
L	L	M	L	4		19	(2)
L	L	M	M	2		7	(3)
L	L	M	H	0		0	
L	L	H	L	0		0	
L	L	H	M	2		0	
L	L	H	H	0		0	
L	M	L	I	0		3	
L	M	L	M	8	(1)	2	
L	M	L	H	0		0	
L	M	M	L	6	(2)	6	(4)
L	M	M	M	13	(3)	8	(5)
L	M	M	H	3		0	
L	M	H	L	0		1	
L	M	H	M	3		3	
L	M	H	H	0		0	
L	H	L	L	0		0	
L	H	L	M	1		2	
L	H	L	H	0		0	
L	H	M	L	0		0	
L	H	M	M	3		1	
L	H	M	H	1		0	
L	H	H	L	0		0	
L	H	H	M	0		0	
L	H	H	H	1		0	
M	L	L	L	1		7	(6)
M	L	L	M	5	(4)	2	
M	L	L	H	1		0	
M	L	M	L	6	(5)	7	(7)
M	L	M	M	14	(6)	5	(8)
M	L	M	H	0		0	
M	L	H	L	0		1	
M	L	H	M	0		0	
M	L	H	H	1		0	
M	M	L	L	3		4	
M	M	L	M	14	(7)	17	(9)
M	M	L	H	2		0	
M	M	M	L	13	(8)	7	(10)
M	M	M	M	83	(9)	68	(11)
M	M	M	H	13	(10)	12	(12)

Right half:

Holzinger: V S F M / MMPI: I B S T				Holzinger Frequency	Holzinger Core No.	MMPI Frequency	MMPI Core No.
M	M	H	L	2		0	
M	M	H	M	18	(11)	15	(13)
M	M	H	H	2		2	
M	H	L	L	0		0	
M	H	L	M	0		3	
M	H	L	H	0		0	
M	H	M	L	2		0	
M	H	M	M	17	(12)	10	(14)
M	H	M	H	6	(13)	12	(15)
M	H	H	L	1		0	
M	H	H	M	2		5	(16)
M	H	H	H	3		4	
H	L	L	L	0		0	
H	L	L	M	0		0	
H	L	L	H	0		0	
H	L	M	L	0		0	
H	L	M	M	1		0	
H	L	M	H	0		0	
H	L	H	L	0		0	
H	L	H	M	1		0	
H	L	H	H	0		0	
H	M	L	L	0		0	
H	M	L	M	1		4	
H	M	L	H	0		0	
H	M	M	L	1		0	
H	M	M	M	18	(14)	27	(17)
H	M	M	H	1		9	(18)
H	M	H	L	1		0	
H	M	H	M	2		4	
H	M	H	H	3		6	(19)
H	H	L	L	0		0	
H	H	L	M	0		0	
H	H	L	H	0		0	
H	H	M	L	0		0	
H	H	M	M	6	(15)	4	
H	H	M	H	0		10	(20)
H	H	H	L	1		0	
H	H	H	M	2		1	
H	H	H	H	3		7	(21)
Total				301		310	

Note that there are either no children or a trivial number in many of the 81 sectors of Table 8.3. One must therefore decide on the minimal number of cases to be required in a sector in order for it to be dignified by the label of a core O-type. In the Holzinger problem the number is arbitrarily set at five children (about 2 percent). In Table 8.3 under Core No. the sectors that meet this criterion are numbered in serial order from 1 to 15. These are the 15 core O-types plotted in Fig. 8.1.

What is to be done with the 61 children that do not easily fit into these 15 types? We assign each of them to that core O-type with which it has its best fit. The procedure for doing so is to compute for each of these cases its distance with the 15 average core O-types and assign it to the one with which it has its smallest distance. Some of the 61 children have such unique profiles that even their smallest distance with the core O-types shows their best fit to be poor. Some criterion must be established in order to decide when an individual will be excluded from any O-type. A convenient uniqueness criterion is objectively set as follows. No object may be a member of a core O-type if the RMS of its cluster scores from those of a core O-type is greater than 1 standard deviation.

To compute the distances of the 61 individuals from the 15 core O-types requires that the mean score profile on V, S, F, and M of each of the 15 core O-types must first be computed. These values are those of an "abstract" individual that defines a core O-type. The profiles of these 15 abstract individuals are plotted as filled circles or dots in each of the graphs of Fig. 8.1. In the BC TRY System the component program OTYPE computed the distances of the 61 children with the 15 abstract types, routinely assigned each child to the O-type with which it had the smallest distance, and set aside the children who did not meet the uniqueness criterion. The OTYPE program permits the analyst to choose other criteria if he wishes.

The final results of these procedures are all graphically shown in Fig. 8.1. Of the 61 unallocated children, 45 are assigned to the 15 core O-types on the basis of small distances. Their scores are represented in the graph by x's. The uniqueness criterion excludes the 16 unique children shown in the bottom graphs.

The analyst can, of course, be content with his typology at this point. He has derived a highly reduced taxonomy according to which all his cases can be classified objectively. The arbitrary sectors may be sufficient. Indeed, in some problems there may be a priori grounds for presetting a specified set of broad categories in each of the dimensions. If so, they will determine the arbitrary sectors of score space and the assigning of individuals to them may be all that the analyst wishes.

Hierarchical structure of core O-types

If there are no a priori grounds for accepting the established sectors, the above procedures may produce more arbitrariness than one wishes to tolerate, for there may be natural clusters that have been sliced through by the arbitrarily defined sectors. Furthermore, some individual members of a core O-type may lie in corners of arbitrary sectors so that in fact they may be closer to core O-types in adjacent sectors than to the objects in the sectors to which they have first been allocated. Therefore additional procedures are needed to free the analysis of such arbitrariness.

What is needed is a display of the overall structure of the relationships among the 15 core O-types. If the slices cut through a tight natural cluster, the core O-types into which it is fractionated should have similar profiles, with small distances between them. If the core O-types are progressively combined, two at a time, into higher-order clusters, those which constitute a natural cluster would be the first to combine. Progressive hierarchical condensation of the 15 core O-types of the Holzinger problem gives the chart shown in Fig. 8.6.

In this chart the vertical scale is the euclidean distance. The first two core O-types with the smallest distance between them are shown at the upper left, namely, types 4 and 6, whose distance is a little below 11. The profiles of type 4 in Fig. 8.1 consist of individuals distinctively low on speed and form, hence the symbols SF are placed above type 4 in the hierarchical chart. Type 6 consists of individuals distinctively low on S alone. This new higher-order type consists of individuals whose profiles are low on S but vary on F from middle down to low. By such a combination we lose the two core types 4 and 6 but gain type 16 in their stead.

The progressive condensation of core O-types is shown as one reads down the chart. On the left of the chart, headed at the top by Lows, these condensations always lead to higher-order types that are low on at least one of the four dimensions: at the bottom opposite a D of 22, the grand higher-order type 26 consists of all individuals that meet this condition of varying from the mid-value of 50 to a very low value on one or more of the dimensions. On the right of the chart are the Highs whose final higher-order type 28 at the bottom right of the chart consists of all individuals that vary in the higher ranges of the cluster on at least one dimension.

Reading the chart from bottom up is like observing a breakdown of the subjects into a taxonomic structure varying, as it were, from genus to species to varieties—all quite objectively and quantitatively determined. Faced by such a chart, one must decide at what level of similarity, as measured by D, one wishes to declare his final typology to be. One decision is not to accept any higher-order types but to retain all 15 core O-types as depicting the final typological structure. But if it is desirable to deal with less than 15, one could reject types 4 and 6, and accept in their stead type 16, etc.

The cutoff point of similarity depends upon the purpose of the typology. If one wants, for example, a grand higher-order type in which all children are at least low in one or more ability and below average in the rest, one would accept type 26 (excluding type 9 children who are in the middle ranges in all abilities). Analogously, all individuals in the grand higher-order type 28 are high in one or more of the abilities. For certain purposes an analyst may wish to deal with two such grand high vs. low person-clusters.

In the present analysis we retain all 15 of the original core types in order to preserve as much differentiation between the types as possible. The reason for doing so is that inspection of the profiles of Fig. 8.1 does not reveal that the arbitrary slices through the score space have cut through any clearly evident natural higher-order cluster. Later we will find a rather different story in the MMPI study. Some of the original core O-types in that study appear to constitute artificial fragmentations of natural clusters.

One difficulty with using this euclidean distance D as an index of similarity is that the D in different problems may not be comparable. If we convert D to RMS, however, these indexes are comparable across different problems. The reason is that RMS describes the average deviation between standard scores of O-types independent of the number of dimensions in the different problems. The RMS values corresponding to the D values are shown in the chart of Fig. 8.6 at the right of the D scale.

Spherical configuration of core O-types

As a further aid in discovering whether the arbitrary sectioning has cut through clusters and in choosing higher-order combinations of core types, we need a more precise display of the structure of the relationship among the core O-types. Such a display is the spherical configuration presented in Fig. 8.2. The 15 core O-types are spread out on the surface of a sphere in such a fashion that the spatial separation between core O-types is a function of their distances. The core O-types linked as similar in the hierarchical chart have been enclosed in dashed lines in the spherical configuration. Types 4 and 6, for example, are close together on the surface at the left. All types designated under Low in the hierarchy are at the left in the configuration, those designated High at the right. The advantage of this display is that the whole configuration can be seen much more easily than in the hierarchy of Fig. 8.6 or in the multiple profiles of Fig. 8.1. The standard score scales of the verbal, speed, form, and memory dimensions are set into the configuration so that the standard scores of the various types can be read from the configuration.

In this display a spherical triangle is plotted by long-dashed lines, the three apexes of which are designated by squares. These "corners" of the triangles are 90° apart as seen from the origin of the sphere. Using these corners as reference points, it can be seen that any two types that exceed 90° will have profiles that are mirror images. For example, type 5 and type 13 generally have standard scores departing in opposite directions from the mean of 50 and presenting mirror images in the patterns of scores across the four abilities.

This configuration of 15 core O-types actually lies on the surface of a hypersphere, i.e., on a sphere of more than three dimensions. However, the procedures for developing the configuration are such as to project as many types as possible on the surface of *one* three-dimensional sphere, leaving a minimal number to be projected into a fourth dimension, or more. How successful the procedure is may be seen from the fact that there is only one O-type, 11, lying in the fourth dimension, denoted by an x in the figure. This fourth dimension is mainly concerned with form, for the form axis lies in the fourth dimension only at its ends, as is the case with core type 11, which has a high standard score on F.

Final types from the profiles, the hierarchy, and the spherical configuration

Information from Figs. 8.1, 8.2, and 8.6 is used in arriving at the final types to retain from an O-analysis. In the Holzinger problem it was finally decided not to deal with any higher-order combinations of the 15 core O-types but to leave them as they stand. The reason is that it did not appear as if any particular cutting through of natural clusters had occurred. The 15 types are listed in Table 8.4, being denoted as the final type by the prefix "T". Thus, in the first column they are symbolized as T1, T2, down to T15. A descriptive term is given to each one based upon its distinctive high and low mean standard scores. The profile level of each O-cluster is given by its four standard scores, listed in the columns headed by Z under Profile Level and Homogeneities. The frequency of cases in each O-type is listed in the column headed Initial. For example, in T1, described as low verbal, low form, is a group of ten children consisting of the original eight in core O-type 1 plus two originally unallocated subjects that are assigned to it later on the basis of their smallest euclidean distances. The standard

TABLE 8.4 CONDENSED INTELLECTUAL TYPES IN THE HOLZINGER PROBLEM

Type	Descriptive Name	Frequency of Cases		Profile Level and Homogeneities								Overall H
		Initial	After D fit	V		S		F		M		
				Z	H	Z	H	Z	H	Z	H	
T1	Low verbal, low form	10	14	36	.93	42	.87	35	.83	45	.90	.88
T2	Low verbal, low memory	10	8	35	.89	48	.85	48	.86	36	.90	.88
T3	Low verbal	22	22	35	.91	48	.86	51	.78	53	.88	.86
T4	Low speed, low form	7	10	50	.75	38	.97	35	.95	48	.82	.88
T5	Low speed, low memory	10	11	49	.80	34	.89	45	.88	36	.93	.88
T6	Low speed	15	17	47	.85	35	.90	48	.93	50	.84	.88
T7	Low form	15	18	50	.88	49	.88	36	.91	45	.84	.88
T8	Low memory	15	23	49	.88	50	.87	48	.76	37	.93	.86
T9	Average	83	44	50	.92	49	.90	49	.88	50	.91	.90
T10	High memory	18	26	52	.80	50	.88	51	.73	65	.91	.83
T11	High form	20	24	52	.89	53	.83	66	.88	50	.81	.85
T12	High speed	22	24	50	.79	65	.94	52	.77	53	.79	.83
T13	High speed, high memory	10	9	53	.68	65	.94	53	.64	67	.86	.79
T14	High verbal	20	22	66	.94	51	.86	53	.88	48	.90	.90
T15	High verbal, high speed	8	12	68	.88	66	.86	58	.66	56	.91	.83
Unique		16	17	58	—.23	53	—.65	59	—.55	54	—.85	—.59

scores of this type are shown in its row of Z values under Profile Level and Homogeneities.

Having made the decision on the final types, we now assign all 301 subjects to those O-types with which they have the smallest distance. Although we assign to O-types individuals that have not been original members of the core O-types, there may remain some members of an original core type that may be closer to core types in adjacent sectors. When all 301 individuals are assigned to those core types with which they have the smallest distance, the frequency in the final types may change from that in initial O-clusters. For example, T1 has 14 individuals in it after adjustment, a gain of 4 individuals from adjacent sectors. Type T9 shows the greatest change. There were 83 subjects originally in this large average core O-type. Obviously, many subjects from outlying edges of it have joined up with core types in adjacent sectors. This process of assignment and reassignment could be reiterated, but the problems involved in this reiterative procedure must be left for later development.

Multivariate selection in the final types

Each of the 15 types is a multivariate selection of individuals from the full supply of 301 children, because it is based on four different abilities, each of which samples several tests. The selection has occurred in two ways. First, the profile levels of mean standard cluster scores reveal how distinctively the selected groups depart from the average value of 50 in the full supply. A further index of selection is the homogeneity coefficient H for the selected group. This H value is a measure of the "tightness" of the profiles of individuals that compose a given type. The H value is a function of the within-variance of the group's cluster scores compared to the total variance of individuals in the full supply. In the case of type T1, its H value is formulated

$$H_{\text{T1}} = \sqrt{1 - \frac{\text{variance of cluster scores of the 10 children in T1}}{\text{variance of cluster scores of all 301 children}}}$$

When the members of an O-type are identical in their profiles, the within-group variance is zero and the homogeneity H is unity. If, however, the O-type is a random selection from the full supply, within-group variance equals the variance in the full supply and H becomes zero.

Type T1 consists of 10 individuals whose cluster scores on V are close to being identical because its H is .93. It is illuminating to look down the column of H values for each of the abilities to see how each of the types has been selected on that particular ability. For a given O-type, the row of H values across the different abilities indicates how homogeneous the multivariate selected group is on the four different abilities. An overall index of homogeneity is given in the last column of Table 8.4; it is merely the average of the H values across the four abilities. The greatest overall selection has occurred in types T9 and the highly verbal type T14. The least selective is T13, with a value of .79, a type with relatively low selection on V and F.

In the bottom row of Table 8.4 are the unique individuals. Their cluster scores reveal that the unique children, as a group, tend to run somewhat above the average in the four abilities. Their homogeneity coefficients are negative. A negative homogeneity coefficient means that the individuals are more heterogeneous (greater variance of scores) than the full supply, a point to be expected with the full array of unique individuals. The square root in the homogeneity equation does not permit evaluation of H when the value under the radical is negative. In practice the square root of the absolute value is found, and the sign of the value is attached to the square root.

It is desirable to estimate the degree to which each of these O-types has profile levels and homogeneities that could be values arising by pure chance selection from the full supply. This matter is treated in a later chapter in a description of the BC TRY component 4CAST, which estimates these probabilities.

The final typology indicates that there are disjunctive, incompatible patterns of scores on the abilities. No compensatory low-high opposite extremes of abilities exist. Furthermore, some patterns of extremes in the same direction are also absent. Whereas in types T1 and T2, low verbal occurs with low form and with low memory, respectively, low verbal does *not* occur with low speed. At the other extreme we find, conversely, that in type 15, high verbal occurs in 12 children with high speed but *not* with high form or with high memory. Just why there should be reciprocal incompatibilities is a fascinating question but beyond the scope of this book.

An example: personality, the MMPI problem

Four general attributes, I, B, S, T, discovered by V-analysis of the MMPI items

The clinical scales that sample the item pool of 566 items of the MMPI suffer from the defect of including items that overlap between scales.

The result is that a considerable but unknown amount of redundancy occurs in the different scales. Tryon and two of his associates, Kenneth Stein and Chen-Lin Chu, made what appears to be the first comprehensive effort to develop item-cluster scales that involve no overlap of items. This newer approach was not possible before the development of the convergence method, by which, under the BIGNV procedures of BC TRY, V-analysis was released from restriction by the number of variables in a problem (Tryon, 1966). The importance of this new development is that subjects can be scored on experimentally independent item pools and a serious effort made to form a taxonomy of person-clusters. This chapter presents such a typology, performed by the new condensation method of O-analysis.

In our own V-analysis of the 556 MMPI items we found seven dimensions defined by psychologically meaningful oblique item-clusters. Four of these seven dimensions accounted for most of the total communality of all the items of the MMPI. About two-thirds of the items were eliminated because of trivial communalities. These four item-clusters are briefly summarized in Sec. A of Table 8.5. Only the item numbers are indicated in these four scales, because the MMPI blank is readily available to readers who wish to look up the individual items.

The example developed here presents a complete typology of individuals, based on their profiles on these four item-cluster scales, identified as introversion (I), body symptoms (B), suspicion (S) and tension (T). Since the MMPI is used in connection with psychiatric examinations, a group expressly selected to be heterogeneous with respect to mental illness was used in this study. The supply of subjects consisted of 70 psychotics (schizophrenics), all with a history of hospitalization within the previous 5 years, 150 diagnosed as "anxieties" but with no history of hospitalization, and 90 normal subjects who were Armed Services officers matched with the 220 psychiatric patients with respect to age and education. In Sec. B of Table 8.5 the cluster scores on the four attributes are shown to correlate higher than the four intellectual abilities in the Holzinger problem. Indeed, the correlation of the tension cluster with the body symptoms cluster is a high level of .75. The reliability coefficients, in Sec. C, run a bit higher, generally in the .90s, compared to those of intellectual abilities, and the generalities of the four MMPI attributes, in Sec. D, are higher than those of the intellectual abilities clusters.

Persons in the cluster score space of I, B, S, T

After the cluster scores of each of the 310 subjects are computed on each of the four attributes, the 310 individuals are represented as points in the

TABLE 8.5   FOUR BASIC ITEM–CLUSTER
ATTRIBUTES, I, B, S, T, OF THE MMPI

A. Defining Variables of I, B, S, T

I: Introversion, defined by the following 26 items:

377	267	52	317	−415
− 57	172	−309	−264	−482
321	86	−479	138	
201	171	509	−353	
180	−547	292	304	
−371	−521	− 79	−449	

B: Body symptoms, defined by the following 33 items:

−243	−230	125	72	−160	− 18
189	114	− 68	− 3	191	−192
108	47	10	− 36	−153	14
−190	44	23	−163	263	
62	− 55	161	− 51	−330	
−175	29	544	−103	− 2	

S: Suspicion and mistrust, defined by the following 25 items:

404	244	447	284	455
507	348	319	438	
383	368	71	89	
390	280	558	112	
436	265	406	426	
136	469	278	316	

T: Tension, worry, and fears, defined by the following 36 items:

555	543	448	182	158	22
431	442	186	32	303	351
337	43	499	439	13	−131
217	−242	166	335	388	365
238	340	338	102	322	494
506	−152	−407	473	360	492

B. Intercorrelations between Cluster Scores I, B, S, T

	I	B	S	T
I		.47	.27	.68
B	.47		.32	.75
S	.27	.32		.48
T	.68	.75	.48	

C. Reliability Coefficients of Cluster Scores

α *reliability*	.93	.92	.85	.92

D. Generality across All Seven Item-clusters, Each Defined by Its Best 17 Items

Mean square of $r's$	.65	.55	.41	.88

cluster score space of four dimensions, and the similarity between score patterns of any two individuals is represented by the euclidean distance D between them.

Core O-types, O-clusters, and unique persons from arbitrary sectors of score space

Each of the scales is sectioned into the same three broad categories as in the Holzinger problem, namely, into low (L), middle (M), and high (H). The result, presented earlier in Table 8.3, is that the same 81 sectors are made in the cluster score space. Since the total number of subjects in the two problems is about the same, it is easily seen that there are relatively more subjects with extreme profiles in the MMPI than in the Holzinger problem. There are, for example, five times as many cases with MMPI patterns of LLLL and LLML than we find in the Holzinger problem. At the other extreme, for the patterns HHHH and HHMH the ratio is 17:3.

When we set a minimal criterion of five subjects in a core type, 21 core types emerge. The profiles of the subjects in these core types are displayed in Fig. 8.7. There are 41 subjects in sectors that have less than five cases; when each of these subjects is assigned to the particular core type with which it has its smallest D, every one of them becomes a member of a core type. The scores of the assignees are plotted as xs in Fig. 8.7. The absence of unique individuals in this MMPI problem, compared with the presence of 16 of them in the Holzinger study, is not due to there being more core types to join. When the 21 core types are condensed to a final set of 14, there are still no subjects who are unique when the same criterion of uniqueness is used.

Hierarchical structure of the 21 core O-types

The hierarchical structure of the 21 core O-types is charted in Fig. 8.8. A dashed line has been drawn across the chart at the similarity level of O-types having a distance of about 11, the smallest distance between any of the 15 core O-types in the Holzinger problem. All the MMPI core types above the line are therefore more similar than any of those in the intelligence problem.

The first question to decide is whether the arbitrary sectioning of the cluster score space has cut directly through natural clusters in MMPI score space. Consider the three O-types 2, 3, and 4 shown to have small D values at the top left of the chart. The profiles of these clusters in Fig. 8.7 indicate that the low-score cutoff of 40 has in fact passed directly through the three sets, and their profiles appear not to be critically distinguishable from each

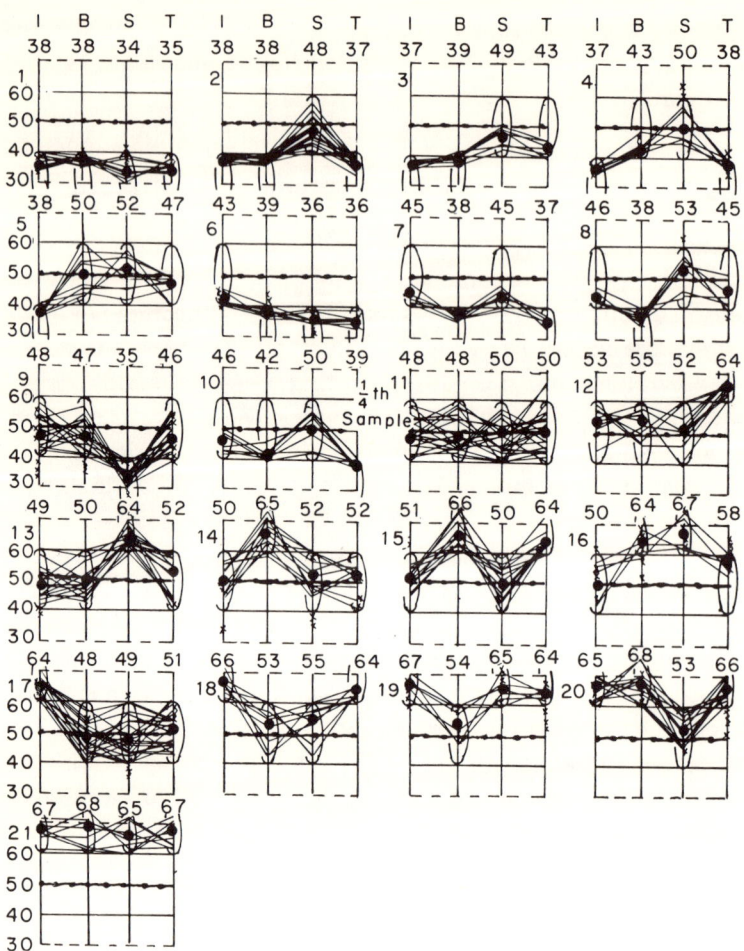

FIGURE 8.7
Profiles of 310 adults in 21 core O-types in the MMPI problem.

other. This is why their D values are so low. Combined, they form the higher-order type 24. This situation is about the same for types 1 and 6 that together form higher-order type 23. Generally, the following additional higher-order combinations appear reasonable: $7+10+8$, $12+15$, and $18+19$. O-type 5 does not, however, seem to be justifiably combinable with type 11.

A striking similarity with the intellectual typological structure appears in the gross differentiation between the lows on the left, where all the types have low cluster scores in at least one attribute and none of them high, and the highs at the right, all of which have at least one attribute with high cluster scores in at least one attribute and none of them low. Recall that the symbol above each of the types refers to that attribute with respect to which it is a high or low extreme. For example, the higher-order type 24 at upper left consists of persons all low in I, introversion. From this hierarchical chart one can quickly read off the distinctive attributes of any of the core types or of higher-order combinations of them. If, for example, one wishes to isolate only persons who are low in at least one attribute and not high in any of the others, then higher-order type 37 is the target group. Conversely, those high in at least one attribute and not low in any other would be a combination of types 35, 36, 13, and 17.

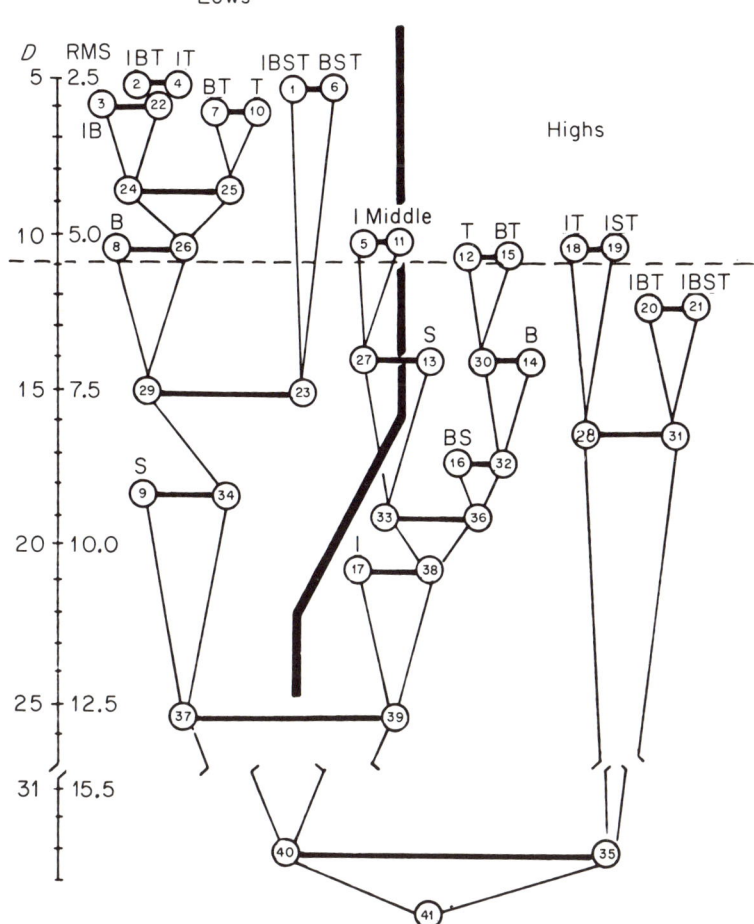

FIGURE 8.8
Hierarchical structure of the 21 core O-types in the MMPI problem.

Spherical configuration of the 21 core O-types

The total configuration of the core types is displayed on the surface of a sphere in Fig. 8.9. The lows are depicted at the left, highs at the right. Higher-order combinations linked in the hierarchical chart are enclosed in dashed lines. The symbol x signifies the fourth dimension which would cover high cluster scores in the body cluster but otherwise average, as is the case with core type 14, or low scores in the suspicion cluster but otherwise average, as is the case with core type 9.

Final types from the profiles, the hierarchy, and the spherical configuration

From a study of Figs. 8.7 to 8.9 those core types which seemed originally to have been produced by arbitrary sectioning of the cluster score space are condensed. From these condensations a final set of 14 types emerged. These types are listed in Table 8.6, where they are symbolized T1 to T14. They are also plotted on the sphere of Fig. 8.9 at their approximate loci.

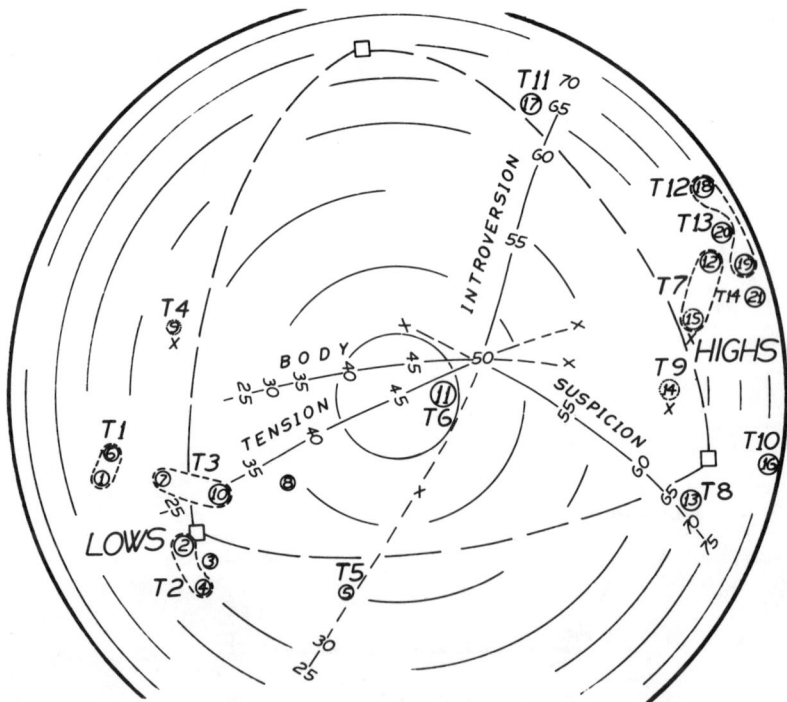

FIGURE 8.9
Spherical structure of the 21 core O-types and of the 14 final condensed types in the MMPI problem.

Those formed by a condensation of O-clusters are clearly evident from the notation under the column headed O-cluster Origin. T1, for example, is such a condensation, being a combination of O-clusters 1 and 6. Others, such as T4, are not a combination, remaining as an O-cluster but being given a new, final type number.

The profile levels of the 14 types are shown in Table 8.6 under the columns headed Profile Level and Homogeneities, these being the rows of cluster score means printed there. The descriptive names given the 14 types are written in the O-cluster Origin column. Type T1, for example, is called extrovert, healthy, trusting, relaxed, these words being based on the attributes in which the type is 1 or more standard deviations from the mean of 50.

The number of cases in each type is shown in the columns with the overall heading Frequency of Cases. For example, there are 68 persons in the initial O-cluster that formed type T6. This number shrinks to 24 when, as the last condensation step in our procedure, we reassign all the 310 persons to the final 14 types with which they have their smallest distances. The reason for this shrinkage is that some of the persons in the periphery of the large initial core O-type of average individuals have, in the reassign-ment, joined other adjacent types with which they have smaller distances than they do with T6.

The homogeneities of the final types after the reassignment are shown in their columns of H values. For example, their overall H values are in most cases on the order of .90, signifying that the final types are very tight person-clusters. The dimension on which each type is most homogeneous is the one with highest H value among the four values of H given in its row under I, B, S, and T.

Some insight into the validity of this typology is provided in the last three columns of Table 8.6. Recall that the supply of 310 persons consisted of normals, anxieties, and schizophrenics. The typology is itself formed without any direct reference to these three classes of persons. Neverthe-less, the 90 Armed Services officers are heavily concentrated in types T1, T2, and T3, and, surprisingly, in T8, the suspicious. The distributions of the two classes of psychiatric patients are quite different from those of the normal subjects. When we compare the two patient groups *inter se*, we find that the anxiety patients appear to be more concentrated in T4, the trusting, and T9, the somatic, whereas the schizophrenics appear relatively more numerous in T8, the suspicious, and T11, the introvert. Interestingly, there also appears to be an excess of schizophrenics in another type, T5, the extrovert. Full confirmation of these differences, however, would, re-quire many more cases than we have in this study.

TABLE 8.6 CONDENSED HIGHER-ORDER TYPES IN THE MMPI PROBLEM

Type	O-cluster Origin	Frequency of Cases		Profile Level and Homogeneities										% of the Criterion Cases		
		Ini-tial	After D fit	I		B		S		T		Over-all H		Normal	Anx-ious	Schizo-phrenic
				Z	H	Z	H	Z	H	Z	H					
T1	1+6: extrovert, healthy, trusting relaxed	24	24	40	.96	38	.98	35	.96	35	.98	.97		20	1	6
T2	2+3+4: extrovert, healthy, relaxed	34	38	37	.99	40	.98	48	.85	38	.96	.94		39	1	3
T3	7+8+10: healthy, relaxed	20	21	46	.92	39	.97	48	.80	40	.94	.91		16	1	7
T4	9: trusting	24	31	48	.82	47	.77	36	.93	46	.91	.86		5	15	6
T5	5: extrovert	8	17	39	.97	50	.87	50	.82	47	.96	.91		3	5	10
T6	11: average	68	24	50	.92	50	.84	50	.92	50	.94	.91		3	11	7
T7	12+15: somatic, tense	24	22	52	.93	62	.77	52	.88	64	.94	.89		0	11	7
T8	13: suspicious	17	26	48	.91	48	.93	64	.93	52	.84	.90		12	3	14
T9	14: somatic	13	16	50	.79	65	.87	52	.79	52	.88	.83		0	10	1
T10	16: somatic, suspicious, tense	10	14	50	.74	64	.90	66	.56	60	.91	.79		0	8	3
T11	17: introvert	27	30	64	.92	48	.91	48	.55	50	.88	.83		2	10	19
T12	18+19: introvert, suspicious, tense	19	20	66	.91	54	.89	60	.89	64	.90	.90		0	9	10
T13	20: introvert, somatic, tense	14	17	65	.96	65	.83	53	.90	66	.90	.90		0	9	4
T14	21: introvert, somatic, suspicious, tense	8	10	67	.92	67	.85	65	.95	67	.94	.92		0	5	2

Multivariate selection

The relatively high frequency of some extreme patterns of cluster scores and the absence of other equally probable patterns reveal how the 14 types are differentially selected in the four MMPI attributes. The most common type with extreme low scores are T2, the extrovert, healthy, relaxed, and T4, the trusting. In the high extremes, the most common are T11, the introverts, T8, the suspicious, and T7, the somatic, tense. Just as we found with the intellectual patterns, there are no types of individuals that have both high *and* low extreme scores in any two of the MMPI dimensions. Such disjunctive patterns would not be expected in view of the generally positive correlations among the four attributes; the values of correlation coefficients do not, however, provide precise statements about multivariate patterns of scores across attributes.

With regard to conjunctive patterns, unlike the intelligence typology, we do find among the 14 MMPI types an extreme type that has high scores in all four attributes, namely, T14, and one with low scores in all four, namely, T1. A particularly potent conjunctive pattern is the combination of extreme low scores in both the body and the tension clusters, namely, the healthy, relaxed person. In Table 8.6 this conjunction appears with relatively high frequency in three types T1, T2, and T3, inhabited largely by Armed Services officers.

One intriguing and baffling type of selection is the discovery that some of the types reveal extremely high homogeneities in some of the attributes but low homogeneities in others. For example, types T10 and T11 both have tight profiles in the body and the tension clusters, their H's in these two attributes being of the order .90, whereas they are simultaneously heterogeneous in the suspicion cluster, with homogeneities of the order .50. This lack of tightness in the suspicion cluster is not a general typological feature, for some of the other types have homogeneities in the suspicion cluster of the order .90. To explain such complex selections in the different types is a challenging research problem.

Screening method of fitting a new person to the typology

The discovery of where a new person fits into this typology is achieved by a quick screening procedure. His responses are coded by keys on the four item-clusters, raw total scores are summed on the four attributes, and these are converted to standard scores on a scale with a mean of 50 and standard deviation of 10 (statistical constants for this conversion are given below). Typing the subject is then simply a matter of discovering with which

A. **Dimension scales showing the 14 types and three subjects' profiles**

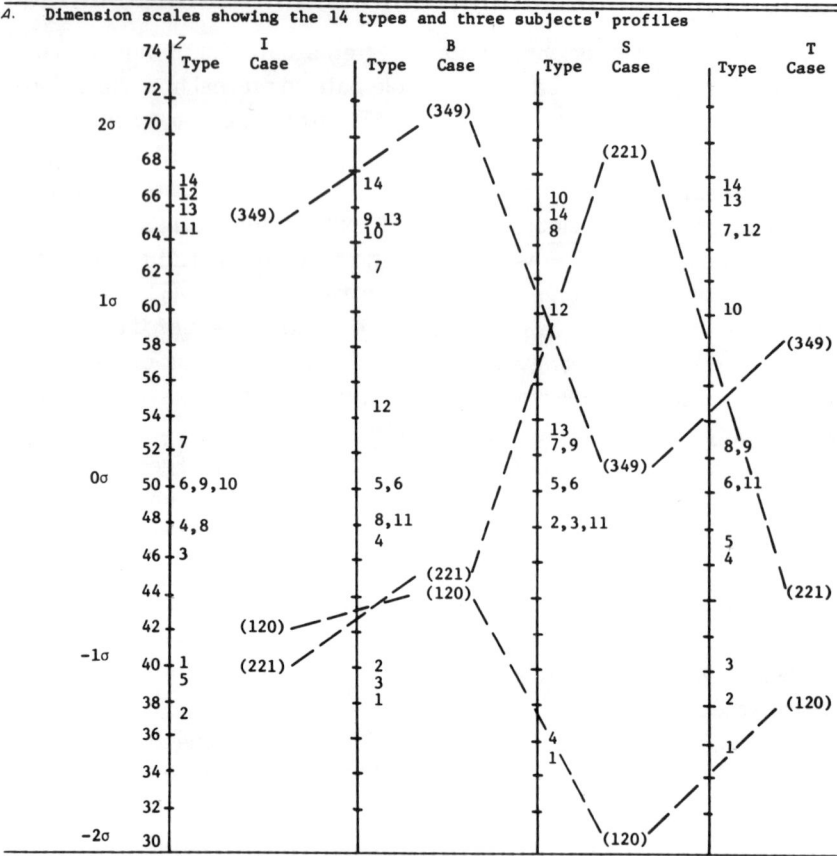

B. Fit of a person to his most similar type.
1. Plot his profile. Then select the dimension scale on which he has his fewest
   $ds<10$. For each type there with a $\underline{d}<10$, count the number of scales with which
   the given type has $\underline{d}<10$ with him, thus:

Type:	1	2	3	4	5	6	7	8	9	10	11	12	13	14
Case 120	4			4										
Case 221						4		1			2		1	
Case 349										3	3		4	3

2. Read off from the scales the $\underline{d}$ values of the types with which he has four
   values of $\underline{d}<10$, and compute their mean $|\underline{d}|$, thus:

	No. 120 With T1	No. 221 With T4	No. 221 With T8	No. 349 With T13
I	+2	-6	-8	0
B	+6	-3	-3	+6
S	-5	-6	+5	-2
T	+3	-8	-8	-8
$\Sigma|d|$	16	23	24	16
Mean $|d|$	4.0	5.8	6.0	4.0

FIGURE 8.10
Finding the MMPI type with which a subject has his best fit.

one of the 14 MMPI types he has his lowest euclidean distance D. This takes only a few minutes by using the screening chart shown in Fig. 8.10. The four attributes are represented by four vertical standard score scales in the figure. On each of the scales the score values of each of the 14 types is represented by its type-number at the right of the scale. The tester writes in the subject's standard scores on this chart, as illustrated by the values in parentheses for three sample subjects in Fig. 8.10. The simple calculations of a subject's best fit to a type are illustrated for the three sample subjects in the lower part of Fig. 8.10b. Since an individual is not to be identified with any type unless his scores lie within 1 standard deviation of the average from those of one of the types, one can quickly locate that particular scale on which there are the fewest number of types lying within 10 score units, from the subject's plotted scores. It is then merely a matter of discovering which of these types is the one with which he has four deviations of not more than 10 units on the four scales. When this type is located, the actual values of the four deviations are written down and the absolute mean of them computed. This mean deviation should not exceed 10. If it does, the subject is unique. If there should be more than one type that yields a mean deviation below 10, the subject is assigned to that type with the lowest value. Since the D of a subject is the square root of the sum of the squares of the deviations, the type with which he has the smallest mean absolute deviation will also be the type with which he has his smallest D. If there is any question about the matter one can compute the actual value of D.

The conversion of raw scores to standard scores requires the means and sigmas of the norm group of 310 subjects of this study. The constants are as follows:

	Mean	σ	Reliability Coefficients	
			α	Split half
I	10.7	7.4	.93	.93
B	10.2	7.8	.92	.93
S	11.4	5.6	.85	.89
T	14.1	8.5	.92	.92

To illustrate, the Z score of a subject with a raw score of 15 on I is

$$Z = 10\,\frac{15 - 10.7}{7.4} + 50 = 5.8 + 50 = 55.8 = 56$$

Problems in the scoring of individuals for typological analysis

In the foregoing treatment we dealt only with the special case in which the observations on individuals are equally weighted cluster (subset) scores from complete data on subjects. We now consider how, for typological purposes, to score individuals on factors instead of clusters, how to handle incomplete (missing) data, how to weight cluster-defined dimensions differentially, and how to correct a typology for certain subgroup biases by means of subgroup standardization of scores. We discuss how these problems are dealt with by programs of the BC TRY System.

Differential weighting of variables in determining cluster or factor scores

The weight matrix

Scores of an individual on any dimension can always be conceptualized as the weighted sum of his standard scores on all the n variables of a study. To make this point clear, the general scoring program of the BC TRY System, FACS, that computes dimension scores of individuals, prints out at the beginning of its output the "nominal weight matrix." An example is given in Table 8.7, the FACS weight matrix in the Holzinger abilities problem for the calculation of five rational composite scores of children on F (form or space), V (verbal), S (speed), M (memory), and N (number or mathe-

TABLE 8.7 ILLUSTRATION OF NOMINAL WEIGHT MATRIX: SIMPLE SUM RATIONAL COMPOSITE SCORES IN THE HOLZINGER PROBLEM

Variables			Scores						Variables			Scores				
			Z_V	Z_S	Z_F	Z_M	Z_N					Z_V	Z_S	Z_F	Z_M	Z_N
1	F1	Vis	.00	.00	1.00	.00	.00		13	S13	Scc	.00	1.00	.00	.00	.00
2	F2	Cub	.00	.00	1.00	.00	.00		14	M14	Wrg	.00	.00	.00	1.00	.00
3	F3	Fbd	.00	.00	1.00	.00	.00		15	M15	Nrg	.00	.00	.00	1.00	.00
4	F4	Loz	.00	.00	1.00	.00	.00		16	M16	Frg	.00	.00	.00	1.00	.00
5	V5	Inf	1.00	.00	.00	.00	.00		17	M17	Wn	.00	.00	.00	1.00	.00
6	V6	Cmp	1.00	.00	.00	.00	.00		18	M18	Nf	.00	.00	.00	1.00	.00
7	V7	Snt	1.00	.00	.00	.00	.00		19	M19	Fw	.00	.00	.00	1.00	.00
8	V8	Wcl	1.00	.00	.00	.00	.00		20	N20	Ded	.00	.00	.00	.00	1.00
9	V9	Wmn	1.00	.00	.00	.00	.00		21	N21	Puz	.00	.00	.00	.00	1.00
10	S10	Add	.00	1.00	.00	.00	.00		22	N22	Rsn	.00	.00	.00	.00	1.00
11	S11	Cod	.00	1.00	.00	.00	.00		23	N23	Ser	.00	.00	.00	.00	1.00
12	S12	Cnt	.00	1.00	.00	.00	.00		24	N24	Ari	.00	.00	.00	.00	1.00

matics). Note, for instance, the first column of numbers gives the weights of *all* 24 tests required to form a composite score on the V (verbal) dimension, these being the value 1.00 for each of the five verbal tests and .00 for all the other tests. These values mean that individual differences in the five verbal tests are being fully utilized in the verbal composite but that individual differences in the remaining 19 tests are not represented at all in the weight pattern.

Individual differences in each test are expressed in standard score form with fixed mean and standard deviation. The raw composite score of a child on V is a simple sum of his 24 standard scores, each multiplied by the weights given in the column headed Z_V in the matrix. The 19 tests with zero weights produce no variation at all in the V composite. The final raw composite score on V is therefore the sum of standard scores over the five verbal tests. As explained in Chap. 2, each child's weighted sum score is restandardized and then rescaled so that the final composite score Z_V has a mean of 50 and a standard deviation of 10.

Table 8.7 shows that, in similar fashion, the columns of weights are so chosen as to provide scores on each of the other four groups of variables by the simple expedient of multiplying the standard scores of the definers of each cluster by 1.00 and the nondefiners by .00.

The weights do not have to be all unities or zeros. They can be any values. Indeed, it is how these weights are chosen that determines the nature of the dimensions that are to be utilized in the typological analysis. Instead, the values of 1.00 could be replaced by the "best weights" of a regression analysis, or they could be values preset by rational considerations. Instead of ones and zeros they could be values between 1.00 and .00, such as those determined by a factoring program, in which case the dimensions would then be called factors. A systematic consideration of these cases is instructive.

Simple sum cluster or rational composite scores

The most meaningful weight matrix is the simple sum type illustrated in Table 8.7, where the standard scores on a subset of variables form a composite score on the dimension; each variable participating in the composite does so with a weight of 1.00; the nondefining remaining variables contribute a weight of .00. On commonsense grounds this form of weighting makes dimensions easier to interpret than the case in which the variables show graded weights.

The weight matrix of the simple sum type is usually derived from the empirical key-cluster solution, which defines each dimension by the best subset of most collinear variables. This matrix is usually improved on by the analyst in a preset solution following the empirical run. This same type of on-off weight matrix may be defined by the investigator who wishes on theoretical or social grounds to form scores of individuals on rational composites of the variables.

Regression-weighted cluster scores

A modified form of the zero-one weight matrix is one in which each dimension is defined by a subset of all the variables but the restricted number of definers are not all given the equal weights of 1.00. The definers of each dimension vary in the size of their factor coefficients, which suggests that they be given differential effective weights according to the degree to which they correlate with the dimension. To illustrate, in the CSA component of the BC TRY System the oblique factor coefficients of the variables assigned to the verbal cluster, ordered by size of coefficients, as calculated by the CSA component of BC TRY are given in Table 8.8. Since the oblique factor coefficients are the correlations of the five predictor variables with a hypothetical score on the "criterion" verbal domain, we can adjoin their column of five correlations with the domain to the matrix of correlations between the five predictor definers (with 1.00 in the diagonal). The results of the orthodox design for computing the multiple regression β weights of the criterion on the five predictors are the varying nominal weights that go in Table 8.7 in place of the values of the constant 1.00 for all five definers as assigned by the simple sum pattern.

Computing this refined set of predicted weights, however, is usually hardly worth the extra effort because in the simple sum weighting of Table 8.7, although all definers have the same nominal weight of 1.00, the

TABLE 8.8 OBLIQUE FACTOR
COEFFICIENTS AND MEAN
CORRELATIONS OF VERBAL CLUSTER
VARIABLES AND THE
VERBAL CLUSTER DIMENSION

	Oblique Factor Coefficients	Mean Correlation with Definers of V
V7 Snt	.84	.68
V9 Wmn	.84	.68
V6 Cmp	.82	.66
V5 Inf	.81	.65
V8 Wcl	.70	.57

definers do in fact have different *effective* weights in determining the full variance of the composite score on the criterion. This effective weight of a predictor is a sheer function of its average correlation with the other four predictors. Thus, V8 appears in the simple sum weight matrix of Table 8.7 to have a weight equal to that of the other four predictors in determining the variance of the scores on V. Actually, however, since its average correlation with the other predictors of V is, as shown in Table 8.8, sensibly lower than that of the other four, its effective weight in determining the variance of the composite score on V is the least.

Regression estimates of factors

An investigator may wish to score individuals on factors, i.e., on dimensions derived by a factoring method that yields factor coefficients between 1.00 and .00 for *all* n variables. There are two kinds of such factors: orthogonal and oblique. Orthogonal factors are usually varimax or quartimax rotations of principal-axes factors, or they are derived by the key-cluster component of BC TRY. Oblique factors are correlated dimensions such as those derived in the CSA program of the BC TRY System. The weighting principle for estimating scores on a factor is identical with that described above for regression weighting of a subset, except that all n variables are predictors of a factor.

There are serious drawbacks to scoring individuals on factors. In the first place there are variables with low weights that introduce unnecessary and therefore undesirable "noise" into the prediction. Second, on general scientific grounds, the theory that all variables play a role in all factors (or vice versa) is probably quite indefensible in most studies. Third, it is usually inevitable that some variables overlap on different factors, i.e., have significantly positive or negative weights on two or more factors. As such, they introduce redundant or "experimentally dependent" sources of variation in the estimates of the different factors on which they overlap. Fourth, if there are missing data generously scattered throughout the original score matrix, the problem is much more serious in replacing missing data when factor scores are to be estimated than when cluster scores are desired.

Differential weighting of the
dimensions of typological space

In the examples of O-analysis given in the foregoing sections of this chapter the scores on different dimensions computed by FACS were put in standard scale form with a mean of 50 and standard deviation of 10. There are distinct advantages in using standard scales because differences between objects in profile level on the different dimensions are commensurate and easy to grasp. Since the typological groupings discovered by O-type analysis are sensitively affected by the scale values of each dimension, however, there is no doubt that the standard scaling embodies the assumption that each dimension is equally important in determining the typology.

Some analysts may not accept the assumption of dimensional equality and prefer a differential weighting. A decision on this matter is not made in the O-type analysis itself but when the scores of the objects on the different dimensions are computed. Options are provided in the program FACS of BC TRY that permit three types of weighting. The first is equality, which is the standard option.

The second option weights each dimension in accordance with the total communality of all variables exhausted by the dimension during the factoring process. Order of the dimensions is important in key-cluster factoring, which has an elasticity to it in this respect not available to principal-axes factoring. In key-cluster factoring, the analyst can determine the meaning of a dimension any way he pleases by selecting its definers and by subjecting them to the CC factoring procedures in any position he wishes relative to that of other dimensions. Thus, if one defines the first dimension by the five verbal tests, it will take out a different portion of the total communality than if the verbal tests are preset as definers of the second, third, or fourth dimensions. The analyst is therefore able to decide the weight to be given to the verbal dimension in typological analysis under the second option by specifying which dimension the verbal tests are to define.

A third option permits the analyst to set the exact weights of the dimensions to any particular values he wishes. In BC TRY he does so by presetting the scaling factors of the dimensions, i.e., by specifying on a detail card in FACS the values of the standard deviations of the scores that FACS computes on each of the dimensions. When cluster scores are to be computed by FACS, one basis on which to specify the weights of the successive dimensions is the generality of each dimension taken as a single factor, as this generality is calculated in CSA. Since these generalities are expressed in terms of communalities or squared correlations, the scaling factors would be proportionate to the square roots of these quantities.

Removing subgroup differences in typological analysis by subgroup standardization in FACS

In some problems, particularly of comparative analysis (Chap. 9), one may wish to control the average score levels of different subgroups before

O-typing proceeds. The adjustments are made when one computes the cluster or factor scores of members of the different subgroups. For example, in the comparative analysis of time and place groups in social-area analysis (Chap. 9) it was decided to compute the typology of all neighborhoods observed in 1940 and in 1950 taken as one large inclusive group. The inclusive group thus consisted of two subgroups, one observed at the beginning of the decade, the other at the end.

We wanted to eliminate differences in average score level of subgroups of neighborhoods separated in time, i.e., in 1940 vs. 1950. The reason was that if some of the variables were in some ways affected by such general trends as inflation, then neighborhoods with the same profiles on such variables within a given year, say 1940, would have different levels of profiles on those variables across periods, say 1940 compared with 1950. Such gross effects are eliminated by standardization of scores within subgroups, i.e., by computing the scores on the subgroup of 1940 neighborhoods separately from computing them on the 1950 subgroup of neighborhoods, and then merging the two sets of scores as an input deck to one inclusive O-analysis by OTYPE.

On the other hand, we wanted to preserve differences in score levels of subgroups of neighborhoods located in different places, i.e., in San Francisco vs. East Bay. In this case the data of San Francisco and East Bay neighborhoods were merged before the scores were computed and thus were not standardized separately.

Before an analyst undertakes a typological analysis, he may wish to give some thought to the subgroup structure of the inclusive groups he is dealing with in order to employ the correct design of cluster score computation that will preserve or eliminate group differences according to his wishes. Decisions on this matter can be troublesome. In the Holzinger problem, for example, where the inclusive group consisted of children from the factory group vs. children from the suburban group, the question arose whether to standardize the V, S, F, M cluster scores in the two groups separated before O-typing. The factory subgroup came extensively from homes where the parental native language was not English; hence their profile level might be depressed as the result of bilingualism. If cluster scores were computed separately in the two subgroups, differences between them due to language would be eliminated. But this within-group standardization would also eliminate *all* selective forces beside language differences that would produce mean differences in scores of the two groups on V, S, F, and M. Such a sweeping equalization seemed undesirable, especially since some of the children from the suburban group also came from bilingual homes and would be thus penalized by separate standardization. Since the matter was obviously complex and not to be solved rationally by assuming differences before they were found or, if found, presuming to know the reason for them, separate standardization was not used. The effects of multivariate selection stemming from the subgroup differences can be assessed in the typology itself.

O-analysis when the number of dimensions is large

Generally one would not seek to place individuals into types on the basis of their scores on a large number of dimensions unless a very large number of subjects were involved. Even with such a crude differentiation between individuals as high and low on each dimension, there are so many possible patterns with many dimensions that it is not likely that there are many persons with similar patterns. For example, Table 8.9 gives the number of possible patterns as the number of dimensions increases: (1) with dichotomous cuts of high and low on each dimension and (2) with trichotomous cuts of high, middle, and low on each dimension. With a crude dichotomous cut, for example, of high and low on 10 dimensions, more than a thousand subjects are needed in order to provide an opportunity for the subjects to fall into the 1,024 possible patterns of highs and lows across the 10 dimensions. With a trichotomous cut in 10 dimensions, still very crude, one would need nearly 60,000 subjects.

The importance of this relationship becomes acute in the use of program OTYPE of the BC TRY System. This program typologizes the N subjects of a group on k cluster score dimensions. The principle employed in a standard run is to start the solution with a set of arbitrary core O-types

TABLE 8.9 NUMBER OF PATTERNS FOR SOME ILLUSTRATIVE NUMBERS OF DIMENSIONS AND NUMBERS OF CUTS

Number of Dimensions	Dichotomous Cuts (High, Low)	Trichotomous Cuts (High, Middle, Low)
1	2	3
2	4	9
3	8	27
4	16	81
5	32	243
6	64	729
7	128	2,187
8	256	6,561
9	512	19,683
10	1,024	59,049

lying in the most populated sectors of the cluster score space formed by trichotomizing each dimension. All the cases are then assigned to the core types with which they have their smallest euclidean distance, thus forming a set of O-types to which all the cases are reassigned. The process is then continuously iterated. Most of the O-types will converge on a stable membership after as few as five iterations.

The critical first step in the solution is to start with trial core O-types that are in highly populated arbitrary sectors of the score space (the arbitrariness is removed as the result of iteration). Under standard options, the selected core O-types are those which occupy sectors that contain at least 2 percent (an optional value) of all the cases.

But if there are too many dimensions, it is possible that *no* sectors will contain more than one case, so that the selection of most-populated sectors is not possible and the iterated solution cannot get under way. For example, in a recent problem with $k = 14$ dimensions and $N = 500$ subjects, virtually all populated sectors had only *one* case in them and none had more than *two*, a not surprising result considering that with 14 trichotomized dimensions there were nearly 5 million sectors into which the 500 cases had to be distributed (the computer took .70 min to do this monumental job!). A solution was finally achieved by rerunning OTYPE with each of the 14 dimensions dichotomously cut at the mean, a procedure that gave only 11 core types, some with as few as three cases in them, a slim frequency on which to select core O-types.

A procedure of determining core O-types
unrestricted by dimensionality

We obviously need some general procedure to follow that will always give a suitable number of trial core O-types by which O-type analyses can initiate the iteration process. Here is such a procedure:

1   If the number of dimensions is not greater than six, employ the standard OTYPE condensation method.
2   If there are more than six dimensions, perform this standard OTYPE analysis only on a selection of the most salient and meaningful dimensions, the number of them being that number in the Table 8.9 which yields a total number of trichotomized sectors less than the total number of subjects in the problem; for example, with 500 cases, the maximal number of dimensions chosen for OTYPE analysis is six, which gives 729 trichotomized sectors into which the 500 cases are cast. In the case of the 14-dimensional problem mentioned above, the dimensions were thus cut down to three sets. Each contained not more than six correlated dimensions that were rationally related, and an OTYPE run was made on each set.

If, however, it is desirable to perform a typology on all the dimensions, here is a procedure for finding suitable core O-types:

1 Run the standard condensation method on all the dimensions. With luck, there may be a sufficient number of core O-types for iteration.

2 If procedure 1 does not work, change the number of partitions to dichotomies and rerun OTYPE under this option.

3 If procedure 2 does not work either, utilize the convergence method to locate core O-types, as described in the next paragraph.

Convergence method of obtaining core O-types

By this method, whatever the dimensionality may be, you can always find a set of trial core O-types which spans the score space and which lies at centers of gravity in the configuration of the objects. The procedure is a EUCO analysis applied to a representative sample of the subjects drawn from the full supply. A simple procedure is to take the cluster scores of every $(N/120)$th case, thus producing a deck of 120 subjects on which to perform the EUCO analysis, as described in the User's Manual. Under this procedure, however, preset the CC5 program so that it factors on $k + 1$ dimensions (if k is 15, preset CC5 to 15). The resulting definers of the $k + 1$ dimensions, i.e., the REFLX1 file, are the members of $k + 1$ core O-types. Input these definers into OSTAT, which sets them up as an OTYPE1 file. Then call OTYPE, which reads the OTYPE1 file and starts the iterative process on the core O-type given there.

If more than $k + 1$ O-types are wanted with which to initiate iteration, a good set of additional core O-types would be dependent clusters displayed in the SPAN diagrams of the EUCO analysis.

Chapter 9

COMPARATIVE CLUSTER ANALYSIS OF VARIABLES, INDIVIDUALS, AND GROUPS

The first objective in comparative cluster analysis is to describe the similarity of the dimensions discovered in different groups. This problem is known as the "comparative dimensional analysis of variables" or "factor matching." In the domain of the intellectual abilities, for example, one may discover in a middle-class suburban group of children that the 24 Holzinger tests of diverse specific abilities (Holzinger and Swineford, 1939) can be accounted for by four "basic" general abilities, or factors, verbal (V), space (form, F), speed (S), and memory (M), as described in previous chapters. Are these dimensions identical with those found in a lower-class school of children of factory workers? Are the seven general MMPI dimensions of introversion, body, suspicion, tension, depression, resentment, and autism found in a group of psychiatric patients the same dimensions discovered in a group of normals?

This problem has a direct, simple solution when approached by the logic and procedures of cluster analysis based upon domain sampling principles and incorporated procedurally in the BC TRY System of cluster and factor analysis (Chap. 3). Dimensional analysis requires as basic data the intercorrelations between the variables in the groups. These correlations are not defined in the usual way in the original data when the comparisons are across different groups, as in comparative analysis. Rather, in comparative dimensional analysis what are needed are the factor coefficients of the dimensions within each group (these are referred to in factor analysis as the "rotated oblique factor coefficients"). These factorial data within the different groups are compared across the groups by the comparative cluster analysis programs called COMP1 and COMP2 of the BC TRY System. The methods are described below.

The second general objective is that of comparing the typologies of two or more groups of individuals. When, for example, the subjects from the factory and the suburban groups are scored on the four general abilities, V, S, F, and M, the children in each group can be sorted Into different types based upon the patterns of the scores. The person clusters (or profile types) in the two groups can differ in two ways: (1) Even though the same kinds of profile types may appear in the two groups, those which occur with high frequency in one group may be rare in the other group. This type of typological comparison across groups is based on the similarity of their "frequency patterns" on a common typology. (2) The kinds of types in the two groups may be different; those which compose one group may not match the types of the other group. In the BC TRY System, the programs expressly designed to perform the comparative typology of groups are the components OTYPE, OSTAT, and EUCO.

The plan of this chapter is as follows. The comparison of the dimensions of different groups (COMP) and of their typologies (OCOMP) is first made for the case of the Holzinger study of the abilities of two groups, the children from the factory and the suburban groups. Under exactly the same format of analysis the COMP and OCOMP analyses of the patient and the normal groups in a study of MMPI item-clusters are then presented. Finally, COMP and OCOMP analyses on the social-area data are presented. Our interest in these three studies is as much substantive as procedural because they refer to important problems in cognitive, personality, and social psychology.

The study of abilities: the Holzinger problem

Basic data structure

In the Holzinger problem, 301 grade school children were given 24 separate tests of specific abilities. These tests are listed in Table 4.1, grouped under the five domains of spatial, verbal, speed, memory, and mathematical abilities. Most of these tests are similar to tests that are included today in test batteries of "intelligence," e.g., the WISC and WAIS batteries from which verbal, performance, and full scale IQs are determined (Anastasi, 1961, chap. 12).

The total group of children, here called the "inclusive group," were children from two Chicago grade schools. Holzinger and Swineford (1939, p. 6) describe them as follows: "The children in the Pasteur School came largely from the homes of workers in nearby factories. Many of the parents were foreign-born . . . using their native language at home . . . . Both parents were American-born in 29 per cent of the cases, while in 48 per cent, both were foreign-born." The second school was the Grant-White school in the suburb of Forest Park, Illinois. In this group ". . . both parents were American-born in 72 per cent of the cases while both were foreign-born in only 15 per cent. Almost 100 per cent of the children were born in the suburb in which the school was located."

The inclusive group can therefore be thought of as being composed of two ecological groups. The 156 from the Pasteur School are called here the "factory children," the 145 from the Grant White School, the "suburban children." The inclusive group has other subgroup structures, notably sex groups and grade groups. Furthermore, the suburban children were organized into two types of classrooms, homogeneous groups and random classes.

Dimensional analysis of the 24 variables in the inclusive group

A direct comparison of the dimensions of the 24 variables in the factory and suburban children and of their separate typological structures can best be made when the definers of their dimensions are the same. The first objective, therefore, is to decide on the number of dimensions on which the subgroup comparisons are to be made and on a common set of definers of each dimension. A full-cycle key-cluster analysis of the 24 variables in the inclusive group, reported in previous chapters, discovered that after four dimensions were extracted from the intercorrelations among the 24 tests, their residuals were trivial. The defining variables of each of the four dimensions are listed in Table 7.1.

Dimensional analysis of the 24 variables in the factory children

To discover the cluster structure of the tests in the factory children, a full-cycle key-cluster solution of this group's intercorrelations among the

24 tests was "preset" on the definers of the four basic dimensions found in the inclusive group. The results are shown pictorially in Fig. 9.1, the bottom spherical plot, which is an annotated tracing of the printout of the diagram in program SPAN of the BC TRY System. The surface separation of any two tests on this sphere is a function of the correlation between them (technically, of their interdomain or common factor correlation). Two tests that correlate 1.00 have superimposed points, two that correlate .00 are 90° apart, represented in Fig. 9.1 by the distances between the three boxes that form the spherical triangle; the boxes represent the subset of three independent dimensions derived by factoring on residuals.

The five verbal tests cluster tightly together at lower left in the configuration, the four speed tests more loosely at lower right, the four form tests at the top. The six memory tests are marked by X, denoting that they all project into a fourth dimension which cannot be shown since it projects at right angles to the three depicted in Fig. 9.1. The five mathematical tests are depicted in these three dimensions; they are all dependent on V, S, F, and M in the sense of being predictable from the four.

Dimensional analysis of the 24 variables in the suburban children

Applying the same dimensional procedure to the correlation matrix of the suburban children gives the configuration shown in the top SPAN diagram of Fig. 9.1. At lower left in the configuration is the same verbal cluster as in the factory group, at lower right is the speed cluster, at the top is the space cluster, and the memory cluster also projects into a fourth dimension; the mathematical abilities once again deploy centrally as dependent variables predictable from the V, S, F, and M dimensions. Clearly, the cluster structure of the suburban children closely resembles that of the factory children. One obvious difference is that although the cluster groups are about the same, they are, as groups, more separated from each other in the factory children than in the suburban children, i.e., less correlated with each other.

Comparison of the dimensions within each group separately (COMP)

A metric description of the within-group structures is provided by a program that computes the correlations between the ability clusters defined as oblique dimensions. The oblique dimensions are computed by the CSA (Cluster Structure Analysis) program of the BC TRY System. The values

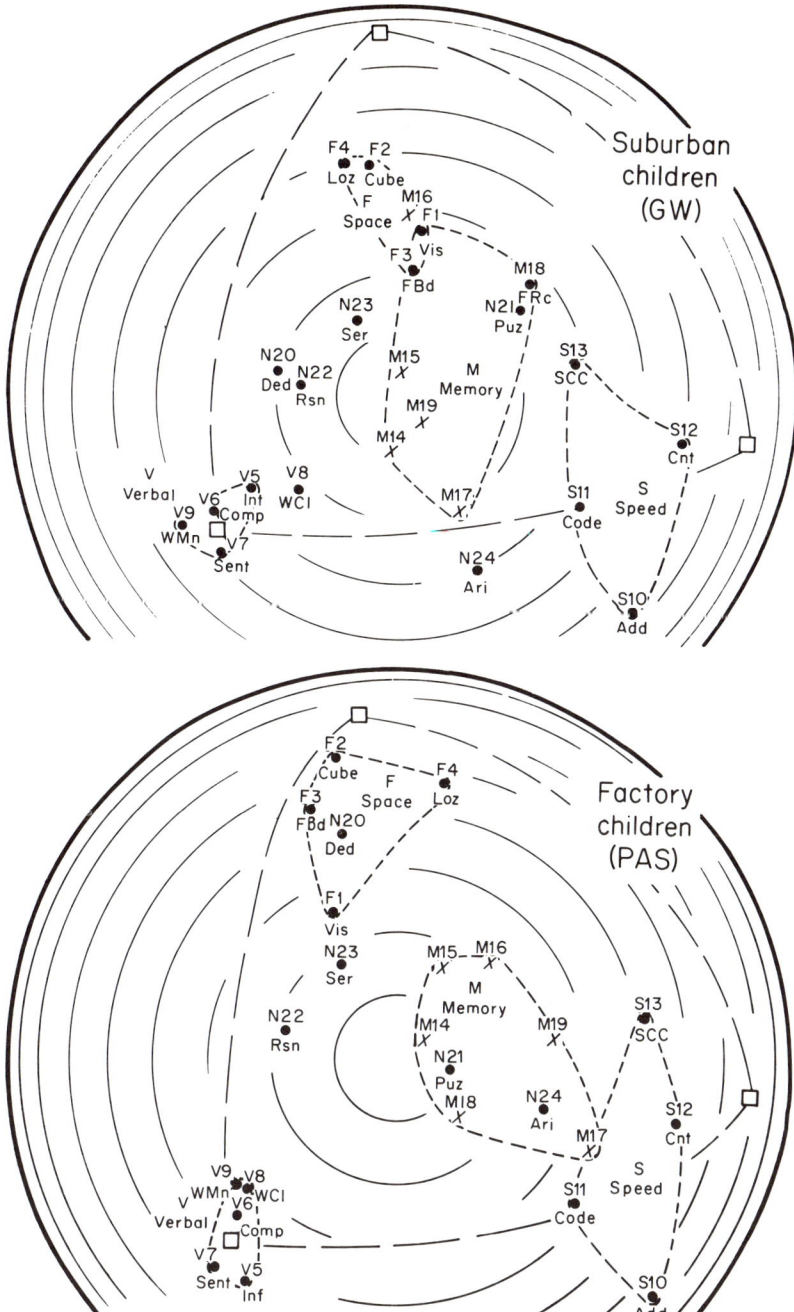

FIGURE 9.1
Cluster structure of abilities within the suburban and factory groups.

of the interdimension correlations are given in Table 9.1, Sec. A. These correlations are known in factor analysis as the "correlations between rotated oblique factors" or the "common factor correlations." In cluster analysis they are called "interdomain correlations," where each cluster is conceptualized as a domain score C_i on many variables collinear with the observed definers of the cluster (Tryon, 1959, eq. 24). Thus, the domain score C_V on the verbal cluster is a hypothetical score on many variables collinear with the observed set, V5, V6, V7, V8, and V9, shown in the SPAN diagram.

The interdomain correlations, listed in Table 9.1 under the columns headed r_{CC}, are computed from the raw correlation matrix using the well-

TABLE 9.1 SIMILARITY OF THE FOUR BASIC HOLZINGER ABILITIES, V, S, F, M, WITHIN AND BETWEEN THE SUBURBAN AND FACTORY GROUPS

A. Similarity of Cluster Dimensions within Each Group

Ability	Group	Verbal, V		Speed, S		Form (Space), F		Memory, M	
		r_{CC}	$\cos \theta$	r_{CC}	$\cos \theta$	r_{CC}	$\cos \theta$	r_{CC}	$\cos \theta$
Verbal, V	Suburban	1.00	1.00	.43	.43	.58	.58	.46	.47
	Factory	1.00	1.00	.42	.43	.35	.37	.14	.14
Speed, S	Suburban	.43	.43	1.00	1.00	.53	.51	.56	.54
	Factory	.42	.43	1.00	1.00	.29	.28	.39	.36
Form (Space), F	Suburban	.58	.58	.53	.51	1.00	1.00	.60	.56
	Factory	.35	.37	.29	.28	1.00	1.00	.27	.26
Memory, M	Suburban	.46	.47	.56	.54	.60	.56	1.00	1.00
	Factory	.14	.14	.39	.36	.27	.26	1.00	1.00

B. Similarity of Cluster Dimensions between Groups (cos θ Only)

	V_s	S_s	F_s	M_s
V_f	.96	.39	.48	.32
S_f	.46	.89	.42	.48
F_f	.46	.36	.92	.39
M_f	.28	.42	.41	.83

C. Generality of Each Dimension (Reproducibility of Correlations)

	V	S	F	M
Suburban	.51	.37	.47	.40
Factory	.50	.27	.28	.18

D. Reliability Coefficient (α) of Cluster Score on each Dimension

Suburban	.90	.83	.70	.76
Factory	.90	.74	.69	.73

known formula for the correlation of sums. The r_{CC} values are precise metric expressions of the degree of similarity of the four basic ability dimensions, V, S, F, and M, in the suburban and factory children. For example, the interdomain correlations between the verbal and speed dimensions in the two groups are .43 and .42, respectively; i.e., the two dimensions have almost exactly the same degree of similarity in the two groups. Between the other dimensions the correlations are generally higher for the suburban children than for the factory children, a fact already seen visually in the SPAN diagrams of Fig. 9.1. The r_{CC} values are a metric statement of similarity that is displayed visually in the diagrams.

In the lower sections of Table 9.1 other metric properties of the four basic ability dimensions are displayed. The "generality" of each, given in Sec. C, is the degree to which each dimension accounts for all the original intercorrelations among the 24 abilities. In both groups the verbal dimension is the most general, but in the factory group the other three dimensions are more specific than in the suburban. Of special interest to the typological analysis is the reliability coefficient of the raw scores on the four dimensions. The α reliablllty (Sec. D) of V is .90, but of the other three, only of the order .70 or .80.

Direct comparative analysis of the dimensions across groups (COMP)

We have assessed the similarity of the V, S, F, and M dimensions of the factory and suburban children by the subjective process of cross-referencing their separate configurations in Fig. 9.1 and by comparing their within-group r_{CC} values in Table 9.1, procedures that are indirect and inferential. We now compare their dimensions objectively and directly.

Figure 9.2 displays the direct comparison achieved by the program COMP2 of the BC TRY System. In this SPAN diagram, traced from the printout, the verbal dimension of the suburban children, labeled V_G (for the Grant-White School), and of the factory children, labeled V_P (for the Pasteur School), are tightly clustered at lower left, meaning that they are quite similar. At the lower right are the two points representing the speed dimensions of the two schools. At the top are the two space dimensions, and extending into the fourth dimension are the two memory dimensions. This cluster structure directly compares, in one diagram, the similarity of the two-dimensional structures that we only indirectly observed above by cross-referencing.

The direct index of the similarity of any two dimensions *across* different groups is the "index of similarity" of the two dimensions (or "factors"), called the cos θ between them. For two dimensions *within* a group cos θ is

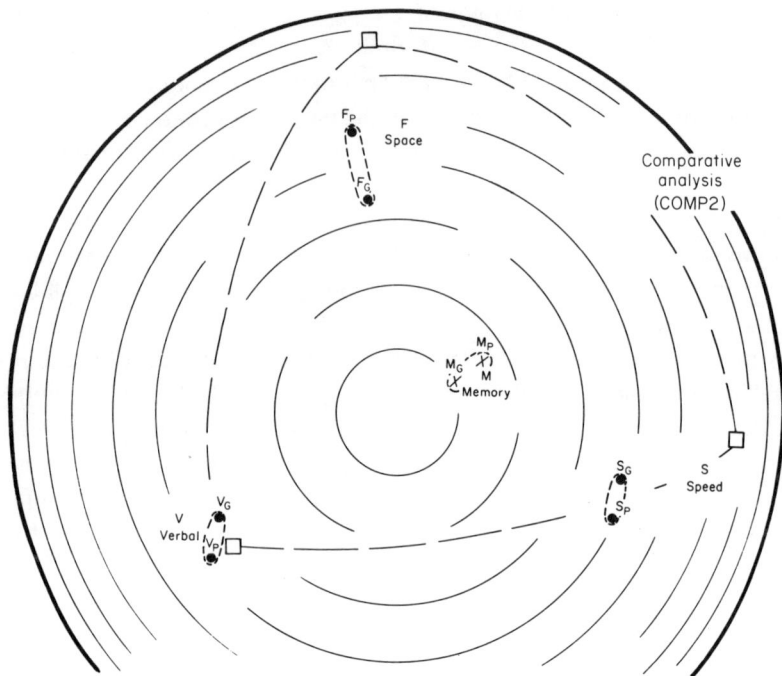

FIGURE 9.2
Cluster structure of abilities across the suburban and factory groups.

equivalent to the interdomain correlation r_{CC}, but it is estimated not from the raw correlation matrix, as r_{CC} is, but from the oblique factor coefficients of the two dimensions. These similarity values are given in Table 9.1, Sec. B. They tell the same story metrically that is shown pictorially in the spherical configuration of Fig. 9.2. On the upper left to lower right diagonal can be seen the index of similarity of V in the factory children and in the suburban children, then of S, F, and M. For example, cos θ between the verbal dimensions in the two groups is .96, between the two speed clusters it is .89, between form clusters .92, between memory clusters .83.

Briefly, the reasoning by which we designate two dimensions as identical is based on the universal logic by which we conceive any two entities as being the same, namely, that they show the same pattern of "observations" in relation to a common set of other "referent entities." For example, the verbal dimensions in the two groups are virtually identical because their patterns of factor coefficients on the constant set of 24 referent abilities are virtually identical. The index of pattern similarity of any two entities on a common set of referent entities is P, called the "index of proportionality" or "index of collinearity," described in detail in Chap. 12 for the case of pattern similarity of the factor coefficients of any two dimensions. The value of the index of similarity cos θ of any two dimensions in different groups is a simple quadratic function of P.

To sum up, we find in Fig. 9.2, and from the metric values in Table 9.1, that the four basic dimensions V, S, F, and M in the two groups are highly similar. But in the factory children they are somewhat more independent of each other than in the suburban children. An environmental explanation of this is that the parents of the suburban children stress scholastic achievement, implementing their ambition by pushing their promising children in all abilities, letting their less promising children fend for themselves. Consistent with this theory, we find that it is precisely in the suburban children that the scholastic institution of "homogeneous" classification is employed, namely, the sorting of sheep and goats into different classrooms. In the factory group, children generally are left to fend for themselves.

There is an alternative genetic explanation. Probably there is more stringent assortative mating on abilities among suburban parents. This sort of sexual selection would generate a higher correlation among all abilities in the suburban group than in the factory, where assortative mating would be more random. A systematic treatment of such environmental vs. genetic correlation-producing agencies in the case of abilities is presented elsewhere (Tryon, 1935, 1939).

Comparative typological analysis in the Holzinger problem (OCOMP)

When we allocate children having the same patterns of scores on the basic abilities, V, S, F, and M, to O-types, do we find the same typological structure of these O-types in the factory and suburban groups? The next two sections are discussions of procedures in the BC TRY System used to answer this question.

Similarity of frequency patterns of the two groups on the common typology of the inclusive group

The first of two ways of determining the typological similarity of two groups is to discover the degree to which they show the same frequency of cases falling in the common typology of the inclusive group. This common set of O-types is shown in Table 9.2 under the general heading Inclusive Typology. The first type, labeled H1, consists of 14 children whose pattern of respective mean cluster scores on basic abilities, V, S, F, and M, is 48, 36, 44, 37. These are mean standard scores on a scale whose mean for the 301 children in the inclusive group is 50 and standard deviation 10. Underlined scores of 40 (1 standard deviation below the mean) or below are termed "low" in the column headed Descriptive Name; those 60 or above are called "high." Type H1 therefore is described in the table as "low speed and memory."

Some types have a high frequency, like H9, the average type, with 38 children in it, others with low frequency, like H2, the low verbal and memory type, with only 8 cases in it. The logic of typological similarity of the factory and suburban children is simple. If both groups show the same frequency pattern on these common 16 inclusive classes, then they have the same typological structure, but to the degree that their frequencies in these 16 classes differ from each other, their typologies differ.

It is a simple matter to count how many children in each group fall into the 16 classes, from which the percentage falling into each class is computed. These percentages are given in Table 9.2 under the heading Factory vs. Suburban. The listed values in the two columns labeled p_f and p_s are the frequency patterns of the two groups, on the basis of which their typological similarity is determined. The overall index of similarity at the bottom of the table is the same general index of proportionality P discussed earlier. If two groups have exactly the same frequency patterns, the index P is 1.00. If their patterns are utterly different, i.e., if the occurrence of each type in one group is matched with the absence in the other group, then the index P is .00. The value of P for the two ecological groups of children is .75, denoting a considerable amount of typological similarity of the two groups.

Of greater interest, however, are the specific type differences between the two groups. These values are listed under Differences in Table 9.2. Because the sampling error of such differences can be large, it is desirable to indicate which of these differences is unlikely to occur by chance. Expressing the percentages as proportions p', we note that the mean proportion in the 16 classes is $1.00/16 = .06$. Since most of the proportions of the types in both groups are not too greatly different from .06, we compute the standard error of a difference σ_d between two true proportions of .06, using the well-known formula for this error, i.e.,

$$\sigma_d = \sqrt{p'(1 - p')\left(\frac{1}{N_f} + \frac{1}{N_s}\right)}$$

TABLE 9.2 SIMILARITY OF FREQUENCY PATTERNS OF FACTORY CHILDREN VS. SUBURBAN CHILDREN ON THE COMMON INCLUSIVE TYPOLOGY IN THE HOLZINGER PROBLEM

Type	Frequency	Inclusive Typology — Z scores				Descriptive Name	Factory vs. Suburban, % in Each Group			Boys vs. Girls, % in Each Group		
		V	S	F	M		Factory p_f	Suburban p_s	Differences	Boys p_b	Girls p_g	Differences
H1	14	48	36	44	37	Low speed and memory	4	6	−2	8	2	6(b)
H2	8	36	47	49	35	Low verbal and memory	5	0	5(f)	3	2	1
H3	21	48	50	48	38	Low memory	4	10	−6(s)	10	5	5(b)
H4	9	50	39	36	48	Low speed and form	2	4	−2	2	4	−2
H5	13	36	42	36	45	Low verbal and form	5	4	1	2	6	−4
H6	20	50	50	38	46	Low form	6	7	−1	5	8	−3
H7	19	47	37	48	50	Low speed	3	10	−7(s)	8	5	3
H8	23	35	47	53	54	Low verbal	13	1	12(f)	9	6	3
H9	38	51	51	48	51	Average	13	12	1	10	15	−5(g)
H10	22	65	50	53	47	High verbal	3	12	−9(s)	7	8	−1
H11	23	47	65	51	52	High speed	11	4	7(f)	6	9	−3
H12	14	64	64	59	58	High verbal and speed	3	6	−3	3	6	−3
H13	27	52	51	63	49	High form	9	9	0	13	5	8(b)
H14	23	52	49	51	63	High memory	9	6	3	5	10	−5(g)
H15	8	57	63	54	67	High speed and memory	3	3	0	2	3	−1
Unique	19					Unique	7	6	1	7	6	1
Total %							100	100		100	100	
N	301						156	145		146	155	

For $N_f = 156$, $N_s = 145$, $p' = .06$ the equation gives $\sigma_d = .027$. Setting the limit for d as 2 standard deviations gives a critical value (significance probability of .045) of approximately 5 percent.

All differences above 5 are indicated by (s) for suburban or (f) for factory, depending on which group has the highest percent. For example, the largest difference between percents is 12 in type H8, low verbal. For this difference the greatest percent frequency is 13 in the factory group. Next is H10, high verbal, most characteristic of the suburban group. These two verbal types therefore represent the greatest typological difference between the two groups. The other significant differences indicate that the suburban group falls more heavily into low memory (H3) and low speed (H7), whereas the factory children occur more frequently in the high speed (H11) type. Verbal, memory, and speed dimensions most markedly differentiate the typological differences between factory and suburban children.

Since sex differences in abilities are of universal interest, we present the data for determining the typological similarity of the boy vs. girl subgroups, in the far right columns of Table 9.2. From the percent columns in the 16 classes, the index of similarity for the sex groups is $P = .85$, somewhat higher than for the factory and suburban groups. The significant differences show that boys more frequently fall into low speed and memory, and low memory, the girls into high memory.

Similarity of empirically derived typologies of the groups

The above analysis informs us of differences between factory and suburban children only on the single common typology of the inclusive group. But for fuller information, we need empirically to discover the typology of each group independently of the other and to compare the two typologies. The procedures for doing so are available in programs of the BC TRY System. On the 156 factory children separately the typology is determined by the OTYPE and OSTAT programs previously described. The resulting 15 classes are shown in Table 9.3, where they are listed as types F1 down through F14 to Unique. Their mean standard score profile values and descriptive names are also given, as are the homogeneity coefficients $\bar{H}$ that describe how "tight" each O-type is in its standard scores on the four dimensions.

In similar fashion the separately worked-out typology of the suburban children is also given in Table 9.3, indicating the 13 classes of these children, listed as S1 through S12 and Unique.

TABLE 9.3 WITHIN-GROUP TYPOLOGIES OF THE FACTORY AND SUBURBAN CHILDREN IN THE HOLZINGER PROBLEM

Factory Children

Type	Frequency	Profile Level and Homogeneity					Descriptive Name
		V	S	F	M	$\bar{H}$	
F1	7	34	36	46	37	.81	Low verbal, speed and memory
F2	5	40	52	52	34	.87	Low verbal and memory
F3	14	57	50	57	40	.86	Low memory
F4	18	41	45	37	45	.87	Low form
F5	5	48	54	40	44	.91	Low form
F6	9	53	46	45	46	.89	
F7	18	42	57	47	51	.87	
F8	11	58	55	51	55	.87	
F9	4	47	66	48	46	.83	High speed
F10	4	52	70	55	58	.91	High speed
F11	15	40	48	62	51	.77	High form
F12	5	63	63	67	53	.75	High verbal, speed, and form
F13	23	43	50	53	59	.82	High memory
F14	9	52	51	48	66	.79	
Unique	9						
N	156						
η		.85	.81	.84	.88		

Suburban Children

Type	Frequency	Profile Level and Homogeneity					Descriptive Name
		V	S	F	M	$\bar{H}$	
S1	8	51	35	48	36	.86	Low speed and memory
S2	13	48	51	43	37	.92	Low memory
S3	14	50	43	36	45	.87	Low form
S4	17	45	36	46	49	.86	Low speed
S5	4	36	48	44	50	.94	Low verbal
S6	21	51	51	49	49	.93	
S7	16	66	51	53	49	.92	High verbal
S8	10	52	62	49	53	.86	High speed
S9	6	66	67	59	57	.85	High verbal and speed
S10	15	52	49	63	50	.86	High form
S11	12	56	53	52	64	.85	High memory
S12	3	68	50	62	69	.94	High verbal, form, and memory
Unique	6						
	145						
		.87	.91	.87	.90		

A general impression of the typological similarity of the two groups can be obtained by comparing the descriptive names of the two and by noting from these names which types are present in both groups and which are present in one but absent in the other.

A more precise comparison of the different typologies is achieved by including all 26 types of both groups (14 F types plus 12 S types) into the same analysis, from which we get exact values of the similarities and differences between them. The procedures for this analysis are called "EUCO analysis" in the BC TRY System. The logic of the analysis is simple. Each type is considered to be an abstract "individual" plotted as a point in the cluster score space of V, S, F, and M, where its locus is determined by its four standard scores listed in Table 9.3. Program EUCO of the BC TRY System computes the euclidean distance between each pair of types and prints these values in a pair-comparison matrix from which one can read off precisely the degree of similarity between any two types.

Space limitations do not permit printing this euclidean distance matrix here. In its stead, however, we present a pictorial representation of the distances between the types in the form of the SPAN diagram given in Fig. 9.3. To secure this diagram, the EUCO matrix is first transformed to a correlation matrix by correlating columns of EUCO values, then running this correlation matrix through a standard key-cluster analysis, ending in the SPAN diagram of Fig. 9.3.

The configuration on the SPAN diagram describes the similarities and differences between the factory and suburban O-types. The circles represent the 14 factory O-types, the squares the 12 suburban O-types. Also included in this analysis are the 15 inclusive H types from Table 9.2. The sizes of the circles and squares and the length of the underline of the H types are proportional to the frequency of each type. The four dimensions, V, S, F, and M, are also plotted, these being secured by inputting abstract model "individuals" whose four standard score values are especially selected to enable one to plot the dimension lines as score axes.

The large supercluster at left center consists of types all in the low region, meaning that generally they have standard scores below the mean on all four dimensions. However, this supercluster breaks off into two general subclusters. The upper subcluster consists largely of suburban types S1, S2, S3, fairly well represented by the inclusive types H1, H4, H3, and H6, whereas the lower subcluster consists largely of F, or factory types, which with S4, are well represented by inclusive types H2, H5, and H7. From these facts the similarities and differences between the types in this general region of low scoring can be ascertained. There are real differences in the typologies of the two groups in this region.

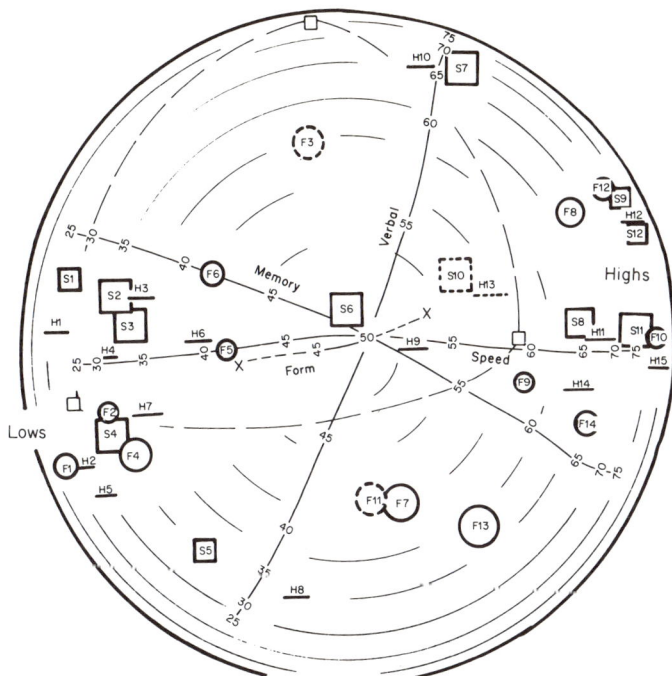

FIGURE 9.3
Spherical representation of the typologies of the factory and suburban
groups.

Generally, a study of the configuration reveals findings similar to those found from the similarity of the frequency patterns, namely, that verbal, memory, and speed dimensions most markedly differentiate the factory and suburban groups. For example, at the top of the diagram, high verbal is represented only by a suburban type, S7. Low verbal through the lower half of the diagram is heavily dominated by factory types.

How well do the 15 inclusive O-types representatively sample the 26 different types in both ecological groups of children? This question is important because in the practical usage of the typology of abilities, these would be the types usually used for the classifications of individuals. The answer is provided by noting whether one or more of the 15 H types lie in all regions occupied by the 26 factory and suburban types. By inspecting the SPAN diagram and by comparing the F and S types of Table 9.3 with the H types of Table 9.2 it becomes clear that the 15 H types fairly cover the ground.

Basic data structures
___

The second study selected for comparative dimensional and typological analysis is that of the responses to the items of the MMPI by groups of normal subjects and psychiatric patients.

The variables are 118 items of the MMPI drawn from the full item supply of 566 to which the subjects responded. The 118 were those tabulated in the previous chapter. The subjects were the inclusive group consisting of the normal and the patient groups. The normal subjects were 90 Armed Services officers matched for age and education against 220 patients. The latter were outpatients of a Veterans Administration mental health clinic, consisting of 70 diagnosed schizophrenics all with a history of hospitalization within the previous 6 years and 150 diagnosed anxiety patients none with a history of any hospitalization for psychiatric disorder.

Dimensional analysis of the 118 item-variables
___
in the inclusive group
___

Recall from the Holzinger study that a comparative dimensional analysis of two groups, here the normal and patient groups, is best performed when the subjects are measured on the same dimensions defined by the same variables, usually those discovered in a dimensional analysis of the inclusive group. This analysis revealed four "basic" MMPI item-clusters, I (introversion), B (body), S (suspicion), and T (tension). The defining items of these four dimensions are those whose item numbers are listed in Table 9.4. Each item-cluster consists of a full form and a short form. The comparative dimensional analysis presented in this section was performed on the scores of subjects on the short forms, and it also includes the short-form items of the other three "dependent" item-clusters, D (depression), R (resentment), and A (autism), whose item numbers are also given in Table 9.4. The full set of items is given in Table 11.2.

The dimensional analysis of the inclusive group from which the four basic and three dependent dimensions were derived was described in some detail in Chap. 8. However, the results of the full analysis are similar to those given below on the patient group. In sum, it was found that seven dimensions are required to account for the intercorrelations among the 118 items but that the first three basic dimensions, introversion, body, and suspicion, were the most nearly independent clusters (see Fig. 9.4). Only four pools of small residuals remained in the matrices of the four D, R, A,

TABLE 9.4 DEFINING ITEMS OF THE SEVEN
ITEM–CLUSTERS OF THE MMPI

A. The Four Basic Item-clusters

I: Introversion (Full form, 26 items, reliability .93; short
form, first[a] 17 items, reliability .91)

377	267	52	317	−415
− 57	172	−309	−264	−482
321	86	−479	138	
201	171	509	−353	
180	−547	292	304	
−371	−521	− 79	−449	

B: Body symptoms (Full form, 33 items, reliability .92;
short form, first 17 items, reliability .89)

−243	−230	125	72	−160	− 18
189	114	− 68	− 3	191	−192
108	47	10	− 36	−153	14
−190	44	23	−163	263	
62	− 55	161	− 51	−330	
−175	29	544	−103	− 2	

S: Suspicion and mistrust (Full form, 25 items, reliability .85; short form, first 17 items, reliability .83)

404	244	447	284	455
507	348	319	438	
383	368	71	89	
390	280	558	112	
436	265	406	426	
136	469	278	316	

T: Tension, worry and fears (Full form, 36 items, reliability .92; short form, first 17 items, reliability .88)

555	543	448	182	158	22
431	442	186	32	303	351
337	43	499	439	13	−131
217	−242	166	335	388	365
238	340	338	102	322	494
506	−152	−407	473	360	492

B. The Three Remaining "Dependent" Item-clusters

D: Depression and apathy (Full form, 28 items, reliability .94; short form, first 17 items, reliability .91)

76	41	414	526	339
−107	259	396	361	− 88
236	418	61	384	− 46
301	− 8	411	84	104
−379	549	142	357	
487	67	397	168	

R: Resentment and aggression (Full form, 21 items,
reliability .87; short form, first 16 items, reliability .82)

94	381	145	106
336	97	148	147
468	536	28	443
−399	139	162	
375	234	416	
39	129	382	

A: Autism and disruptive thoughts (Full form, 23 items,
reliability .86; short form, first 17 items, reliability .81)

559	545	342	389
241	358	374	356
15	560	459	40
349	−329	297	31
425	100	33	134
511	345	359	

---

[a] Reading by columns from the left.

and T clusters. Since the last of these, T (tension), had the greatest generality of the remaining four, it was decided to add T to I, B, and S as the final set of basic four dimensions of the MMPI.

Dimensional analysis of the 118 item-variables in the patient group

A full-cycle key-cluster solution of the intercorrelations between the 118 items in the patient group resulted in the cluster structure depicted in Fig. 9.4 (top diagram). This factoring process was "preset" on the four basic dimensions defined by the items of I, B, S, and T. In the tight cluster at lower left in the configuration the symbols plotted as I and enclosed in a broken line are 15 of the 17 introversion items that define this cluster. The remaining two lie nearby in the direction of the two arrows. In another tight cluster at lower right are 16 body, or B, items; the seventeenth item was dropped from the analysis because of trivial communality ($h^2 < .10$). The suspicion cluster is at the top. The remaining four clusters, depression, resentment, autism, and tension, lie within the framework of the three I, B, S clusters. Clearly the total configuration for the patient group shows an excellent cluster structure; it is virtually the same as that found previously in the total inclusive group (Tryon, 1966, fig. 1).

Dimensional analysis of the 118 item-variables in the normal group

A radically different dimensional structure emerges in the normal group, shown in the lower portion of the SPAN diagram of Fig. 9.4. The dramatic change is in the body cluster, which was so sharply evident in the patient group. It is *absent* as a distinct cluster among normal subjects, and so are the depression and autism clusters. But the introversion and suspicion clusters do appear as fairly independent item groups. Tension and resentment clusters also remain but move into a grand arc bounded by the introversion and suspicion clusters. It appears that only introversion and suspicion are the dominant and distinctive dimensions of normal subjects in the MMPI item-clusters.

Comparison of the dimensions within each group separately

Precise numerical statements about the seven item-clusters in each of the two groups are given in Table 9.5, Sec. A (analogous to Table 9.1 in the Holzinger problem) and Secs. C and D. The relationships between the seven

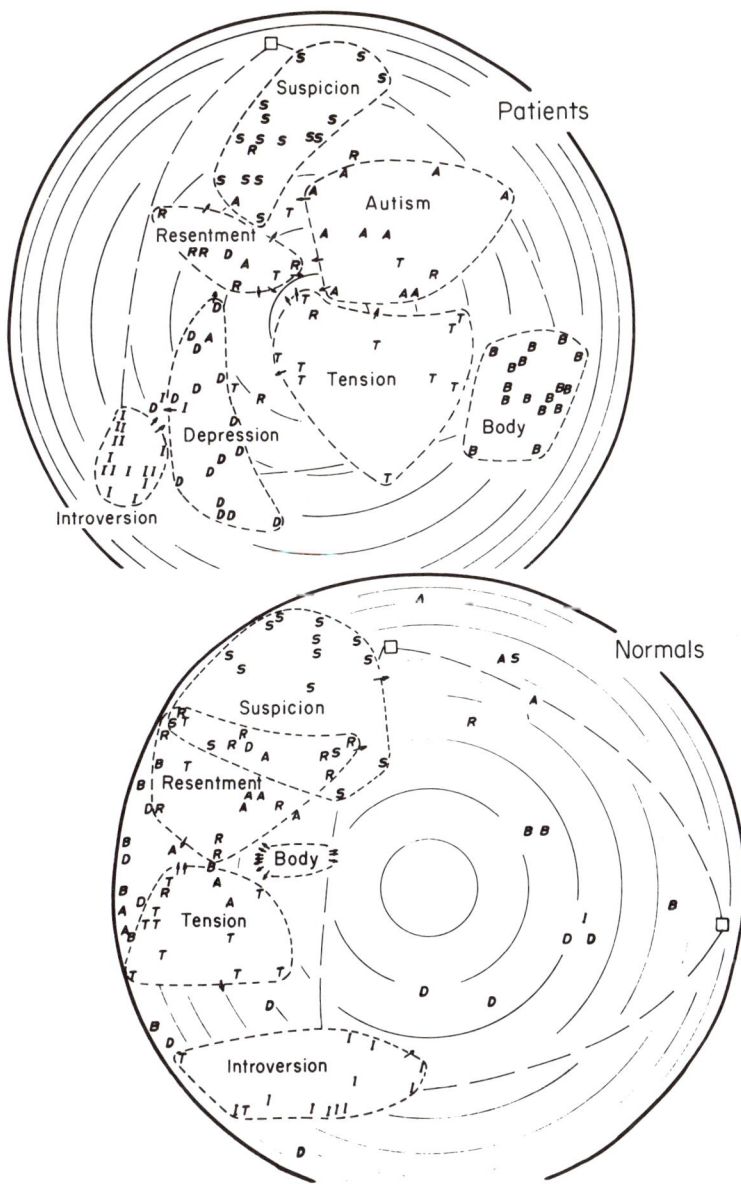

FIGURE 9.4
Cluster structure of 118 MMPI items within the patient and normal groups.

domains represented as dimensions (or oblique factors) are given in Sec. A by the interdomain r_{cc} values; those for the patients are above the lined-off diagonal, those for normal subjects below. These "correlations between oblique factors" are merely abstract metric descriptions of the complex relationships depicted in the SPAN diagram, and though they are more precise numerical statements compared with the verbal statements about the configuration, they are more difficult to organize conceptually. We must leave to the reader a detailed examination of this complex table of relationships, suggesting that he cross-reference his study of it by simultaneously referring to the visual configuration in Fig. 9.4.

Several obvious points need mention here. In both groups the introversion and suspicion dimensions are the most independent, and tension is

TABLE 9.5 SIMILARITY OF THE SEVEN MMPI ITEM-CLUSTER DIMENSIONS WITHIN AND BETWEEN THE NORMAL AND PATIENT GROUPS

A. Similarity of Item-cluster Dimensions within Each Group[a]

	I		B		S		D		R		A		T	
	r_{cc}	cos θ	r_{cc}	cos θ	r_{cc}	cos θ	r_{cc}	cos θ	r_{cc}	cos θ	r_{cc}	cos θ	r_{cc}	cos θ
I			.12	.13	.31	.31	.71	.69	.47	.46	.33	.38	.50	.49
B	.52	.46			.34	.34	.32	.31	.37	.37	.50	.49	.63	.60
S	.07	.06	.61	.52			.38	.37	.66	.62	.65	.61	.59	.57
D	.76	.60	.56	.43	.43	.38			.66	.65	.57	.55	.78	.76
R	.37	.32	.64	.52	.76	.69	.76	.65			.64	.62	.79	.77
A	.51	.45	.90	.69	.77	.70	.72	.59	.74	.68			.77	.74
T	.59	.53	.72	.53	.50	.48	.72	.59	.71	.63	.67	.61		

B. Similarity of Item-cluster Dimensions between Groups (cos θ only)

	I_P	B_P	S_P	D_P	R_P	A_P	T_P
I_N	.71	.20	.22	.53	.42	.38	.49
B_N	.33	.49	.47	.32	.49	.51	.50
S_N	.15	.30	.74	.28	.60	.57	.46
D_N	.56	.31	.43	.61	.65	.58	.59
R_N	.33	.31	.64	.48	.80	.57	.62
A_N	.35	.42	.61	.45	.60	.69	.60
T_N	.48	.41	.53	.52	.70	.57	.73

C. Generality of Each Dimension (Reproducibility of Correlations)

	I	B	S.	D	R	A	T
Normals	.38	.51	.43	.53	.52	.52	.48
Patients	.31	.19	.24	.52	.42	.36	.61

D. Reliability Coefficient of Cluster Score on Each Dimension

	I	B	S	D	R	A	T
Normals	.81	.54	.83	.72	.80	.76	.75
Patients	.90	.87	.83	.88	.79	.79	.81

[a] Values for patients lie above the ruled diagonal and those for normals below.

most positively correlated with all the other dimensions. The body dimension is radically different in the two groups, fairly specific in the patients but rather general in the normal subjects, correlating .90 with autism. This generality of the body dimension is misleading in the normal group, because we know from the SPAN configuration that the body dimension is not a cluster defined dimension in normal subjects but a mere sampling of heterogeneous items from their whole sphere of items. It is a grab bag of items in the normal group, just as autism is, so that their high correlation is merely due to both being similar hodgepodges.

Direct comparative analysis of the dimensions across groups

When we project the dimensions of the two groups into the same COMP2 analysis, the relations among the dimensions not only within but especially across the two groups are clearly displayed. They are pictorially displayed in the single SPAN diagram of Fig. 9.5 (analogous to Fig. 9.1 of the Holzinger

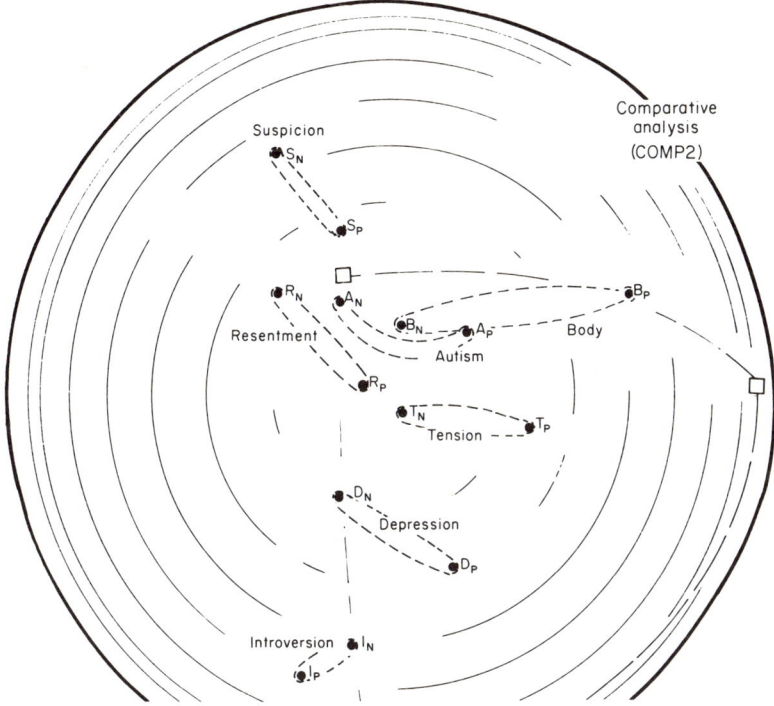

FIGURE 9.5
Cluster structure of 118 MMPI items across the patient and normal groups.

problem). The sharply differentiated and spread-out dimensions of the patients, denoted by the subscript P attached to the seven dimensions, I, B, S, D, R, A, T, confirm the within-group cluster structure of their items as previously depicted in the upper sphere of Fig. 9.4. In contrast, the within-group structure of the dimensions of the normal subjects, indicated by the subscript N, confirms the narrow, essentially two-dimensional band ranging from the introversion dimension to the suspicion dimension.

Consider, now, the similarity of the dimensions across the two groups as objectively measured by the cos θ values, given in Table 9.5, Sec. B, especially those on the diagonal. The most similar dimensions across the groups are introversion (.71), suspicion (.74), resentment (.80), and tension (.73). The least similar is the body dimension (.49), a different kind of dimension in the two groups.

The index of dimensional similarity cos θ and the interdomain (common factor) correlations r_{cc} given as paired values in Sec. A of Table 9.5 show a close correspondence only for tight clusters I, S, R, and T.

Comparative typological analysis in the MMPI problem

The comparative typological objective is to discover the degree to which O-types of individuals, formed by classifying together individuals having the same pattern of standard scores on the four basic MMPI dimensions, I, B, S, T, have the same structure in the patient and normal groups. In this analysis, each person was scored by his full-form scores on I, B, S, and T.

Similarity of frequency patterns of the two groups on the common typology of the inclusive group

In the typological analysis of the inclusive group by program OTYPE 14 O-types emerged. These are listed as types M1 to M14 in Table 9.6, under Inclusive Typology, with the frequencies, standard scores on I, B, S, T, and descriptive names. When the normal and patient subjects are sorted separately into these 14 inclusive O-types, the percentages falling into them are the values listed in the columns labeled "% in Each Group." As a point of special interest, the patient group is separated into its two component diagnostic groups, anxieties and schizophrenics.

The overall similarity of the typology of the three groups in relation to each other is given by their P values: normal vs. anxiety groups show a $P = .17$, indicating virtually no similarity in their typological structures;

TABLE 9.6 SIMILARITY OF THE FREQUENCY PATTERNS OF NORMALS AND PATIENTS ON THE COMMON INCLUSIVE TYPOLOGY IN THE MMPI PROBLEM

| Type | Fre-quency | Z Scores | | | | Descriptive Name | % in Each Group | | | Differences between | | |
		I	B	S	T		Normals P_N	Anxieties P_A	Schizo-phrenics P_S	Normals and Anxieties	Normals and Schizo-phrenics	Anxieties and Schizo-phrenics
M1	24	40	38	35	35	Extrovert, healthy, trusting, relaxed	20	1	6	19(N)	14(N)	−5
M2	38	37	40	48	38	Extrovert, healthy, relaxed	39	1	3	38(N)	36(N)	−2
M3	21	46	39	48	40	Healthy, relaxed	16	1	7	15(N)	9(N)	−6
M4	31	48	47	36	46	Trusting	5	15	6	−10(A)	−1	9(A)
M5	17	39	50	50	47	Extrovert	3	5	10	−2	−7(S)	−5
M6	24	50	50	50	50	Average	3	11	7	−8(A)	−4	4
M7	22	52	62	52	64	Somatic, tense	0	11	7	−11(A)	−7	4
M8	26	48	48	64	52	Suspicious	12	3	14	9(N)	−2	−11(S)
M9	16	50	65	52	52	Somatic	0	10	1	−10(A)	−1	9(A)
M10	14	50	64	66	60	Somatic, suspicion, tense	0	8	3	−8(A)	−3	5
M11	30	64	48	48	50	Introvert	2	10	19	−8(A)	−17(S)	−9(S)
M12	20	66	54	60	64	Introvert, suspicion, tense	0	9	10	−9(A)	−10(S)	−1
M13	17	65	65	53	66	Introvert, somatic, tense	0	9	4	−9(A)	−4	5
M14	10	67	67	65	67	Introvert, somatic, suspicion, tense	0	5	3	−5	−3	2
Unique	0					Unique						
N	310						90	150	70			

there is a mild typological similarity of normal subjects and schizophrenics, with $P = .41$; the anxiety and schizophrenic patients, in contrast, bear considerable resemblance, having $P = .73$.

The details of the group differences, given in the columns headed Differences, are of great interest. The normal subjects are almost exclusively concentrated in types M1, M2, and M3, described generally as extrovert, healthy, and relaxed, with a few in M8, the suspicious. The anxiety patients excel in the somatic types, M7, M9, M10, M13, M14, indicating persons most preoccupied by body disturbances. The schizophrenic patients, compared to the anxiety patients, behave typologically somewhat like normal subjects, except that they fall heavily in the introvert type, M11. The standard errors of the differences are .034, .041, and .037 for the normals-anxieties, normals-schizophrenics, and anxieties-schizophrenics differences, respectively.

Similarity of empirically derived typologies of the groups

Fuller information on the differences between the O-types of the normal and patient groups comes from a direct comparison of their typologies, as these are empirically derived separately by the OTYPE and OSTAT programs but projected then into the same comparative analysis. Table 9.7 gives the basic data: 14 types of normals and 13 types of patients.

Looking through the descriptive names of the normal and patient O-types reveals, perhaps astonishingly, that there is no overlap of their 27 types except for the average and the trusting O-types but that even in these the normal subjects have only a handful of cases whereas they are abundant in the patient group.

In sum, patients are clearly distinguished from normal subjects in their objectively derived patterns of MMPI scores. This finding goes directly to the question of the validity of the MMPI in distinguishing patients from normal persons. Our finding here definitely demonstrates the validity of the MMPI items in differentiating normal subjects from patients, *provided* the item-cluster scores on I, B, S, and T are used (and not the hodgepodge in the usual unclustered scales) and *provided* the objective typology described in these pages is used as the classificatory scheme.

When the 27 types are projected into the same EUCO analysis along with the 14 inclusive O-types, the grossly different typological structure of the normal subjects and patients stands out boldly. This fact is clearly evident in the spherical representation of the types given in Fig. 9.6. The normal object types (abbreviated N and placed in circles) are virtually all located in a supercluster at the left in the low score ranges on all dimensions.

TABLE 9.7 WITHIN-GROUP MMPI ITEM-CLUSTER TYPOLOGIES OF THE NORMALS AND PATIENTS

Normals

Type	Frequency	Profile Level and Homogeneity					Descriptive Name
		I	B	S	T	$\bar{H}$	
N1	8	38	39	34	35	.99	Extrovert, healthy, trusting, relaxed
N2	10	41	39	38	37	.98	Healthy, trusting, relaxed
N3	3	38	42	37	37	.99	Extrovert, trusting, relaxed
N4	16	39	38	47	37	.98	Extrovert, healthy, relaxed
N5	3	45	39	44	36	.98	Healthy, relaxed
N6	4	39	45	48	39	.97	Extrovert, relaxed
N7	4	45	43	52	39	.97	Relaxed
N8	4	53	41	38	41	.95	Trusting
N9	6	39	39	45	42	.98	Extrovert, healthy
N10	3	44	42	52	45	.97	Average 1
N11	12	39	40	57	39	.96	Extrovert, healthy, relaxed
N12	6	53	45	52	48	.91	Average 2
N13	5	39	46	66	45	.96	Extrovert, suspicious
N14	6	44	47	60	46	.98	Suspicious
Unique	0						
N	90						
η		.97	.98	.97	.97		

Patients

Type	Frequency	Profile Level and Homogeneity					Descriptive Name
		I	B	S	T	$\bar{H}$	
P1	36	45	45	37	44	.85	Trusting
P2	15	41	51	51	48	.90	Average 1
P3	28	49	47	53	50	.89	Average 2
P4	28	62	47	44	51	.88	Introvert
P5	15	49	63	46	50	.84	Somatic
P6	10	51	50	65	57	.93	Suspicious
P7	12	49	64	64	59	.89	Somatic, suspicious
P8	15	54	57	52	62	.93	Tense
P9	11	64	50	57	60	.90	Introvert, tense
P10	12	49	66	49	64	.90	Somatic, tense
P11	14	65	67	50	64	.90	Introvert, somatic, tense
P12	14	65	54	64	60	.90	Introvert, suspicious, tense
P13	10	65	68	64	66	.92	Introvert, somatic, suspicious, tense
Unique	0						
N	220						
η		.91	.88	.88	.90		

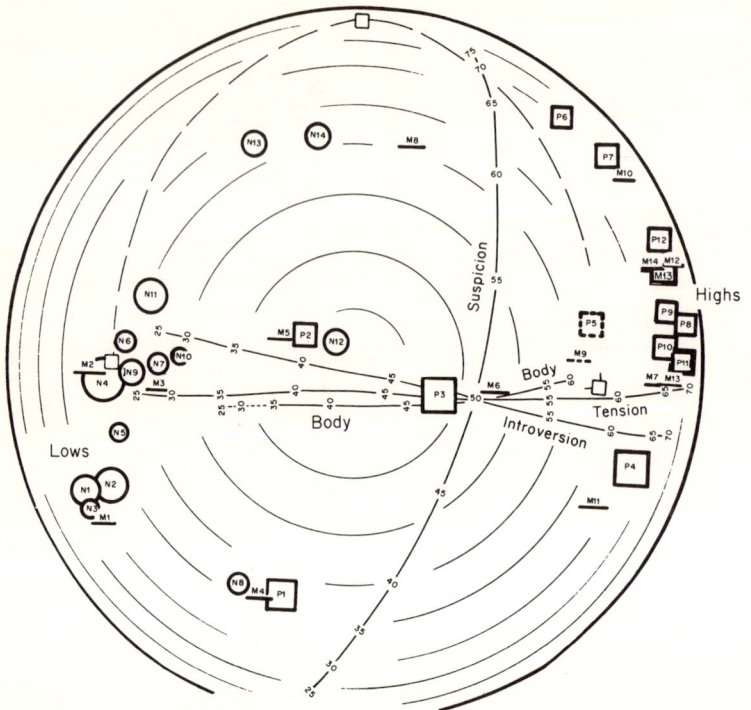

FIGURE 9.6
Spherical representation of the typologies of the normals and patients.

The patient types (abbreviated P and placed in squares) are largely in the cluster at the right or high region of the configuration. This separation confirms, of course, the finding of the previous section, but the SPAN configuration provides a more differentiated description.

The locus of the 14 inclusive types (abbreviated M and underlined) are located in all regions of this typological space where there are normal and patient types. This fact means that as a system of classifying individuals, normal or mentally ill, the 14 inclusive types satisfactorily cover the ground.

The study of social structure: the social-area problem

Basic data structure

In previous chapters we described V-analysis and O-analysis of the social areas of the San Francisco Bay region. In this chapter we present a comparative analysis of the dimensions and of the typology described in the previous chapter. There are 225 census tracts in the analysis, each "observed" in 1940 (prewar) and in 1950 (postwar), of which 105 are from San Francisco and 120 from the East Bay communities. Each census tract was treated as two units, one for the 1940 data and one for the 1950 data. Thus, there are 450 objects in the analysis. Of the 33 census variables of the study 14 were selected in V-analysis to represent the three dimensions F, family life; A, assimilation; and S, socioeconomic independence.

Dimensional analyses

The four subsets of data were submitted to preset V-analysis through the BC TRY programs CC5, CSA, and SPAN. The metric descriptions of the results of these analyses are given in Table 9.8, which also shows the defining variables of each of the clusters, the oblique factor coefficients for the definers in each analysis (on the definers' dimension only), the cluster reliability (α) of each cluster in each of the four analyses, and the generality of each cluster in the four analyses (reproducibility of correlations). The actual structure of the different cluster analyses, time and place, are highly similar. The degree of similarity and the specific differences involved are described below in the comparative analyses. Since the analyses are generally similar to the analysis reported in greater detail earlier, no additional data are reported here.

Comparison of the dimensions
within each group separately

The metric comparison of the cluster structures within each of the four groups is given in Table 9.9, both in terms of the domain intercorrelations and the cos θ measures. The four 3 by 3 tables for the two measures are merged, cell by cell, in order to facilitate comparison. For example, the correlation between the S and F cluster dimensions for San Francisco is .08 in 1940 and .27 in 1950. Comparable correlations in the East Bay communities are .14 and .15, as shown in the table. The data reported in Table 9.9 indicate a high degree of cluster structure consistency in the four sets of data. Clusters S and F define dimensions that are nearly independent, while clusters S and A have a moderately high degree of correlation and A and F are in the intermediate correlation range. The pictorial comparison of these four analyses can be made by inspecting the spherical diagrams of the 14 variables defining the three demographic dimensions, shown in Fig. 9.7. The four upper diagrams in Fig. 9.7 are marked according to the place and time for the respective data represented.

TABLE 9.8 DEFINING VARIABLES OF THE F, A, S DIMENSIONS OF METROPOLITAN NEIGHBORHOODS AND THEIR RELIABILITY AND GENERALITY BY SITE BEFORE AND AFTER WORLD WAR II

Cluster-defined Dimensions

Socioeconomic Independence S		Family Life F		Assimilation A	

A. Defining Variables

Socioeconomic Independence S	Family Life F	Assimilation A
Mm Managerial-profession males Df Domestics (females) Om Own-account males Co College-educated	Oo Owner-occupied Fl Large families Fd Family-detached Uf Housewives (unemployed females) Am Young children (males)	Sm Skilled males Nw Native-born whites F Females Fe Foreign from Protestant Europe Wf White-collar females

B. Oblique Factor Coefficient of Definers of Each Dimension

Definer	San Francisco		East Bay		Definer	San Francisco		East Bay		Definer	San Francisco		East Bay	
	'40	'50	'40	'50		'40	'50	'40	'50		'40	'50	'40	'50
Mm	.98	.94	.97	1.00	Oo	.92	.94	.92	.85	Sm	.78	.88	.81	.88
Df	.87	.92	.94	.91	Fl	.96	.97	.97	.92	Nw	.82	.78	.76	.76
Om	.90	.86	.69	.79	Fd	.81	.85	.95	.85	F	.69	.68	.58	.52
Co	.81	.84	.87	.83	Uf	.83	.87	.85	.83	Fe	.68	.51	.78	.73
					Am	.81	.83	.73	.46	Wf	.81	.81	.91	.97

C. Reliability of Cluster Scores on Each Dimension

.95	.94	.93	.94		.95	.96	.96	.91		.89	.88	.89	.90

D. Generality of Each Dimension (Reproducibility of Correlations Between Definers)

.32	.35	.41	.48		.44	.45	.42	.34		.33	.34	.44	.52

Direct comparative analysis across groups

The metric comparative analysis of the cluster dimensions across the place and time groups produced by the BC TRY program COMP2 are given in Table 9.10. The similarities of the three dimensions for a given place but spanning the war years are given in the first two rows of the matrix. These cos θ values are all very high, indicating a very small change in the social cluster structure across the intervening 10 years even though there was a major global upheaval and a dramatic change in the populace of the areas (see Chap. 10). Similar remarks about comparisons between the two localities for each time are justified by the next two rows of figures in the table.

The direct comparative analysis involves comparisons of all twelve

TABLE 9.9 SIMILARITY OF CLUSTER DIMENSIONS WITHIN EACH
PLACE–TIME GROUP

Place-Time Group	Socioeconomic, S		Family Life, F		Assimilation, A	
	r_{CC}	$\cos \theta$	r_{CC}	$\cos \theta$	r_{CC}	$\cos \theta$
Socioeconomic, S						
San Francisco '40	1.00	1.00	.08	.08	.30	.29
San Francisco '50	1.00	1.00	.27	.10	.59	.43
East Bay '40	1.00	1.00	.14	.15	.57	.56
East Bay '50	1.00	1.00	.15	.27	.57	.57
Family life, F						
San Francisco '40	.08	.08	1.00	1.00	.27	.27
San Francisco '50	.27	.10	1.00	1.00	.33	.23
East Bay '40	.14	.15	1.00	1.00	.31	.30
East Bay '50	.15	.27	1.00	1.00	.31	.32
Assimilation, A						
San Francisco '40	.30	.29	.27	.27	1.00	1.00
San Francisco '50	.59	.43	.33	.23	1.00	1.00
East Bay '40	.57	.56	.31	.30	1.00	1.00
East Bay '50	.57	.57	.31	.32	1.00	1.00

TABLE 9.10 SIMILARITY OF CLUSTER–DEFINED DIMENSIONS
BETWEEN THE PLACE–TIME GROUPS (COS θ ONLY)

	Socio-economic, S	Family Life, F	Assimilation, A
Same place but spanning World War II:			
San Francisco, '40 vs. '50	.97	.98	.89
East Bay, '40 vs. '50	.94	.89	.94
Same time but different place:			
San Francisco '40 vs. East Bay '40	.88	.92	.90
San Francisco '50 vs. East Bay '50	.92	.84	.84
Mean	.93	.91	.89

dimensions, three from each of four analyses. The matrix of cos θ values
was subjected to a single full-cycle cluster analysis, yielding the cluster
structure graphically displayed in the lower left portion of Fig. 9.7.

The configuration on the sphere (marked COMP2) dramatically
reveals the identity of socioeconomic, family life, and assimilation dimensions in the four groups. All four S dimensions virtually occupy the same
locus on the sphere, as do the four F dimensions and the four A dimensions, demonstrating that the basic tridimensionality of metropolitan
neighborhoods is undisturbed by differences in locale or time or of course
by difference in the people that occupy them (the people were largely
different individuals in the groups).

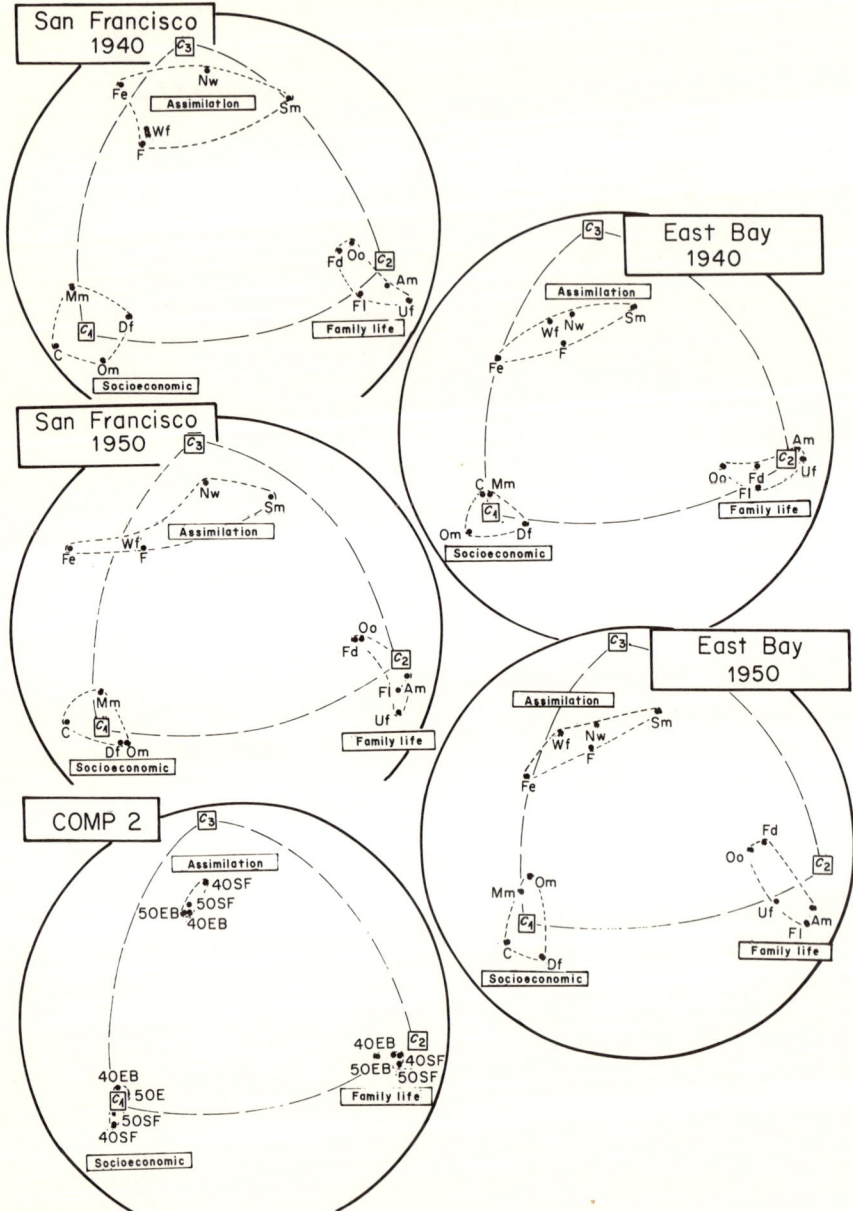

FIGURE 9.7
Comparative cluster structure of the 14 variables that define the three demographic dimensions of neighborhoods.

Similarity of frequency pattern on a common typology

Using the typology methods of OTYPE, 11 types of social areas are discovered in the full inclusive group of 450 neighborhoods. The basic metric data on these 11 neighborhood O-types are presented in Table 9.11. The first column lists the 11 O-types, S1 to S11. Under Common Inclusive Typology are listed the mean standard scores of each O-type on the three cluster dimensions, the overall homogeneity value for each O-type, descriptive names with indications of the high and low cluster score means in the O-type score profile, and the percent of the 450 neighborhoods contained in each O-type. The high values of the homogeneities indicate very tightly defined O-types.

The relative frequencies of the inclusive typologies include all 450 neighborhoods. If the four communities separately have the same pattern of frequencies as the inclusive group, they have the same typology; this is the situation in prewar vs. postwar communities, where the frequencies are very nearly identical in the time groups. The index of similarity of the frequency patterns for the 1940 data and the 1950 data is $P^2 = .99$.

The story is different with respect to locale. Because of the inconsequential effect of time on the typology, we have combined the frequency of neighborhoods on the common typology of San Francisco in 1940 and 1950 and have compared the San Francisco pattern with the East Bay, similarly combined. These frequency patterns are in the columns headed San Francisco vs. East Bay.

Specific differences in the typologies are quickly discerned from the values in the Difference column. Positive differences refer to types that have high frequencies in San Francisco but low frequencies in East Bay. These are, in order of size, the downtowners, minorities, and the fancy apartmentiers object clusters. Negative differences in the Difference column refer to types of higher frequency in East Bay than San Francisco; these are suburbians 2 cluster and workers 1 cluster. An overall statement of the degree of similarity of the San Francisco vs. East Bay is the index of similarity $P^2 = .70$, a high value but considerably less than the degree of similarity between the two time periods.

Similarity of empirically derived typologies

A drawback to comparing the social-area structure of different metropolitan sites on a common typology is that it sets a common mold for them to fit into and does not provide an opportunity for types unique to a particular site to be discovered. It is necessary to perform an empirical typological

TABLE 9.11 SIMILARITY OF FREQUENCY PATTERNS ON A COMMON TYPOLOGY OF PREWAR VS. POSTWAR NEIGHBORHOODS AND OF SAN FRANCISCO VS. EAST BAY NEIGHBORHOODS

Social-area Type	Common Inclusive Typology: San Francisco and East Bay; 1940 and 1950						Prewar vs. Postwar (San Francisco and East Bay), %			San Francisco vs. East Bay (1940 and 1950), %		
	Z-Scores			Overall H	Descriptive Name	%	1940	1950	Difference	SF	EB	Difference
	F	A	S									
S1	38	28	45	.86	Minorities; low F, A	7	7	7	0	12	2	10
S2	51	37	43	.88	Workers 2, low A	11	12	10	2	10	12	−2
S3	35	50	48	.85	Downtowners, low F	15	16	15	1	25	6	19
S4	49	51	45	.94	Mid 3	16	15	16	−1	16	16	0
S5	59	50	43	.94	Workers 1	16	16	16	0	13	19	−6
S6	50	57	52	.95	Mid 2	9	8	11	−3	8	11	−3
S7	61	60	51	.96	Suburbians 2, high F, A	9	9	9	0	3	15	−12
S8	43	51	64	.89	Fancy apartmentiers, high S	4	4	4	0	8	1	7
S9	61	57	81	.92	Exclusives, high F, S	4	4	4	0	2	6	−4
S10	50	60	61	.93	Mid 1, high A, S	4	5	4	1	2	7	−5
S11	62	61	68	.95	Suburbians, high F, A, S	4	4	4	0	2	6	−4
N						450	225	225		210	240	

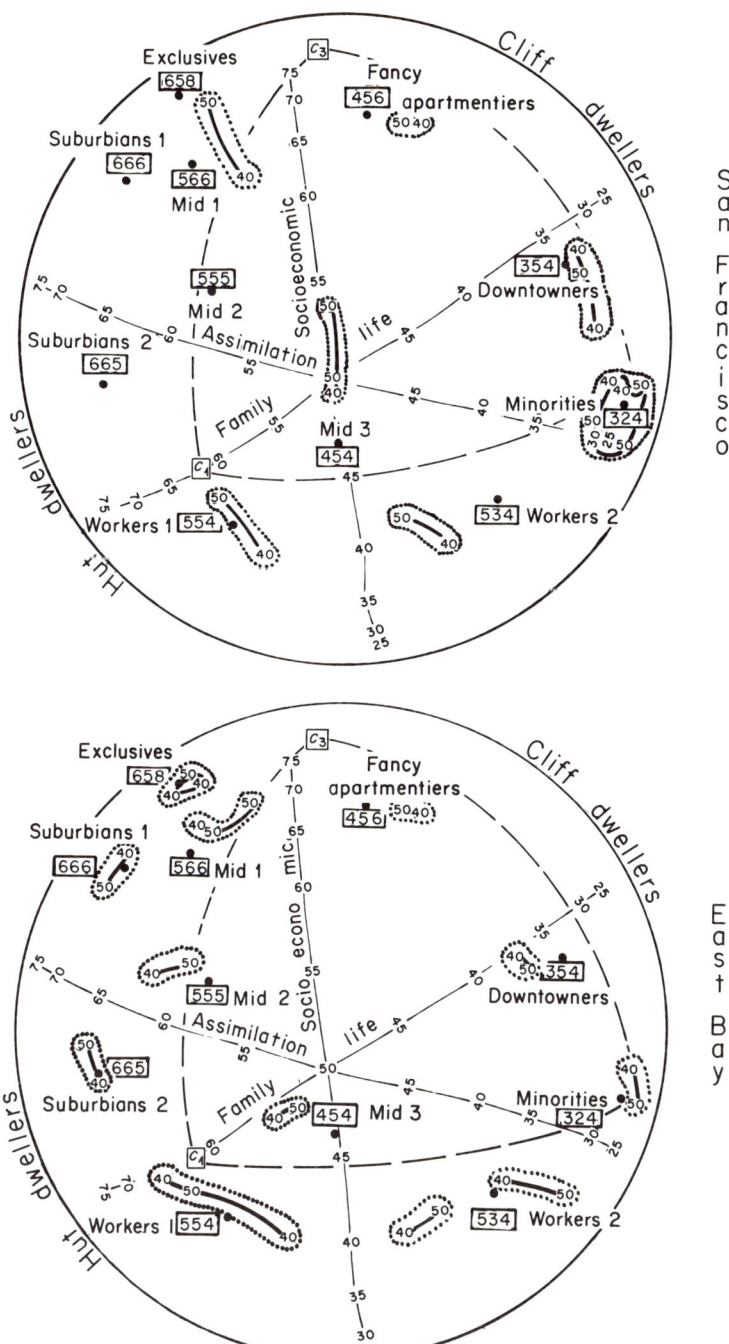

FIGURE 9.8
Comparative social area typologies of prewar and postwar San Francisco and East Bay.

analysis of each metropolitan group separately and then to project all the discovered types of all the groups into the same master analysis by which one can compare the similarities and differences in the subgroup structures. This procedure was followed in the social-area problem. A single "abstract" neighborhood was made up for each discovered type, and these were then projected into a EUCO analysis which terminates in a SPAN diagram showing the relations of all types to each other in one grand configuration. Because this configuration has too much in it for a single presentation, the data are presented in two parts of Fig. 9.8. The top SPAN diagram includes as points the 18 types found in San Francisco, 9 in 1940 (labeled 40) and 9 in 1950 (labeled 50). The bottom sphere displays the configuration of the 25 types in the East Bay, 13 in 1940 and 12 in 1950, similarly labeled. We have also included in each configuration the 11 inclusive types as reference points (black dots), as well as reference points by which to identify the three score axes on the spheres. We have also connected, with lines, the types found in 1940 and 1950 that have the greatest similarity and have encircled them in dotted lines.

The salient conclusions from this comparative analysis seem to be as follows. Though the computer derived the 1940 typology quite independently of the 1950, and though few persons lived in the same neighborhoods in 1950 and 1940, nevertheless almost precisely the same typology was recovered on these two occasions, and this despite the social disruptions of the war. The "reality" of the 11 social areas that compose the common typology seems to be confirmed, though quite obviously the types found in San Francisco are heavily composed of "cliff dwellers" and those of East Bay of "hut dwellers." Nevertheless, there are some common overlapping types that are equally differentiated in the separate computer runs on San Francisco and East Bay. The total configuration consists of a swarm of points representing the 450 neighborhoods, between which there are doubtless no clean dividing lines. Still, there seem to be 11 "natural" areas of concentration which our typological method objectively discovers. One could doubtless further divide the 11 types, or for that matter could combine them into larger social areas. But such possibilities do not deny the "existence" of the 11 areas that have been discovered. There are obvious places in the configuration where no Bay Area neighborhoods existed either in 1940 or 1950. Whether they are truly zones of disjunction, signifying kinds of neighborhoods that *cannot* exist because such patterns of FAS scores are genuine incompatabilities, is not clear from our analysis. Perhaps they exist in Boston, Minneapolis, or New Orleans or will appear in the Bay Area in later decades. Only further comparative social-area analyses will tell us.

Chapter 10
PREDICTING INDIVIDUAL AND GROUP DIFFERENCES IN CLUSTER ANALYSIS

This chapter deals with the problem of predicting differences between *individuals* in mental abilities (the Holzinger problem) and in self-conception (the MMPI) and with the prediction of differences among *groups*, namely, the neighborhoods of the San Francisco Bay Area.

When observing differences among objects in one domain of behavior, how can the observations best be organized to achieve maximal prediction of differences among them in other attributes? For example, in the Holzinger problem, where 301 school children were observed on 24 tests measuring verbal abilities, mental speed, form or space perception, and memory abilities, how can these data be cluster-analyzed in order to predict differences optimally, say, in the mathematical abilities of these children?

It is shown in this chapter that the highest level of prediction is multi-variate and that the best multivariate prediction is achieved when the predictor is a series of object-clusters. In the Holzinger problem, for example, the children are cast into person-clusters, so selected that in each such object-cluster the children are homogeneous in their pattern of scores on the predictor abilities. Prediction from object-clusters is called "differential prediction." For one typological group of children, having a particular pattern of scores on the predictor abilities, we can predict mathematical capability with a high level of accuracy, whereas for children with another pattern of predictor abilities we can predict mathematical ability no better than chance.

Three developments have made differential typological prediction possible: (1) an objective means of determining different types that compose a group, (2) a means of objectively describing the degree of predictability of other attributes by each type, and (3) the capability of assessing the probability that predictions from each O-type, however small the number of persons that compose it, can arise by chance. These developments are now embodied in components of the BC TRY computer system. The component OTYPE quickly and efficiently discovers the typological breakdown within a given group. The objective facts on the degree of predictability in each of the types are computed in the component OSTAT. The component 4CAST calculates the probability that the predictions by each O-type are those of a mere Monte Carlo (equal probability) sampling of subjects from the total supply.

Individual prediction from four "basic" mental abilities

In this first problem, multivariate prediction is of mathematical abilities in children from four cluster-defined "basic" mental abilities, V (verbal), S (speed), F (form or space), and M (memory). The subjects are the 301 seventh- and eighth-grade children who were observed in their responses to 24 ability tests of the Holzinger problem. V-analysis and O-analysis of these data have been described extensively in previous chapters.

The predicted mathematical abilities

Among the 24 tests were 5 different tests of mathematical ability, none of which are definers of the four basic abilities, V, S, F, and M. These five mathematical abilities are listed in Table 10.1, labeled from N20 to N24 in the original series of tests and defined respectively as tests of deduction, numerical puzzles, problem reasoning, number series, and arithmetic.

Special interest in these mathematical abilities inheres in their being considered by many factorists as being different sample measures of a "general factor g" (e.g., Holzinger and Swineford, 1939, p. 8). The tests

TABLE 10.1 MULTIVARIATE PREDICTION OF FIVE MATHEMATICAL
ABILITIES FROM V, S, F, AND M IN THE HOLZINGER PROBLEM

Mathematical Ability	Reliability	Multiple Correlation with V, S, F, M	
		Raw Scores	Domain (Factor) Scores (Theoretical)
N20 Ded	.69	.59	.63
N21 Puz	.80	.60	.65
N22 Rsn	.73	.57	.60
N23 Ser	.92	.63	.68
N24 Ari	.81	.64	.69

were designed to measure general mathematical or deductive reasoning
rather than specific arithmetic computational operations. The structural
organization of the 24 abilities is shown pictorially in Figs. 7.3 and 7.4, which
give the graphic position of the mathematical abilities in relation to the
predictor tests.

Multiple linear prediction of the mathematical abilities

The multiple correlation between the scores of the children on each of the
five mathematical abilities tests and the four predictor clusters, V, S, F,
and M, are listed in Table 10.1, headed Raw Scores. These multiple correlations cannot, in general, exceed the reliability coefficients of the tests,
listed under Reliability. Predicting the mathematics abilities from "domain"
or "factor" scores on V, S, F, and M produces the estimates of multiple
correlation given in the last column. The obtained multiple correlations are
not sensibly lower than those from theoretical factors measuring V, S, F,
and M.

A multivariate prediction of each mathematical ability from the four
basic predictor abilities, V, F, S, and M, yields a multiple correlation of the
order .65. This value is rather low for prediction, not providing much support for the view that a person's general intelligence g, if measured by
ability in mathematics, can be estimated from his standings in verbal,
speed, spatial, and memory abilities. But multiple correlation, like all
correlations, is an average statistic. While average prediction may be poor,
for some *types* of persons it is much better than for others.

Differential prediction of mathematical abilities from person-clusters (O-types) based on patterns of V, S, F, and M scores

Figure 10.1 shows the general results of differential typological prediction, details of which are presented later. In the top section headed Deduction is the distribution of the mathematical deduction scores of two subgroups of children out of the total of 301, each child being represented as a dot. The top histogram is that of the O-type called "T1," consisting of 13 children that have a particularly distinctive pattern of standard scores on the four predictor variables, V, S, F, and M. Next, below, is the histogram of 22 children of O-type T14, that has a different pattern of scores on V, S, F, and M. The common standard score scale at the top is the standard score transformation of the raw test scores to a mean of 50 and a standard

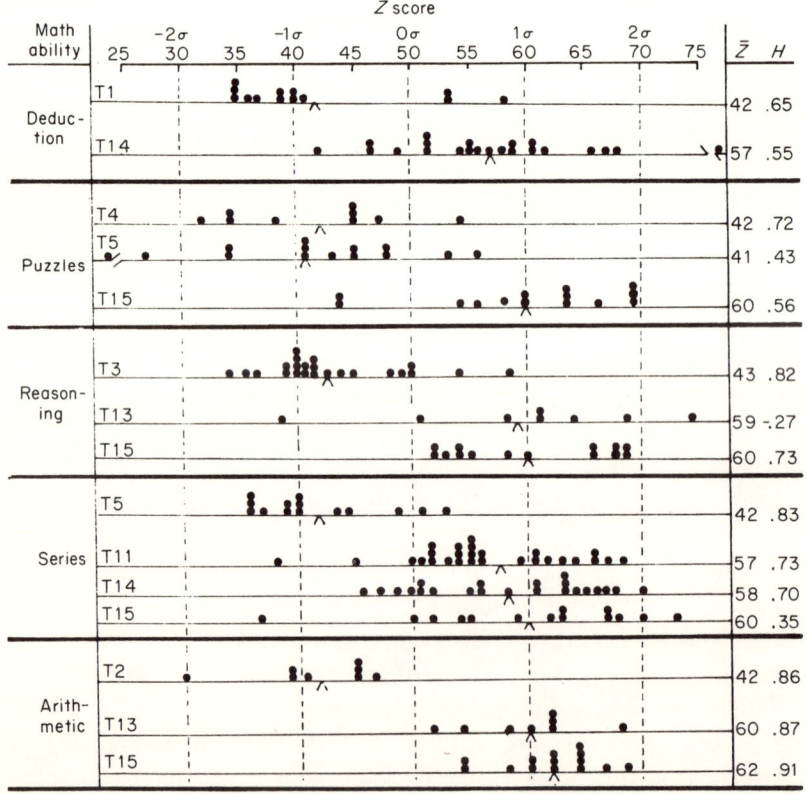

FIGURE 10.1
Distribution of standard scores on mathematical abilities of members of those Holzinger V,S,F,M O-types whose predictions are significant beyond the .001 level.

deviation of 10. The dashed lines down the figure give the standard scores lying at 0, ± 1, and ± 2 standard deviations from the mean.

At the far right of each histogram is given the mean standard score of the children of the O-type; for T1 it is 42 and for T14 it is 57. Thus, these two types differ in their mean deduction scores by $1\frac{1}{2}$ standard deviations. The degree of homogeneity of the deduction scores of each O-type is reflected by the value H, listed at the far right, .65 and .55, respectively, for the two groups.

Further down Fig. 10.1 are found the mathematics scores of different O-types on puzzles, reasoning, series, and arithmetic. The three O-types in the bottom sector of the figure are those for which the prediction of arithmetic scores is the highest. O-type T2 has arithmetic scores that do not overlap in any degree on those of the two types, T13 and T15. All three groups are highly homogeneous, having homogeneity coefficients of the order .90.

The predictions displayed in Fig. 10.1 are of 15 types out of a possible 75 such O-type predictions. These particular 15 are those which, despite the small number of cases in each type, are nonchance, at the significance level lower than .001. That is, in each case the probability of recovering each of the 15 histograms in Fig. 10.1 by random sampling from the full supply of 301 scores is considerably less than .001. As is shown later, more predictions than the 15 shown in Fig. 10.1 would have been included if the level of significance had been changed to .01 and considerably more if .05 had been chosen.

The predictor typology

The first step in the analysis just outlined is to cast the children into homogeneous O-types of their patterns of scores on the four predictor variables V, S, F, and M. The method of doing so has been fully described in previous chapters, where it was shown that these 301 children fall into 15 O-types. These 15 O-types, with a summary description of the distinctive scores on V, S, F, and M, are given in Table 10.2 in the Predictor Abilities columns. For example, T1 consists of those 13 children whose scores are *low* on V, low on S, *low* on F, and low on M. The italicized terms *low* mean that the O-type is 1 standard deviation or more below the mean in standard scores; no italicization means that the score is from $\frac{1}{2}$ to 1 standard deviation below the mean. The type contrasting most to O-type T1 is at the bottom of the table, labeled T15; the scores on the 14 children of T15 are all high on V, S, F, and M, and greater than 1 standard deviation above the mean on V and S. The distinctive patterns of scores of the other types are shown in the first columns. When there is no entry, it means that the

TABLE 10.2 PREDICTED MATHEMATICAL ABILITIES OF VSFM O-TYPES IN THE HOLZINGER PROBLEM

Type	Predictor Abilities				Fre-quency	$\bar{H}$	Predicted Mathematical Abilities														
							Level and Homogeneity										Significant				
	V	S	F	M			Ded		Puz		Rsn		Ser		Ari		Ded	Puz	Rsn	Ser	Ari
							$\bar{Z}$	H	$\bar{Z}$	H	$\bar{Z}$	H	$\bar{Z}$	H	$\bar{Z}$	H					
T1	*Low*	Low	*Low*	Low	13	.87	42	.65	43[a]	−.17	44[a]	.15	42[a]	.30	44	.17	Low	Low	Low	Low	Low
T2	*Low*			*Low*	8	.86	43	.81	43	.57	40[a]	.96	42	.80	42[a]	.86			*Low*		Low
T3	*Low*				23	.88	48	.44	49	.78	43	.82	47	.67	45[a]	.62		Low	Low	Low	Low
T4		*Low*	Low	*Low*	9	.88	46	.67	42	.72	41[a]	.74	44	.65	46	.63		Low			
T5		*Low*	Low		14	.89	45	.61	41	.43	47	−.34	42	.83	41[a]	.57		Low		Low	
T6		*Low*			19	.88	48	.48	49	.60	50	.57	49	.39	49	.59					
T7			*Low*		20	.88	45	.78	47	.40	48	.74	46	.64	49	.46					
T8					21	.86	45[a]	.72	49	.26	47	.57	47	.62	50	.62	Low				
T9				Low	38	.91	50	.20	50	.68	49	.57	47	.60	51	.43					
T10				*High*	23	.86	56[a]	.45	51	.50	52	−.35	50	.52	51	.29	High				
T11			*High*		27	.83	52	−.52	53	.45	54	.63	57	.73	50	.40				High	
T12		*High*		*High*	23	.84	49	.41	50	.41	49	.61	51	.42	52	.22			High		
T13	High	*High*		*High*	8	.85	55	.35	54	−1.13	59	−.27	54	.67	60	.87	High		High	High	*High*
T14	*High*	High			22	.86	57	.55	54	.62	57[a]	.50	58	.70	56[a]	.54			High	High	High
T15	*High*	High	High	High	14	.85	56	.46	60	.56	60	.73	60	.35	62	.91		*High*	*High*	*High*	*High*

Notes: Underlined values are significant at $p < .001$ except those with superscript a at $p < .01$. Descriptive term of a predicted variable is absent if neither $\bar{Z}$ nor H is significant at $< .01$ or $.001$. The terms high and low are italicized if the score deviates from 50 by one sigma or more.

individual's score on a particular ability is indistinguishable from a mean of 50, that is, within $\pm \frac{1}{2}$ standard deviation from the mean.

The 15 O-types are the same as those given in the previous chapter, except that in the use of the computer component OTYPE the iteration procedure of the program executed one more trial for the typology of Table 10.2. The frequencies of cases in the 15 O-types, listed in the column headed Frequency, are about the same, and so are the average homo-geneity coefficients on the predictor variables, in the column headed $\bar{H}$. In addition to the 282 children in these 15 types, there were 19 "unique" children, excluded from O-types because their patterns of scores on V, S, F, and M did not fit satisfactorily the profiles of the 15.

Predicted mathematical abilities
of the 15 V, S, F, and M O-types

The pattern of mathematical abilities of the O-types can be seen in the last five columns to the right under the general heading Significant. O-type T1, for example, is in the low class on four of the five mathematical abilities, earning low mean standard scores in four of them. The descriptive terms of both predictor and predicted abilities of this group of 13 children reveal it to be rather generally low. The numbers in Table 10.2 under the columns titled Level and Homogeneity show in detail, for each of the five mathematical abilities, the mean standard scores $\bar{Z}$ and their associated homo-geneities H. It is from these values that the descriptive terms of the column Significant originate (see the notes to Table 10.2).

Space limitations do not permit discussion of the mathematical abilities of each of the 15 predictor O-types, but specialists in human abilities may find a detailed study of the results in Table 10.2 of interest. One overall conclusion is apparent, namely, that the number of predictor basic abilities in which a given O-type is extreme is associated with the number of mathematical abilities in which it is extreme: both O-types T1 and T15 are extreme in the four predictor abilities and in the four mathematical abilities; O-type T5 is extreme in three predictors and three mathematical abilities; O-types extreme in two predictors generally tend to be extreme in about two mathematical abilities; and in the large across-the-board average group, T9, the 38 children are average in all abilities.

Level of confidence in predicted abilities of O-types

How confident can we be that the predicted level and homogeneity of the O-types are significant values, i.e., could not arise by chance? For example, with respect to the top O-type in Table 10.2, T1 is a selection of 13 children whose score in the first mathematics test, deduction, is 42, almost 1 standard deviation below the mean of 50 in the full supply of 301 children. What is the probability of recovering such a mean value in a sample of 13 scores if we randomly drew 13 from the full supply of scores on the 301 children? The component 4CAST in the BC TRY System computes such a probability by drawing a large number of samples, one at a time, from the full supply. From these samples it computes the relative frequency (probability) of earning mean scores as low as, or lower than, the observed score. When 4CAST was actually applied to the full supply of 301 deduction scores and drew a very large number of samples of 13 from it, computing the mean deduction score of each sample, the program found from the full array of means that the number of samples having a mean of 42 or less was less than 1 in 1,000, that is, the observed mean is significant at the .001 level. Actually, 3,000 such drawings were made (see below), so that we know this result is quite firm.

In Table 10.2 each mean significant at the .001 level is shown by underlining its value; means significant at the .01 level carry a superscript a. In the Significant columns at far right, we have entered the descriptive term high or low only for those cases of O-types whose mean mathematical ability is significant either at the .01 or the .001 level.

Which of the O-types have significant, distinctive mathematical abilities? The answer, of course, depends upon the level of confidence one accepts as significant. In Table 10.3 we have listed the number of O-types shown in Table 10.2 as significant at the .001 level, at the .01 (but not .001) level, and at the .05 (but not .01 level). If we accept .05 in the bottom sector as significant, then the number of O-types that have distinctive mathematical abilities at least at this level are shown in the bottom row headed Total; 38 of the total number of 75 predictions are significant.

This result means, of course, that for about half of the O-types no predictions can be confidently made about their mathematical abilities. There is something about the nature of the mathematical abilities of such types that appears to be unrelated to their standing in the four basic abilities. The very best and strongest cases for prediction are those listed in the top sector of Table 10.3, where there seems to be no question that the mathematical capabilities of the type are very distinctive and predictable. These are the types whose mathematical achievements are clearly distinctive at the .001 level and whose histograms in the five abilities are graphed in Fig. 10.2. That figure therefore presents the best case for prediction in the Holzinger problem.

To this point we have emphasized only the cases where distinctive

TABLE 10.3   BY CONFIDENCE LEVEL, THE O–TYPES
THAT PREDICT MATHEMATICAL ABILITIES IN THE
HOLZINGER PROBLEM

Confidence Level	Mathematical Abilities				
	Ded	Puz	Rsn	Ser	Ari
At $p < .001$					
	T1	T4	T3	T5	T13
	T14	T5	T13	T11	T15
		T15	T15	T14	
				T15	
At $p < .01$					
	T8	T1	T1	T1	T2
	T10		T2		T3
			T4		T5
			T14		T14
At $p < .05$					
	T2	T2	T8	T2	T1
	T7	T14	T11	T4	
	T15			T7	
				T9	
$Total$	7	6	9	9	7

$$\Sigma = 38\ (51\%)$$

mathematical abilities are predictable, i.e., those types whose mean mathematical abilities deviate significantly from the mean of 50. Prediction, however, is not only a matter of finding instances of those who deviate from
the average. It is as important to know that individuals are at the average
as it is to know that they deviate from it, types whose mean value is *not*
significantly different from the average but whose homogeneity at the
average is extremely high. These would be O-types in Table 10.2 with non-
significant means but with significant H values. There are no such O-types
in the Holzinger problem, but we shall later find many instances of this
type of prediction both in the MMPI and in the social-area problem. The
ambiguous cases are those without descriptive terms under Significant
in Table 10.2. For example, type T1 shows no term under Arithmetic, signifying that the mean arithmetic score of these 13 children does not deviate
significantly from 50 or have a homogeneity coefficient significantly different from .00. This result means that we cannot confidently say anything
about these children with respect to their arithmetic ability, either that
they are below average or at the average in arithmetic ability. In this
Holzinger problem there are many blank spaces in the far right columns,

FACS01 (N20)

OTYPE 07

N = 0020

Axis values: 105 90 75 60 45 30 15

Bottom axis: -2.5   -2.0   -1.5   -1.0   -0.5   0.   0.5   1.0   1.5   2.0   2.5

MEAN SCORE: 44.47**** 45.56**** 46.66**** 47.75**** 48.84**** 49.94**** 51.03**** 52.12**** 53.22**** 54.31**** 55.40

FREQ.:
1   11 1111223123322334353635364645557765775657754436545555444332533324221222211111112   1   2
5122327744359459524440830356621850238077700711246320025887377203892408838242174310970026088616964 70

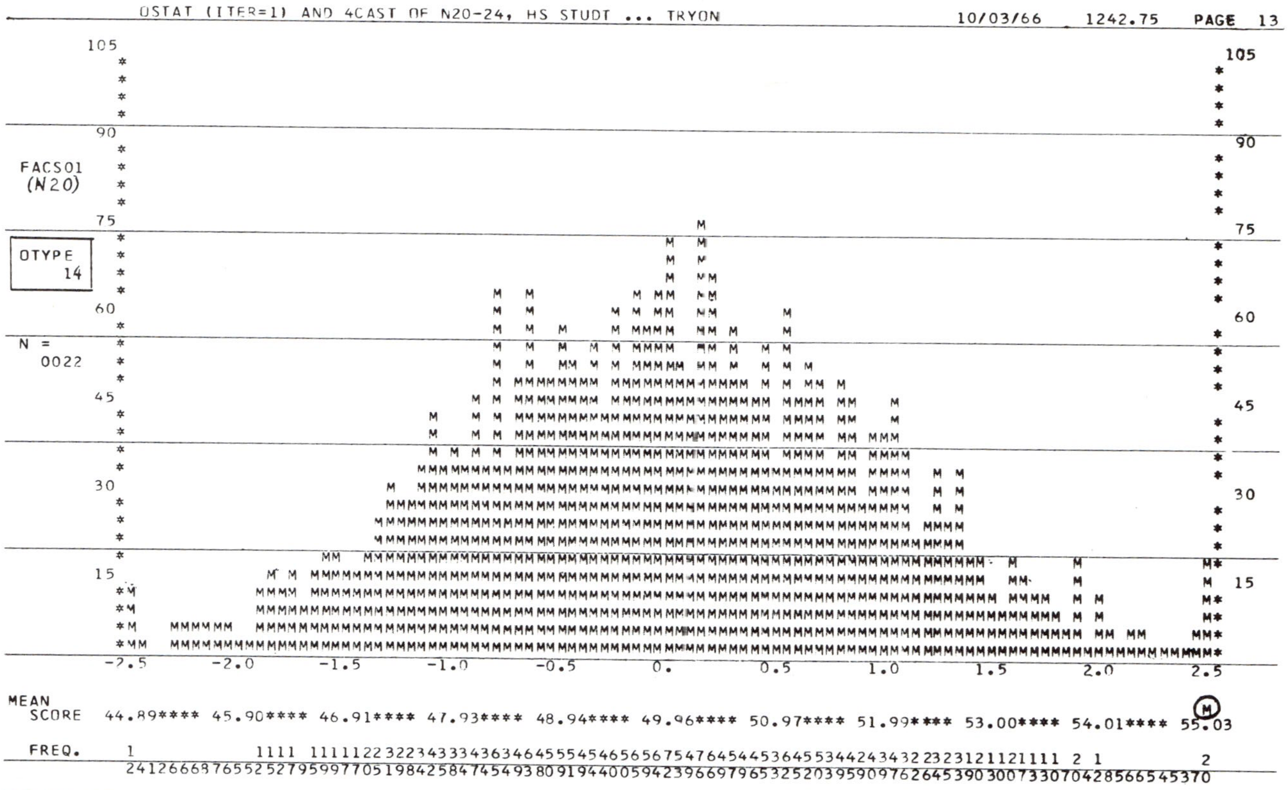

```
105                                                                        105
      *                                                                    *
      *                                                                    *
      *                                                                    *
      *                                                                    *
 90   *                                                                90  *
FACS01 *                                                                   *
(N20)  *                                                                   *
      *                                                                    *
 75   *                                                             M  75  *
+--------+ *                                                           *
| OTYPE  | *                                                           *
|  14    | *                                                           *
+--------+ *                                                           *
 60   *                                                      M      60  *
N =   *                                                      M          *
0022  *                                                                 *
 45   *                                             M                45 *
      *                                             MM                  *
      *                                             MM                  *
      *                                             MMM                 *
 30   *                                   M M     M M    M M        30 *
      *                                   M M     M M    M M           *
      *                                   MMM     MMM    MMM           *
      *                                                                *
 15   M M      MM            M        M   M M          M  M     M   15 *
      ** M     MMM          MMM      MMM  MMM         MMMM   MM      *
      ** M     MMM          MMM      MMM  MMM         MMMM   MM      *
      ** M     MMM          MMM      MMM  MMM         MMMM   MM MMM  *

     -2.0   -1.5   -1.0   -0.5    0.    0.5    1.0    1.5    2.0    2.5

MEAN
SCORE 44.89**** 45.90**** 46.91**** 47.93**** 48.94**** 49.96**** 50.97**** 51.99**** 53.00**** 54.01**** 55.03

FREQ.    1    1111 1111122322343343636465554654556576547645453645534424343223121121121111  2 1  2
         2412666876552527959977051984258474549380919440059423966979653252203590976266539030073070428566545370
```

FIGURE 10.2
Placement of the observed mean in a distribution of means of 3,000 random samples drawn from the supply of 301 cases by program 4CAST for each of two V,S,F,M O-types in the Holzinger problem.

a fact which supports the contention that, generally speaking, predicta-bility of mathematical capabilities from the four basic abilities is rather poor.

Determination of confidence level

Determining the level of confidence in O-type prediction is so critically important that the procedure should be described. The usual means of estimating significance in prediction problems is to resort to formulas that estimate "population values" in the predicted scores. These estimates usually require a belief that sample values are distributed normally. If nonparametric estimations are made, the error of prediction is likely to be so large that for the small numbers of cases found in many O-types it would be difficult to establish significance. Thanks to the computer, we are no longer hemmed in by such crude estimation tests. The BC TRY component 4CAST actually draws random samples from the full supply and shows quite clearly, unbound by any assumptions about normality, the probability of recovering either mean values or homogeneities above and below those actually observed in the particular O-type. Illustrations of how 4CAST works are given in Fig. 10.2. In the top figure is the distribution of the means of 3,000 samplings of 20 cases each from the full supply of 301 deduction scores. This computation was made to discover the chance probability of recovering for O-type T7 a mean value equal to or less than its observed mean of 45. The frequency histogram of these 3,000 means is printed in the top section of the figure. The abscissal scale is the stand-ard score of each mean from the mean of all means. Below the standard score scale is the raw score scale of the means. Thus, the distribution is a pictorialization of the "standard deviation of the mean." It is a real dis-tribution, not a fictitious one. The standard deviation of the distribution of means is, of course, the deviation of a mean at 1 standard deviation from the mean of all means, having in this example the value of

$$52.12 - 49.94 = 2.18$$

The important matter here is where the observed mean of O-type T7 is located in this distribution. Its value, 45.42, is plotted by the program as an M encircled just above the Mean Score Scale at the left. 4CAST thus presents visually and prints in a table (not reproduced here) the propor-tion of means below the observed mean, which in this case turns out to be .014 of all 3,000 sample means. This is a probability less than .05 but not less than .01. Therefore, type T7 is placed in Table 10.3 in the bottom category of .05 under Deduction.

The bottom graph in Fig. 10.2 shows pictorially the sampling means for T14, that is, the probability by random sampling of recovering a mean deduction score greater than T14's observed mean of 55.63. It can be seen that the observed mean, the encircled M at the far right on the mean score scale, lies in the category 2.5 standard deviations or greater. The tabled value in 4CAST shows that this observed mean is way off the scale of sample means. There are *no* values among the 3,000 means that are higher than 55.63. We therefore know that the probability of T14 earning its observed mean by random sampling is $p < .001$. For this reason, type T14 is placed in Table 10.3 in the top confidence level of $p < .001$.

One of the important properties of 4CAST is that no assumptions are made about the shape of the distribution or about the existence of an imaginary "infinitely large population" from which the O-types are drawn. Such assumptions are, in fact, sometimes unrealistic in biological and social data.

Camparison of univariate, multivariate, and O-type prediction

We are now in a position to compare the relative effectiveness of univariate, multivariate, and O-type predictions in the Holzinger problem. The comparisons can be made for each of the five predicted mathematical abilities, as shown in Table 10.4. In rows 1 and 2 is the best prediction in the univariate case. For example, under Deduction the highest correlation of the

TABLE 10.4 INCREASE OF PREDICTION IN PROGRESSING FROM UNIVARIATE TO MULTIVARIATE TO O-TYPE PREDICTION IN THE HOLZINGER PROBLEM

	Ded	Puz	Rsn	Ser	Ari
Univariate prediction:					
1. Best prediction from any of the 19 individual variables of V, S, F, and M	.38 from V9	.37 from M16	.51 from V9	.48 from F1	.44 from V9
2. Best prediction from a single composite, V, S, F, or M	.50 from V	.50 from S	.49 from V	.53 from F	.54 from S
Multivariate prediction:					
3. Linear multiple prediction from all four composites, V, S, F, and M	.59	.60	.57	.63	.64
O-type prediction:					
4. Best prediction from O-types: the three highest H values of O-types	.81 .78 .72	.78 .72 .68	.96 .82 .74	.83 .80 .73	.91 .87 .86

scores of the children on the deduction test is with the scores on the vocabulary test V9. The correlation with this best individual variable is .38. The next highest level of prediction is from single cluster-defined composites, either V or S or F or M. In the case of the deduction test, the highest correlation is .50 with the verbal composite, V. The next highest is linear multivariate prediction from all four composites; the multiple correlation of deduction scores with predicted scores of V, S, F, and M is .59.

But multiple correlation does not represent the highest possible prediction. For some of the O-types the prediction is a great deal better than that signified by a multiple correlation of .59. In the bottom part of Table 10.4 are listed H values for the three O-types that have the highest H values given in Table 10.2. For the deduction test the three highest are of O-types T2, T7, and T8, these being .81, .78, and .72, respectively. Six of the 15 O-types have H values greater than the multiple correlation value of .59. (We show below that the H values are directly comparable to the values of correlation coefficients.)

The fact that about half of the O-types have better and half poorer predictions than that indicated by the multiple correlation is not a matter of chance. The reason is that the value of the multiple correlation is an average of the H values, and, therefore, about half of the O-types will do a better and half a poorer job of prediction than that revealed by correlation.

Here is the relationship between correlation and H. First, express the correlation between any two variables by the more general form, namely, the curvilinear correlation or correlation ratio η

$$\eta_{yx}^2 = \sum_{i=1}^{m} p_i H_i^2$$

where y is the predicted variable and x the predictor value, i runs from 1 to m (the total number of classes, or O-types), and p_i is the percent of cases in the ith class (or type). What this all means is that η^2 is an average of the squared H values, each H^2 being weighted by the proportion of cases in a type. The relationship between η and r is, of course, well known, namely, when the two variables are continuous and their relationship linear, $\eta = r$.

Thus, the power of O-type prediction inheres in the ability to separate the types of individuals for which the prediction of a criterion is best from the types for which the prediction is worst. Handling the problem of prediction *only* in terms of multiple correlation and regression deprives us of the information on which that correlation is based. O-type prediction tells where to look in the group for the good and the poor predictors.

Prediction from individual differences in four MMPI item-clusters

We turn now to the use of cluster analysis in the prediction of individual differences from scores on such a "personality test" as the MMPI.

The question in this section is this: From certain "basic" scores on the MMPI, how well can one predict other attributes of the individuals? The predictor variables and the attributes predicted from them have been described in a previous chapter and will be summarized here only briefly in order to provide an adequate picture of the prediction design. The individuals in this study were 310 subjects: 70 psychotics, 150 anxiety cases, and 90 Armed Services officers matched with the psychiatric cases for age and education. Applying the BC TRY procedures called BIGNV analysis (see Chap. 11), which permit a key-cluster analysis in problems with up to 5,000 variables, the 566 items of the MMPI are found to have all of their generality accounted for by seven dimensions. The first three of these dimensions account for a large percent of the communalities of the items. A fourth dimension Is accepted in order to account for as much of the additional variation as possible.

The scores of the subjects on these four dimensions, symbolized as I, B, S, and T, provide the four-dimensional cluster score space in which the typological structure of the 310 subjects is determined. There result 14 types of individuals whose nature and structure have already been described. The prediction design of this analysis aims to reveal the degree to which scores on the four "basic" MMPI variables, I, B, S, and T, predict scores on the other three scaled attributes, D, R, and A that were not involved directly in the four-dimensional solution. We finally compare univariate, multivariate, and O-type prediction in this problem.

The predictor item-clusters I, B, S, and T

The MMPI item-cluster scales that were the predictors in this study may be briefly summarized as follows:

I, introversion: 26 items with a score reliability of .93
B, body symptoms: 33 items with a score reliability of .92
S, suspicion and mistrust: 25 items with a reliability of .85
T, tension, worry, and fears: 36 items with a reliability of .92

The meaning of each scale inferred from item content is sharply clear. There is no item overlap between any of the scales. The reliability coefficients of total scores on the scales, being about .90, are much higher than those usually reported for other MMPI scales.

The predicted item-clusters D, R, and A

The three predicted item-cluster scales are as follows:

D, depression and apathy: 28 items with score reliability of .94
R, resentment and aggression: 21 items with score reliability of .87
A, autism and disruptive thoughts: 23 items with score reliability of .86

The meaning of these scales, inferred from item content, is also quite evident. None of the scales have item overlaps on the others or upon the predictor scales, and total score reliabilities are also about .90.

Multiple linear prediction of D, R, and A

The MMPI prediction problem can be restated. If we compute only the scores of the subjects on the four basic item-clusters, I, B, S, and T, can we predict the subject's scores on D, R, A from I, B, S, T, and thus avoid computing scores on D, R, A?

The answer, in terms of traditional linear multiple correlation, is as follows:

	Depression	Resentment	Autism
Multiple correlation with I, B, S, T	.84	.75	.73

These multiple correlations are rather high as multiple correlations go, but they are certainly not 1.00, signifying that some information about D, R, A is lost. And this lost information is to some extent reliable information, because the reliability coefficients both of the predictor and the predicted variables are of the order of approximately .90. Nevertheless, prediction in this MMPI problem is better than in the Holzinger problem, where, it may be recalled, the multiple correlation of the predicted variables with their predictors was of the order of .60.

Differential prediction of D, R, and A
from person-clusters (O-types)

Before going into details, we look at a pictorial representation of O-type prediction in this MMPI problem. Figure 10.3 shows the histograms of

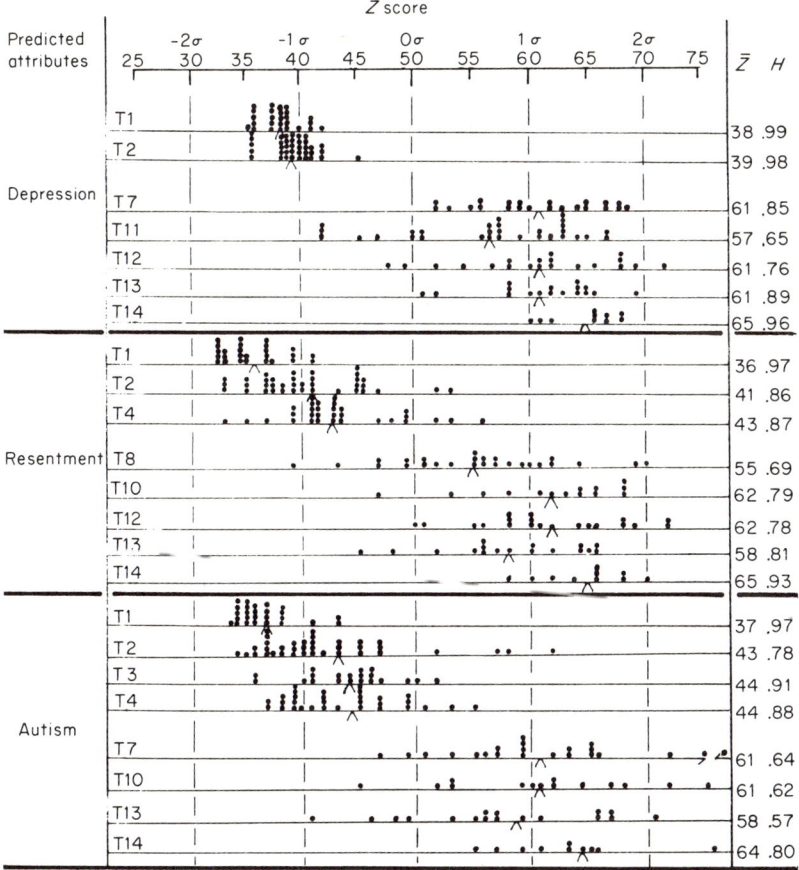

FIGURE 10.3
Distribution of standard scores on depression, resentment, and autism of members of those MMPI I,B,S,T O-types whose predictions are significant beond the .001 level.

standard scroes on the depression, resentment, and autism clusters of those MMPI O-types whose predictions are significant beyond the $p < .001$ level. There is a very high level of prediction in this MMPI problem compared with that of the Holzinger intellectual abilities depicted earlier in Fig. 10.1. For example, types T1 and T2 are both below 1 standard deviation in their depression scores and are visibly quite homogeneous, and metrically so, as indicated by their high homogeneity coefficients, .99 and .98. Plotted below these two types are five others, T7, T11, T12, T13, T14, with high depression scores, four of them having mean standard scores at least 1 standard deviation above the mean of 50. In short, the two lows vs. the five highs are separated from each other by 2 standard deviations. A similar structure is repeated both for the predicted resentment variable in the middle sector of Fig. 10.3 and for the autism variable in the bottom sector.

Certain general conclusions can be drawn from these histograms: (1) It is obvious that prediction of low scores on D, R, A is much more accurate than of high. (2) For some of the O-types, the prediction of these three attributes is considerably higher than that revealed by linear multiple correlation. For example, in the depression group, five of the seven have H values greater than the multiple correlation of .84; in the resentment group seven of the eight have H values in excess of their multiple correlation of .75; and in the autism group five of the eight have H values that exceed the multiple correlation of .73.

The predictor typology

In Table 10.5 the 14 predictor MMPI O-types are listed in the columns to the left, each defined by the descriptive terms that indicate the degree to which it is distinctive relative to the average performance of individuals on the four predictor variables, I, B, S, and T. For example, O-type T1 consists of 26 individuals; it is *low* in all four attributes. At the opposite extreme in the bottom row, T14 consists of 10 individuals that are *high* on the four predictor attributes. In general, types T1 to T5 are *low* in at least one of the four predictor attributes. Low scores on these four item-clusters may be thought to represent "positive mental health." These five types belong to this category. Type T6 has average scores across the board. But from T7 down to T14 are types that are high in one or more of the predictors, i.e., these eight types suffer from one or more "symptoms" of emotional disorder.

In the predictor attributes (the column headed $\bar{H}$ in Table 10.5), the $\bar{H}$ values are all of the order about .90 or higher; not a particularly surprising fact since, of course, they are selected for their homogeneity on these attributes. These findings on the typology of the MMPI are substantially what was presented in the previous chapter. The typology shown in Table 10.5 is a little sharper than that given in the previous chapter, being based on one additional iteration for typological structure by the OTYPE computer component.

Predicted D, R, A attributes from the MMPI O-types

At the right in Table 10.5 are the predicted D, R, A attributes of the 14 O-types. Comparing the descriptive terms in the last three columns with the words in the left column describing the status of each O-type in its predictor attributes gives a simple, clear picture of the predictor-predicted

TABLE 10.5 PREDICTED ATTRIBUTES FROM IBST O-TYPES IN THE MMPI PROBLEM

Type	Predictor Attributes				Frequency	$\bar{H}$	Predicted Attributes — Level and Homogeneity						Predicted Attributes — Significant		
	I	B	S	T			D $\bar{Z}$	D H	R $\bar{Z}$	R H	A $\bar{Z}$	A H	D	R	A
T1	Low	Low		Low	26	.97	38	.99	36	.97	37	.97	Low	Low	Low
T2	Low	Low	Low	Low	37	.95	39	.98	41	.86	43	.78	Low	Low	Low
T3		Low		Low	20	.91	42[a]	.94	44[a]	.88	44	.91	Low	Low	Low
T4			Low		31	.86	45	.82	43	.87	44	.88	Av	Low	Low
T5	Low				17	.91	46	.82	48	.71	49	.68	Av	Av	Av
T6					24	.91	51	.73	51	.76	49	.86	Av	High	High
T7		High	High	High	22	.89	61	.85	56[a]	.64	61	.64	High	High	
T8					26	.90	50	.71	55	.69	54	.65	Av	Av	
T9		High	High		16	.83	47	.82	50	.85	50	.49	Av	Av	
T10	High	High	High	High	14	.87	56[a]	.77	62	.79	61	.62	High	High	High
T11	High				30	.87	57	.65	53	.71	50	.68	High	Av	Av
T12	High	High	High	High	20	.90	61	.76	62	.78	59[a]	.68	High	High	High
T13	High	High		High	17	.90	61	.89	58	.81	58	.57	High	High	High
T14	High	High	High	High	10	.92	65	.96	65	.93	64	.80	High	High	High

Notes: Underlined values are significant at $p < .001$ except those with superscript a at $p < .01$. Descriptive term of a predicted variable is absent if neither $\bar{Z}$ nor H is significant at $<.01$ or $.001$. The terms high and low are italicized if the score deviates from 50 by one sigma or more.

pattern of each O-type. For example, type T1, being low, or "healthy," in all four predictor attributes, is low, or "healthy," in the three predicted attributes of depression, resentment, and autism. The reverse situation is the case with the "sick" group, T14, at the bottom of the table.

The value of this chart in the clinical use of the MMPI should not be underestimated. Once an individual has been located in his typological group by the methods described earlier, a "total" picture of the pattern of his attributes on the MMPI can be achieved by observing the status of his O-type both in the predictor attributes to the left of the table and the predicted attributes to the right. For example, if a given patient falls into type T11, one can say that he is distinctively highly introverted, with a good probability of being highly depressed, though in the other particulars of resentment and autism he is liable to have average status.

As stated earlier, it is as important to know that an individual is *average* in a predicted attribute as it is to know where he distinctively differs from the average. Take the case of type T9, for example. The 16 individuals of this type are preoccupied with body symptoms, confessing to a large number of disturbed body functions. But their predicted status in D, R, and A is quite average in these three particulars, as shown by the values under D, R, A in the columns headed Level and Homogeneity. They are *at* the average in D and R because their homogeneities are of the order .82 to .85, significant at $p < .001$. With respect to the autism dimension, however, nothing can be said about the status of these individuals, for their autism score is not significant. Ten of the predictions are at an average status in the predicted attributes, as indicated by the abbreviation Av. That is, in one-fourth of the O-type predictions the individuals are average.

Finally, the degree of predictability of the O-types in comparison to linear multiple correlation is reported in the columns of H values under Predicted Attributes. For example, in predicting the depression scores of individuals in the 14 O-types, comparing their H values with the multiple correlation between depression and the four predictor variables, shown below to be of the order .84, one finds that six of the O-types have homogeneities above this value, some near 1.00. For *them*, depression scores can be predicted to a higher degree of accuracy than that indicated by a multiple correlation of .84.

Comparison of univariate, multivariate and O-type predictions

The increase in predictability in going from univariate through multivariate to O-type prediction is even more sharply evident in the MMPI problem than in the Holzinger problem, as shown in Table 10.6. Take first the case

TABLE 10.6 INCREASE OF PREDICTION IN PROGRESSING FROM UNIVARIATE TO MULTIVARIATE TO O–TYPE PREDICTION IN THE MMPI PROBLEM

	D	R	A
Univariate prediction:			
1. Best prediction from any of the best individual items of I, B, S, or T	.64 from 431[a]	.59 from 555[b]	.50 from 431[a]
2. Best prediction from a single composite, I, B, S, or T	.78 from T	.69 from T	.66 from T
Multivariate prediction:			
3. Linear multiple prediction from four composites, I, B, S, T	.84	.75	.73
O-type prediction:			
4. Best prediction from O-types: the three highest H values of O-types	.99 .98 .96	.97 .93 .88	.97 .91 .88

[a] Worry over possible misfortunes.
[b] Feel about to go to pieces.

of predicting depression scores of individuals. Under univariate prediction one would certainly not make any serious effort to predict an individual's depression score from any of the individual items on these MMPI scales, for even the best item in all four scales correlates no higher than .64 with the depression scores. Prediction is improved a bit if one takes a composite of items, either I, B, S, or T. The T, or tension, score correlation with the depression score is .78, for example. Multiple prediction from all four scores on I, B, S, and T is a little better; the multiple correlation is .84. But more accurate prediction can be made for *some* O-types isolated by typological techniques.

Conclusions on individual prediction

What are the main conclusions from this systematic study of prediction in the two areas of intellectual ability and the MMPI type of self-report? First, it is evident that as one moves from the very specific responses of individuals in situations, such as to a particular item or stimulus pattern, through increasing composites of his performance across a variety of situations, then prediction to other "outside" attributes improves. This fact can be understood on domain sampling principles, because the more extensive the compositing of the predictor behaviors or the more multivariate the design of the battery of predictors, the greater the likelihood of sampling determinants of individual differences that would operate in other "outside" situations. This finding is of little encouragement to those psychologists who seek to find highly controlled specific experimental situations in which to get a "pure" measure of a given behavior from which to make predictions. Behavior in highly controlled settings is likely to elicit only the most specific kinds of determinants that are unlikely to be elicited in other "outside" situations.

The second major conclusion is that the predictability of an individual's behavior depends upon who the individual is. Domain sampling works out, as it were, differently in different individuals. For some individuals, knowing their pattern of basic intellectual abilities may enable one to make accurate predictions of other abilities or cognitive aspects of their behavior. With other individuals, however, complete knowledge of their basic abilities may be of no earthly use in predicting how the individual will behave in other kinds of situations. Differential O-type prediction is a fact the behavioral scientist must live with. It is unlikely that he will discover "laws" of prediction that will apply to everybody.

Predicting group differences: the social-area problem

This section differs from earlier sections in this chapter by focusing on objects that are *groups* of individuals rather than single individuals. The objects in this case are neighborhoods (census tracts) of individuals in a metropolitan area consisting of the population of San Francisco and the East Bay communities.

Knowing the prewar 1940 scores of a given neighborhood on its three basic demographic FAS dimensions, to what degree can one predict other characteristics of it, e.g., the same FAS characteristics observed a decade later in 1950 or such wholly different attributes as the subjective attitudes of the people in the neighborhood as revealed by their voting behavior? If the forecast is discovered to be accurate, it leads to a second question. May there not be a more basic set of dimensions of social structure than those defined by demographic features, one representing determinants of both the demographic and the attitudinal characteristics of the people, a set enduring over a long period of time?

The general design of this analysis is (1) to determine the FAS predictor values of each of the 243 neighborhoods of the Bay Area in 1940,

before World War II. (2) Predictions are made from them to other concurrent demographic and attitudinal features of the neighborhoods. (3) Postwar characteristics observed up to 15 years later are predicted, spanning a
socially disrupting war and postwar period, at the end of which it is doubt-
ful that many individuals who originally lived in the neighborhood still lived
there. Do neighborhoods remain ecologically constant despite the turnover
of specific inhabitants?

The predictor prewar dimensions, F40, A40, S40

The 33 demographic characteristics of the 243 neighborhoods of the Bay
Area have a cluster structure already described in previous chapters. In
order to distinguish the 1940 variable clusters from the 1950 variable
clusters we denote the respective clusters F40, A40, S40, for 1940 and F50,
A50, S50, for 1950.

Concurrent prediction of social achievement
(Ach40) and vote for Roosevelt (P40)

To discover whether the basic three FAS dimensions predict scores on
other demographic characteristics concurrently observed in 1940, a subset
of six variables not included as definers of the three basic FAS dimensions
was selected as another independently observed demographic cluster.
This achievement cluster, Table 10.7, Sec. A, labeled Ach40, is defined by
six characteristics that generally stand for white-collar middle-class
achievement in American metropolitan society.

How well do the three dimensions predict an utterly different domain
of behavior, namely, the voting attitudes of individuals in these neighborhoods? To find out, P40, the vote for Roosevelt, was computed from the
election tabulations. It seems clear that such a vote in a neighborhood in
1940 reflected the people's deep-rooted beliefs and biases about the role
of the government in their personal lives.

Forecasting postwar demographic
and voting attitudes

A socially disrupting great war uprooted the people from 1940 to 1945. The
lives of most people were deeply affected. It was not uncommon for persons, even families, to move from one place to another. The physical
neighborhoods as defined by the census did not change, but the bodies in

TABLE 10.7 MULTIVARIATE PREDICTION OF
CONCURRENT AND POSTWAR DEMOGRAPHIC AND
VOTING ATTRIBUTES FROM PREWAR FAS SCORES
IN THE SOCIAL–AREA PROBLEM

A. Concurrent Prediction
(1940, N = 243 Tracts of Bay Area)

	Reliability	Multiple Correlation with FAS40
Ach40, achievement:	.96	.92
Wm		
Sc		
Ch		
Em		
Rn		
Re		
P40, political 40:	.88	.93
Vote for Roosevelt, 1940		

B. Postwar Prediction
(1947, 1950, 1954; N = 105 Tracts in San Francisco)

	Reliability	Multiple Correlation
F50, family life[a]	.94[b]	.98
A50, assimilation[a]	.85[b]	.95
S50, socioeconomic[a]	.92[b]	.96
P47, political 47		
Vote for mayor, 1947	.93[c]	.90
P54, political 54:	.94	.85
NdAg		
HsEm		
DmGv		
ExFl		
T54, taxation 54:	.88	.82
HsBn		
AgBn		
ExBn		
VtBn		
E54, ethnic-religious 54:	.87	.66
WlEx		
ChEx		
CoEx		

[a] For definers see Table 9.8.
[b] Reliabilities are for 225 tracts of the Bay Area.
[c] Reliability estimate is highest correlation with any other variable.

them frequently did. But did the changing people in these neighborhoods also change in their demographic and attitudinal characteristics? We can find out by predicting scores of these neighborhoods on the FAS dimensions as observed a decade later in 1950 and on their postwar voting in elections.

The important postwar predicted variables used in this study are shown in Table 10.7, Sec. B. The FAS scores in 1950 are based on the same defining variables used in 1940. In 1947 scores on the political dimension P47 were calculated from the vote for mayor in San Francisco. An analogous political dimension score, P54, was determined in the 1954 election and two other voting-attitude cluster scores were also secured, namely, T54, taxation, and E54, ethnic-religious.

The 1954 voting-attitude cluster scores were calculated only for the neighborhoods of San Francisco. There were 31 state and city propositions and candidates on the ballot. The results of the analysis of these data were reported in Chap. 7. The voting-attitude clusters are listed in Table 10.7, Sec. B.

Multiple linear prediction of demographic and attitudinal attributes

As in the prediction study of Holzinger mental abilities and of the MMPI item-clusters, the first prediction performed is by the standard linear multiple regression method. The results take the form of multiple correlations between the three predictor FAS scores in 1940 and the two concurrent and the seven postwar attributes. These multiple correlations are given in Table 10.7, extreme right column, .92 for Ach40, and .93 for P40. These are high multiple correlations, meaning that if one knew only what a neighborhood's three FAS40 scores were, then by properly weighting those scores by regression coefficients of the predicted variables on the three predictors, one could forecast scores on the achievement variable and the vote for Roosevelt variable with great accuracy.

Forecasting postwar from prewar characteristics is also astonishingly high. The results, in terms of multiple correlation, are in Sec. B of Table 10.7. How well can one forecast a neighborhood's postwar demographic characteristics in 1950 from its prewar FAS values? Almost perfectly for F50, family life, since the multiple correlation with F50 is .98! And the multiple correlations with the other two demographic characteristics, A50, and S50, are almost as high. But the voting attitudes of a neighborhood after the war are also highly predictable from its prewar demography. The multiple correlations between FAS40 and the political attitude vote in 1947 and 14 years later in 1954 are of the high order of .90 and .85. There is a decrease in forecasting accuracy as one moves from demographic to attitudinal features, especially of the ethnic-religious attitude, whose multiple correlation with FAS40 is only .66.

Differential prediction from neighborhood O-types (social areas)

Since multiple correlation is a statement of average prediction across all individual objects, we now turn to the more powerful kind of prediction, the differential prediction for different social-area types of neighborhoods. The basic histograms are given in Fig. 10.4, which shows how three social-area types are arrayed on the three predictor and seven predicted variables. Each of the 39 neighborhoods of the whole Bay Area has scores on the three 1940 FAS clusters about the same as the others, as shown in the three top histograms. The scale across the top is the standard score scale into which scores on all of the variables have been converted. The mean standard score of all census tracts in the supply is 50, and the standard deviation is 10. For 39 neighborhoods of type T1 the mean F40 score in the very top histogram clusters around a value of 56, about $\frac{1}{2}$ a standard deviation above the mean of the whole supply. In the second histogram the A40 scores center around 43, and in the third the S40 score clusters around 41.

It is not surprising that the neighborhoods of type T1 are so homogeneous in the three predictor variables because they were selected that way. But what *is* surprising is the discovery that in most of the nine predicted variables on which they were *not* selected they are just about as homogeneous. This fact is evident in the nine histograms of the predicted variables just below the top three.

Contrast the findings on type T1 with those of the combined social-area types T10 and T12 in the bottom sector of Fig. 10.4. The neighborhoods of this type are very high and homogeneous in scores on FAS40, and they are nearly as extreme in their scores on the nine predicted variables.

It is this general correspondence of homogeneity in predicted variables of individual neighborhoods purposely selected as homogeneous in the predictor variables that is represented by the high multiple correlations (in the .90s). What is dramatic in the findings here is that the predictability is about as high in attributes observed a decade after 1940 as in those observed concurrently in 1940.

In order not to overstate the case for prediction in terms of such extremes, Fig. 10.4 includes a third neighborhood type called type T5; its histograms are in the middle sector of Fig. 10.4. At the top of the sector for

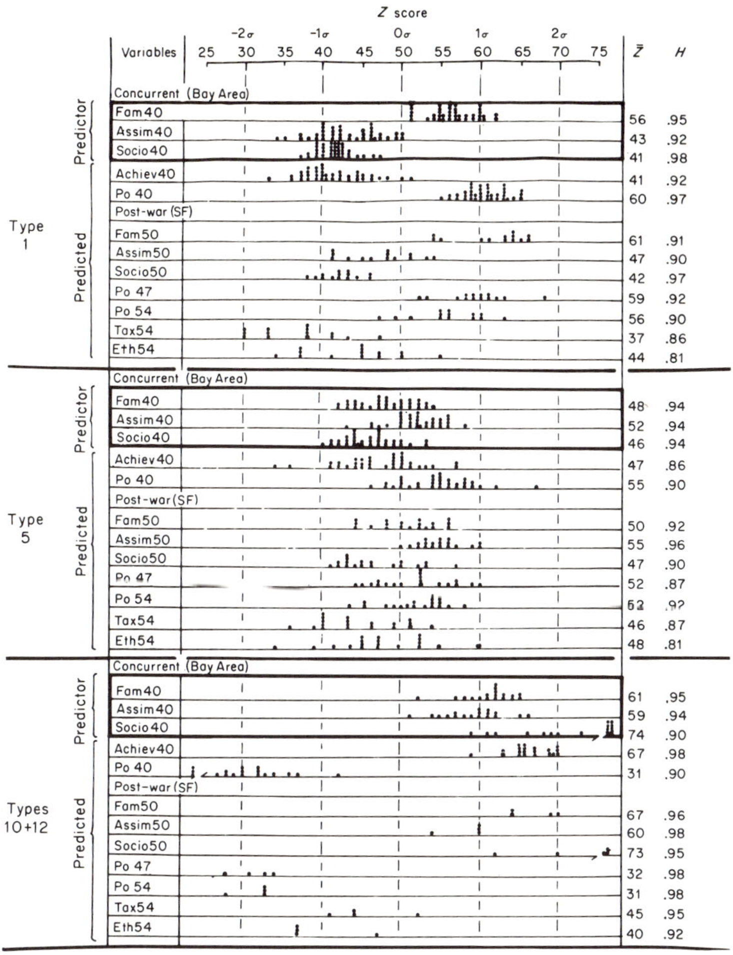

FIGURE 10.4
Distribution of standard scores on predictor and predicted variables of members of four FAS40 O-types.

type T5 are the three histograms on the 1940 predictor variables of FAS40. These 33 neighborhoods cling quite homogeneously around the middle standard score of 50. In the nine histograms of predicted variables, both concurrent and postwar, this average social-area O-type tends to have values on the predicted characteristics that are also near the average standard score of 50. This fact does not mean that the neighborhoods of type T5 behave like a random sample of neighborhoods drawn from the full supply. They behave homogeneously at the average, as revealed in the last column of numbers headed H in Fig. 10.4, showing the homogeneity coefficients of the O-type on the respective variables. If all the neighborhoods of a given type have exactly the same score, the H value is 1.00, but if the scores behave like a random selection from the full supply, H is .00. The H values for the predicted variables of type T5 are of the order .90 or higher, signifying a high degree of prediction (or low error). Indeed, the homogeneity of the neighborhoods of type T5 in the predicted variables is about as high as that of the extreme social area composed of combined types T10 and T12 at the bottom of the figure.

Predicted attributes of the 12 social areas

The complete findings on predicted characteristics, both concurrent and postwar, are given in Table 10.8. Since this table is complex, consider first the top row, for type T1. Across the row in the 1940 concurrent predicted variables, T1's high family score is accompanied by high political, the vote for Roosevelt, and its low assimilation and low socioeconomic is accompanied by low achievement. For a contrast, compare this pattern with the predictor-predicted pattern of T12 at the bottom opposite extreme. The across-the-board average social-area type is T5. When a given characteristic deviates from the mean by 1 standard deviation or more, its descriptive term is italicized in the table, as in the case of T1's *high* under P40, vote for Roosevelt.

The main findings of Table 10.8 are (1) that for all 12 types the standard score level of postwar attributes is generally predictable at a very high level of significance and (2) that the homogeneities, usually in the .90s, are nearly as high on the predicted as on the predictor attributes.

Space limitations do not permit a detailed exploration of disjunctive patterns that do *not* occur, thus disclosing the operation of social and biological forces that prevent certain combinations of demographic and attitudinal characteristics from appearing in any neighborhood. For example, where are neighborhoods that are low on the assimilation variable and high on the socioeconomic? There are none. Social discrimination in our society apparently so intimidates members of minority groups that the well-off individuals among them who can "pass" seem to avoid forming habitat groups that would identify them as a homesite body. Conversely, the opposite combination, high scores on the assimilation variable, and low on the socioeconomic variable, does occur in the conjunctive type T6. Social selection does not prevent the grouping together of poor whites who, when also characterized by high scores on the family life variable as is T6, become the respectable "poor but decent."

TABLE 10.8   PREDICTED CONCURRENT AND POSTWAR ATTRIBUTES OF 1940 DEMOGRAPHIC FAS O-TYPES

Column groups: **1940 Predictor Demographic Variable** (F40, A40, S40); **Concurrent 1940 Predicted Variable** (Frequency; Ach40, Demo40, Vote P40); **Postwar Predicted Variables (San Francisco only, 1947–1954)** — Frequency; **Demographic** (F50, A50, S50, P47); **Voting attitudes** (P54, T54, E54). Each variable cell gives $\bar{Z}$, H, and descriptive term.

1940 Types	F40	A40	S40	Freq.	Ach40	Demo40	Vote P40	Freq.	F50	A50	S50	P47	P54	T54	E54
T1	56 .95 High	43 .92 Low	41 .98 Low	39	41 .92 Low		60 .97 *High*	14	61 .91 *High*	47 .90 Av	42 .97 Low	59 .92 High	56 .90 High	37 .86 *Low*	44[a] .81 Low
T2	33 .86 *Low*	34 .77 *Low*	46 .92 Low	12	39 .55 *Low*		55 .87 High	10	40 .82 *Low*	36 .69 *Low*	42 .90 Low	61 .68 *High*	63 .78 *High*	58[a] .88 High	57 .50
T3	44 .94 Low	29 .70 *Low*	47 .85 Low	12	36 .79 *Low*		54 .57	9	48 .98 Av	30 .65 *Low*	44[a] .64 Low	58[a] .71 High	55 −.25 High	56 −.24	63 −.83 *High*
T4	35 .83 *Low*	50 .87 Low	48 .89 Low	29	49 .52		49 .67 Av	21	39 .78 *Low*	52 .76 Low	49 .81 Low	47 .63 High	51 .49	59 .79 High	55[a] .73 High
T5	48 .94 Low	52 .94	46 .94 Low	33	47 .86 Av		55[a] .90 High	19	50 .92 Av	55[a] .96 High	47 .90 Av	52 .87 Av	52 .92 Av	46 .87 Av	48 .81 Av
T6	59 .97 High	55 .97 High	45 .97 Low	25	50 .95 Av		55[a] .95 High	6	64 .99 *High*	57 .90	49 .82 Av	49 .92 Av	48 .90	41 .90	42 .93
T7	52 .94 High	59 .95 High	55 .92 High	22	57 .91 High		43 .91 Low	7	56 .92	58 .97 Av	59 .90	42 .98 Low	41[a] .96 Low	42 .92 Low	39 .87 *Low*
T8	61 .98 *High*	60 .99 *High*	50 .95 High	16	57[a] .91 High		46 .89 Av	2	63 .99+ *High*	63 .99+ *High*	58 .97	40 .95 Low	38 .99	47 .84	43 .75
T9	43 .89 Low	52 .89	62 .89 *High*	14	60 .84 *High*		40[a] .76 Low	12	46 .89 Av	55 .90	64 .64 *High*	38 .74 *Low*	38 .81 *Low*	57[a] .81 High	51 .77
T10	61 .92 *High*	58 .94 High	81 .93 *High*	9	68 .99 *High*		28 .91 *Low*	2	67[a] .97 *High*	57 .96	80 .98 *High*	31 .96 *Low*	30 .97 *Low*	48 .92	37[a] 1.00 *Low*
T11	51 .92 High	61 .99+ *High*	66 .93 *High*	4	64 .97 *High*		32 .97 *Low*	0							
T12	62 .98 *High*	61 .94 *High*	67 .86 *High*	9	65 .97 *High*		34 .88 *Low*	2	67 .95 *High*	60 .99+ *High*	66 .91	32 .99 *Low*	33[a] 1.00 *Low*	42 .99	42 .85

Notes: Underscored values are significant at $p < .001$ except those with superscript a at $p < .01$. The descriptive term of a predicted variable is absent if neither $\bar{Z}$ nor H is significant at .01 or .001. The terms high and low are italicized if the score deviates from 50 by one sigma or more.

Constancy of social areas despite extensive turnover

The best prediction is of the *same* characteristic over time. There is virtually no change in each neighborhood's distinctive demographic characteristics from 1940 to 1950, despite the fact that the people in most neighborhoods underwent much war dislocation. Here are the facts, expressed as correlations between scores on the neighborhoods' three demographic characteristics observed in 1940 and the same ones 10 years later in 1950:

Demography:

$$r_{(F40)(F50)} = .98$$
$$r_{(A40)(A50)} = .95$$
$$r_{(S40)(S50)} = .95$$

The constancy of voting attributes is nearly as great, despite the fact that the attitude objects voted on were not identical at the different times:

Attitude:

$$r_{(P40)(P47)} = .98$$
$$r_{(P40)(P54)} = .84$$
$$r_{(P47)(P54)} = .93$$

Urban social areas are not only stable but appear to persist in their biosocial characteristics unaffected by the particular persons who inhabit them. A very substantial proportion of the people living in the neighborhoods in 1940 moved out during the war decade and were replaced by a new crop. This fact is revealed in the annual percentage turnover as reported in the 1950 census. The tract statistics include, for each tract, the percent of persons *not* in the same house in 1950, compared with 1949. For the 105 neighborhoods of San Francisco the annual turnover was

Annual turnover, %	5–9	10–14	15–19	20–24	25–29	30–34
Number of tracts	1	21	39	29	10	5

The lowest turnover is about 10 percent, the highest about 33 percent. The average tract has an annual turnover rate of about 20 percent. If this rate did not sensibly change for each one of the 10 years from 1950 back to 1940, and if selection were random, a little simple arithmetic reveals that for every 100 individuals in an average neighborhood in 1940, 80 would be left in 1941, 64 in 1942, and so on up to 1950, at which time, when the census man came to enumerate them, only 11 out of the original 100 would be present. Even for those neighborhoods with the lowest rate of turnover of around, say, 10 percent, after 10 years two out of every three persons would have been replaced.

Despite this copious turnover, the characteristics of the neighborhood remained relatively unchanged. The demographic and attitudinal characteristics of social areas are, as it were, supra-individual. They appear to be invariant with respect to who moves in and moves out. To illustrate, the attitudes and values of Skid Row seem to go on forever despite the fact that during the course of a 10-year period most of the original inhabitants will have died. Only an urban renewal scheme, which bulldozes it away, can change it. An exclusive, conservative neighborhood seems to remain that way whoever moves in or out, especially in a "contained" city with fixed limits.

Comparison of univariate, multivariate, and O-type prediction

We also find here, just as in individual prediction, that as one proceeds from univariate through multivariate to O-type prediction, there is a systematic increase in the predictability of the characteristics of neighborhoods. The facts are shown in Table 10.9. The first row, devoted to univariate prediction, lists the highest correlation of each predicted variable with any one of the three single predictors, F40, A40, or S40. Thus, for P40, vote for Roosevelt, its highest correlation is $-.87$ with the S40 socioeconomic score. The second row gives linear multiple correlations with all three pre-

TABLE 10.9 INCREASE OF PREDICTION IN PROGRESSING FROM UNIVARIATE TO MULTIVARIATE TO O–TYPE PREDICTION IN THE SOCIAL–AREA PROBLEM

	Concurrent		Postwar						
	Ach40	P40	F50	A50	S50	P47	P54	T54	E54
Univariate prediction:									
1. Best prediction from		$-.87$	.98	.96	.95	$-.81$	$-.81$	$-.76$	$-.53$
a single composite,		from	from	from	from	from	from	from	from
F40, A40, S40		S40	F40	A40	S40	S40	S40	F40	F40
Multivariate prediction:									
2. Linear multiple									
correlation from									
F40, A40, S40	.92	.93	.98	.95	.96	.90	.85	.82	.66
O-type prediction:									
3. Best prediction from	.99	.97	.99	.99	.98	.99	1.00	.99	1.00
O-types: three highest	.97	.97	.99	.99	.97	.98	.99	.92	.93
H values of O-types	.97	.95	.98	.97	.97	.96	.97	.92	.87

dictors. Finally, the bottom rows of the table give the three highest H values taken from Table 10.8. Here is the best prediction of all. Prediction for neighborhoods of those types that have H values near unity is nearly perfect. But for neighborhoods of other types that have lower H values it is, of course, poorer. The power of differential prediction from knowledge of the social areas to which neighborhoods belong is that it tells one *where* prediction is nearly perfect. Multivariate prediction by multiple correlation does not yield such a discriminating forecast.

Basic tridimensionality of social areas

Summarizing, we have discovered (1) that 33 demographic characteristics of neighborhoods can be described without loss of generality by only three cluster defined composites drawn from them and (2) that 31 voting attitudes of the neighborhoods also can be accounted for by only three cluster defined attitude composites. In differentiating the neighborhoods, therefore, nothing general is lost if we ignore all the original 64 measures taken on them and in their place describe variation among them by scores on only the six clusters. (3) We have discovered that the three prewar demographic cluster scores predict attributes of neighborhoods from 8 to 15 years later, despite a great turnover of people in the intervening time. Putting these findings all together suggests that there may exist not six but only *three* basic dimensions that differentiate the neighborhoods.

The way to discover how many basic dimensions exist is to project the scores on demographic *and* attitudinal clusters into one single crucial inclusive key-cluster analysis. If the demographic three are different from the attitudinal three, then a six-dimensional solution will result. Indeed, perhaps at least eight may be required if *time* is an additional dimension, for it is possible that all measures observed in 1940 may show special correlation among themselves different from a special association among those observed 10 years later. The crucial test consists of a key-cluster analysis of the following 16 scores in the neighborhoods of San Francisco: the three FAS measures observed in 1940 and 1950; the three political attitude variables, P40, P47, P54; the other two voting dimensions, T54 and E54; and five additional variables defined especially for this inclusive analysis. The additional five were these:

> *Income*
> I49 Income (per capita) 1949
> *Turnout*
> V40 Turnout of registered voters 1940
> V47 Turnout of registered voters 1947

Movement of home (turnover of people in the neighborhood)
  M40 Moved home in 1940
  M50 Moved home in 1950

This factorial analysis of the 16 variables is called the DAT study, since it is a simultaneous dimensional analysis of Demographic, Attitudinal, and Time variables. When the matrix of correlations among the 16 DAT variables is factored by the key-cluster method, a three-dimensional solution *completely* exhausts the initial estimates of communalities of all 16 variables. To check this extraordinary finding a principal-axes solution on the matrix with 1.00 in the diagonal was run. Three principal-axes dimensions account for 94 percent of the total variance of all 16 variables. Crediting the remaining 6 percent of the total variance to unique or trivial common variance, we are compelled to accept this conclusion: *Only three dimensions account for nearly all the variance among neighborhoods in demographic and attitudinal characteristics covering a span of* 15 *years!*

What is the nature of the basic three dimensions that determine variation among neighborhoods? The answer seems clear. The structure of the relationships among the 16 variables depicted in the spherical configuration given in Fig. 10.5 gives the clue. Recall that in such a spherical display, the points represent the 16 variables and that the spatial separations between them are functions of their intercorrelations, as described earlier. The configuration of the purely demographic characteristics at or near the corners of the spherical triangle, the S's, F's, and A's, is exactly the same as shown in Fig. 9.7, but now each of the three clusters also includes the attitudinal characteristics.

We designate the three inclusive basic clusters as "conservatism," "territoriality," and "exclusiveness." Observe, first, the conservatism cluster at the lower left: there is almost perfect collinearity between the socioeconomic dimensions, S40 and S50, and the political dimensions, P40, P47, and P54. The political dimensions have been "reflected," as signified by the minus sign in front of them, meaning that they are associated with the socioeconomic variables in a reflected state. For example, a high P40 score, the vote for Roosevelt, of a neighborhood is associated with a low socioeconomic score, S40 or S50. If we were to form an inclusive composite score on each neighborhood consisting of these five variables, one with a high score on it would therefore refer to a neighborhood in a high socioeconomic independence, anti-Roosevelt condition, generally recognized as that of conservatives.

In analogous fashion, the second dimension at the right, labeled territoriality, embraces the family life characteristics, F40 and F50, coupled

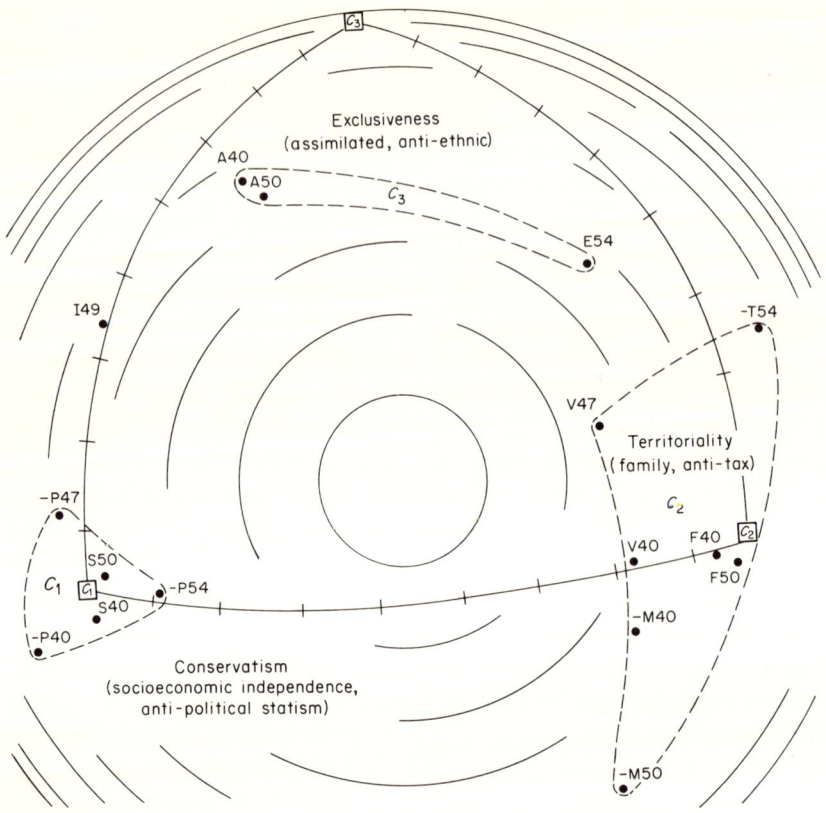

FIGURE 10.5
Spherical configuration showing the relationships among the 16 demographic and voting-attitude clusters of 105 San Francisco neighborhoods spanning 1940 to 1954.

with the anti-tax attitude, −T54. Associated with it are the four new indexes of community involvement, voting turnout, V47 and V40, and nonmoving or physical stability, −M40 and −M50. A high score on a composite of these seven variables seems quite clearly to signify a neighborhood of people highly identified with their families and homes, a characteristic that has been generally designated in sociobiology as territoriality.

The third cluster, exclusiveness, is rather obviously centered on the two assimilation characteristics, A40 and A50, that are somewhat loosely associated with the anti-ethnic vote, −E54. This exclusiveness dimension is the majority-minority dimension, a high score on which would characterize a neighborhood consisting of native white Protestants of northwest European origin holding negative attitudes toward minority groups.

A final point of interest is the locus of *money* in the configuration, as measured by family income, I49. Note that I49, not surprisingly, is associated both with the conservatism dimension and with the exclusiveness dimension but that it is independent of the territoriality dimension.

Clearly, the conservatism and territoriality dimensions are relatively independent. But the exclusiveness dimension is positively related to both. The purely demographic characteristics of exclusiveness, namely, A40 and A50 are, however, more highly associated with the conservatism dimension than with the territoriality dimension, but the attitudinal component, $-$E54, which is tinctured with a reluctance to spend money on tax relief for minorities, leans toward territoriality with its anti-tax component, $-$T54.

It seems quite likely that these three basic dimensions are determinants of the social structure of most urban neighborhoods of modern America. The correlations among the three may not be the same from city to city, but the dimensions will probably be there. Other investigators have observed one or more of the three *demographic* aspects of these three basic dimensions. In the objective measure of Warner's social class dimension, some of the definers of it are those that define our S dimension (Warner, Meeker, and Eels, 1949). The intuitively derived urbanization, segregation, and social rank variables of Shevky include some of the same definers of our F, A, and S, respectively (Shevky and Bell, 1954). Particularly relevant are the Borgatta-Hadden factors recently reported by Cartwright and Howard (1966) and used by them in their study of urban gangs. Their socioeconomic status factor I has similar definers to our S cluster, their suburb and stable family factors II and III include definers of our F dimension, and the disorganization-deprivation factor IV is probably close to our A dimension.

The findings here clearly establish that scores on only the three higher-order clusters, conservatism, territoriality, and exclusiveness, are necessary to predict all the significant variations among the neighborhoods. From a practical measurement viewpoint, however, it is not necessary to include voting behavior when we wish to describe a given neighborhood's characteristics with regard to territoriality, exclusiveness, and conservatism. All we need are the easily obtained, objective demographic characteristics, F, A, and S, for this analysis has shown that the voting attitude of these neighborhoods is almost perfectly predictable from the demographic features.

Chapter 11

UNRESTRICTED CLUSTER AND FACTOR ANALYSIS

Before digital computers were commonly available to the multivariate researcher, a major problem was the clerical difficulty of multivariate analysis. Even in relatively small problems such as the 24-variable Holzinger study, the amount of labor required to execute only the first step in the analysis of individual differences was staggering. Many weeks of hard work could be involved in factoring a correlation matrix. Now that digital computers are available, such studies can be executed in a matter of seconds of computer time. Computer programs, however, are generally written with a definite limit to the number of variables and objects that can be entered into an analysis. In the BC TRY System the limit on the number of variables is 120 on the IBM 7094 and 90 on the CDC 6400. Limitations on the number of subjects for V-analysis are not so stringent, 9,999 on the 7094 and 5,000 on the CDC 6400. In O-analysis the number of objects that can be handled by EUCO analysis is the same as the number of variables, 120 or 90. In OTYPE analysis this restriction is not present, and the limit on the number of objects is 9,999 or 5,000, depending on the program. One might suspect from these statements that the computer has a very definite limitation in the analysis of data. The limitation is not a logical restriction but a practical one. When the number of variables is very large, special techniques in computing must be used in order to handle the large numbers of data involved. These special techniques are more difficult to write into a computer program, and they take a longer time to execute for a given number of variables. Generally, programs written for a smaller number of variables (like 120) cost less to write and to execute than programs written for a larger number of variables (like 500) even if they are both applied to the same data. A major cause of the difficulty is the necessity of assuming, in the large-scale program, that the memory of the computer will not hold all the data at one time. Consequently, more or less elaborate schemes are necessary to store and retrieve segments of the data as the need arises to operate on the data segments. Such programs, to store and recover data, are both difficult to write and costly to operate. As a consequence, we have addressed ourselves to the task of devising methods and procedures to utilize the BC TRY procedures of factor and cluster analysis on large-scale theoretically unrestricted data sets, all without changing the basic programs. In order to fully implement these procedures several other programs have been written, particularly to sample sets of variables and objects.

The procedure for unrestricted cluster and factor analysis in the BC TRY System is called BIGNV analysis. When applied to successive samples of 120 or 90 variables, BIGNV converges on the salient cluster or factor structure of the full supply of variables. This convergent method permits an inverse analysis of individuals (O-analysis, in which objects play the role variables play in V-analysis), also unrestricted in principle by N, the number of subjects. The expression "in principle" means that the theory and logic of the procedures remove restrictions on n and N. In reality, however, hardware constraints of the computer and the costs of running problems with very large numbers of subjects and variables put practical limits on the studies that can be done.

The domain sampling theory and steps of BIGNV are described and illustrated in this chapter. One set of illustrations involves two independent V-analyses of the 566 items that compose the MMPI. The other set involves two O-analyses. The first is on the item responses of 310 adult subjects who took the MMPI, 70 psychotic and 150 anxiety outpatients of a Veterans Administration clinic, matched for age and education with 90 Armed Services officers. The second is on the original scores of the Grant White school sample of 145 children of the Holzinger study. The studies involved have been described in some detail in previous chapters.

BIGNV in the MMPI study

The formal question is as follows: How can one solve for the cluster or factor structure of a full supply of 566 items when the computer program is designed for no more than 120 variables? The cluster or factor structure in the unrestricted analysis should be that structure which would have been found if one did have a superprogram on a large computer that could handle the full supply directly. The solution to the problem, as implemented in BIGNV, is taken from the general ideas of sampling theory in statistics. The structure is estimated in each of a sequence of samples from the full supply, and the cluster structures of the samples are merged into a single structure. The full supply of variables is divided into manageable random samples. The BC TRY procedure that selects these samples is called SAMPLER. The cluster structure of each of the samples is determined by the existing BC TRY programs. The BC TRY procedure called MERGER is then used to combine the structures of the random samples to form an estimate of the structure in the full supply. In short, SAMPLER breaks down the supply into samples, and MERGER builds it up again. In the process the cluster structure of the full supply becomes revealed. It should be noted from this illustration that by using this domain sampling procedure we are in principle no longer constrained from tackling any problem no matter how large the full domain of variables from which the samples are to be drawn.

The detailed procedures of applying BIGNV procedures to the MMPI are not spelled out here. Rather, the general outline of the four primary steps in the analysis are discussed along with the results of the analysis. The BC TRY User's Manual gives the details. Table 11.1 shows the main stages of the analysis as applied in the 120-variable system on the IBM 7094. The implementation of these stages in an initial V-analysis of the 566 MMPI items is illustrated in the subordinate steps under the four stages. In stage 1 the SAMPLER procedures break down the 566 items into five random samples of items, each sample containing 120 items. The communality of each variable in a sample is calculated within that sample. A new sample is drawn, composed of the 120 items, out of all 566 items, that have the highest communalities. This sample of items is submitted to a standard key-cluster analysis. The cluster solution reveals nearly a dozen cluster-defined dimensions. Some of the dimensions are highly correlated, and others are defined by narrow, specific doublets. Combining the highly correlated clusters and deleting the trivial doublets in the analysis produce a hierarchical condensation that yields a salient set of three pivotal dimensions. These dimensions are defined by three clusters of items readily identified as the introversion (I), body symptoms (B), and suspicion (S) clusters introduced in earlier chapters.

In stage 2 a concern is the further elimination of items having trivial communality. Items with the highest communalities are the ones most likely to form clusters (or to be saturated by factors). On the other hand,

TABLE 11.1 BIGNV PROCEDURES OF BC TRY APPLIED TO V–ANALYSIS
(FIRST TRIAL ON 566 MMPI ITEMS)

Preliminary

Input by the data processor of raw scores on the full supply of 566 items

Stage 1. Sampler: Breakdown of the Supply of n Items into Samples of 120; Determination of the Pivotal V-dimensions from Them

Selection of five random samples of items
Calculation of communalities of all items in their samples
Selection of the 120 most general items for pivot dimension analysis
Discovery of pivotal cluster-defined dimensions on the most general samples, after hierarchical condensation: pivotal I, B, S

Stage 2. Preset Cluster Analyses on the Samples: on Item Samples of Decreasing Generality (Size of Communality)

Selection of five item samples of decreasing communality; rejection of items of trivial communality (200)
Preset factoring and structure analysis of the four most general samples on the three pivotal dimensions
Testing the sufficiency of the pivotal dimensions in the samples

Stage 3. Merger: Synthesis of the Cluster Structure of the Supply of n Items from the Samples

Rejection of additional specific items of trivial communality (117)
Formation of a composite statistical structure and geometric configuration of the supply from the samples (249)

Stage 4. Description: Nature of the Cluster Structure of the Full Supply of n Items

Final decision on three pivotal and four dependent oblique item-clusters from the composites and from auxiliary sector analysis
Rejection of rationally ambiguous items (57)
An abridged representative structure of 118 of the "best" items of the seven oblique clusters
Conceptualization of the seven MMPI item-domains: pivotal I, B, S and dependent D, R, A, T (retained items: 192 or 34%)

an item that shares little or no variation with any other item cannot possibly be a defining variable of any cluster or enter into a factored general dimension. Accordingly, in stage 2 of Table 11.1, 200 items of trivial communality are first deleted, including the 16 duplicate items of the MMPI. Remaining items are then regrouped into samples of decreasing communality. Defining items for the three pivotal dimensions are included in each of the samples as markers. Cluster analyses are carried out on these samples, with solutions preset on the common set of three pivotal dimensions. The sufficiency of the three pivotal dimensions is determined by inspecting the residual correlations.

In stage 3, an additional search for items with trivial communality (now with respect to the common dimensions) is carried out. Another 117 items are removed as a result of this further search. Since the item samples are factored on common dimensions in stage 2, in stage 3 the MERGER procedures consist of simply forming a single composite cluster structure. A surviving pool of 249 items (all with salient communality) is involved in this procedure.

With the composite structure of the full supply revealed, the stage 4 phase consists of making a final decision on the salient clusters in the structure. Careful study leads to the selection of seven clusters; the three pivotal dimension definers and four additional clusters, depression (D), resentment (R), autism (A), and tension (T). In this process 57 rationally ambiguous items are lost. As a final abridged sample, the 118 items that best define the seven clusters are chosen. A single dimensional analysis of these items is then carried out, thus yielding, in one computer run, a representative structure of the full supply of MMPI items. The seven item-clusters are listed in Table 11.2.

The first three clusters, labeled I introversion, B body symptoms, and S suspicion and mistrust, are the pivotal three that defined the tridimensionality of the analysis. For each cluster the contents of the best 17 items only, paraphrased for simplicity, are shown. They are best in the sense that they have the highest factor coefficients, i.e., the highest correlations with their respective clusters. The 17 contents in the introversion cluster rather compellingly reveal that they are all symptoms of social withdrawal. These 17 are the abridged set of items that, in stage 4, most clearly reveal conceptually what the cluster means. At the foot of the list, under Other, are shown the item numbers and factor coefficients of nine additional items in the introversion cluster that fill out the full cluster. All are also introversive symptoms. The nine other items are less distinctive than the 17 above them in the special sense that they show a bit more correlation with some of the other six clusters than the 17 do. The 17 are sharply differentiated in terms of correlation from the items that compose the other six clusters. Finally, the reliability coefficients are marked Reliability at the foot of the introversion listing. The full cluster of 26 items yields a composite cluster score that has an α reliability coefficient of .925. Even for the lesser number of 17 abridged items, a composite score has a reliability with the high value of .911.

The reader may review the other six clusters in detail. Generally, they have sharp meanings rather close to the titles assigned them in Table

11.2. Cluster scores on all have high α reliabilities, of the order of approximately .90. The reliabilities are much higher in general than those found for the standard nine clinical scales, despite the fact that the scales are composed of many more items than are these clusters.

A graphical picture of the cluster structure of the MMPI items is provided in the geometric diagram of Fig. 11.1, which is a map of the correlations among the items. It is a tracing of the printout of the component SPAN of BC TRY, produced as the final outcome of a single computer run of a full cycle key-cluster analysis of the 118 best items for the seven clusters. It should also be noted that the items selected to define the axes in the figure are the three pivotal clusters, which are introversion, body, and suspicion, circled as shaded domains of correlation. Though the three factored dimensions are perforce orthogonal (the second and third being

TABLE 11.2 MOST DISTINCTIVE DEFINING ITEMS OF MMPI CLUSTERS (FIRST TRIAL) RANKED BY OBLIQUE FACTOR COEFFICIENTS Fc

No.	Content	Fc
C_1, I, introversion (pivot):		
377	Sit alone at parties	.70
− 57	Poor mixer	.66
321	Easily embarrassed	.66
201	Shy	.65
180	Poor conversationalist	.65
−371	Self-conscious	.65
267	Hard to talk in group	.62
172	Bashful	.61
86	Lack self-confidence	.61
171	Poor at party stunts	.61
−547	Do not like parties	.60
−521	Hard to talk in group	.59
52	Pass friends without speaking	.58
−309	Don't make friends quickly	.58
−479	Mind meeting strangers	.56
509	Can't stick up for self	.53
292	Don't speak first	.46

Other:

No.	−79	317	−264	138	−353	304	−449	−415	−482
Fc	.54	.54	.51	.51	.51	.50	.47	.46	.43

Reliability: For $n = 26$, .925; $n = 17^a$, .911

[a] Denotes abridged set of best defining items.

No.	Content	F_c
C_2, B, body symptoms (pivot):		
-243	Many pains	.66
189	Weak all over	.63
108	Fullness in my head	.62
-190	Many headaches	.61
62	Body tingle, burn, crawl, sleep	.60
-175	Dizzy spells	.60
-230	Heart pounds; breath short	.57
114	Tight band around head	.57
47	Hot all over	.55
44	Head hurts	.54
-55	Heart, chest pains	.54
29	Acid stomach	.54
125	Stomach trouble	.49
-68	Pain in neck	.47
10	Lump in throat	.45
23	Nausea, vomiting	.41
161	Top of head feels tender	.37

Other:

No.	544	72	-3	-36	-163	-51	-103	-160
F_c	.65	.58	.56	.56	.53	.52	.50	.50
No.	191	-153	263	-330	-2	-18	-192	14
F_c	.49	.49	.42	.36	.33	.33	.33	.32

Reliability: For $n = 33$, .919; $n = 17^a$, .886

No.	Content	F_c
C_3, S, suspicion and mistrust (pivot)		
404	People misunderstand me	.63
507	People take credit, shift blame	.58
383	People disappoint me	.57
390	I'm misunderstood when I help	.52
436	People want respect they don't give others	.50
136	People have reasons when nice	.49
244	My ways are misunderstood	.48
348	On guard when people friendly	.48
368	Avoid people I may abuse	.46
280	People make friends for use	.45
265	Trust nobody	.44
469	People jealous of my ideas	.43
447	When opposed, I must win	.40
319	People don't put out to help	.40
7-	People exaggerate for sympathy	.39
553	People are guilty of bad sex	.36
405	"Experts" no better than I	.35

Other:

No.	278	284	438	89	112	426	316	455
F_c	.45	.40	.39	.37	.34	.34	.34	.32

Reliability: For $n = 25$, .854; $n = 17^a$, .830

a Denotes abridged set of best defining terms.

No.	Content	F_c
C$_4$, D, depression and apathy:		
76	Most of time feel blue	.78
-107	Not happy	.75
236	Brood	.74
301	Life is a strain	.74
-379	Have spells of blues	.67
487	Give up quickly	.67
41	Just can't get going	.63
259	Can't start to do things	.62
418	Am no good at all	.62
-8	Few things interest me	.61
549	Shrink from facing crises	.61
67	Wish I could be happy	.61
414	Can't forget disappointments	.60
396	When things fine, don't care	.57
61	Have not lived right	.56
411	Feel a failure when others succeed	.56
142	Feel useless	.55

Other:

No.	397	526	361	384	84	357
F_c	.61	.53	.53	.53	.52	.52
No.	168	339	-88	-46	104	
F_c	.49	.47	.44	.43	.39	

Reliability: For $n = 28$, .935; $n = 17^a$, .910

No.	Content	F_c
C$_5$, R, resentment and aggression:		
94	Do many things I regret	.66
336	Impatient with people	.60
468	Sorry to be cross, grouchy	.59
-399	Easily angered	.56
375	Someone spoils my feeling good	.51
39	Feel like smashing things	.50
381	Said to be hotheaded	.50
97	Urge to harm or shock	.49
536	Get angry when people hurry me	.48
139	Urge to injure self or others	.47
234	Easily angered but recover soon	.46
129	Why am I cross and grouchy?	.45
145	Feel like picking a fist fight	.43
148	Can't stand interruptions	.41
28	When wronged, I pay back	.34
162	Resent admitting being taken in	.32

Other:

No.	416	382	106	147	443
F_c	.59	.47	.46	.45	.39

Reliability: For $n = 21$, .865; $n = 16^a$, .820

[a] Denotes abridged set of best defining items.

(Continued)

C6, A, autism and disruptive thoughts

No.	Content	F_c
559	Frightened in middle of night	.61
241	Dream things should keep secret	.55
15	Think unmentionable things	.55
349	Strange, peculiar thoughts	.54
425	Dream frequently	.52
511	Secret daydream life	.47
545	Recurrent dreams	.47
358	Can't stop thinking bad words	.46
560	Bothered by forgetfulness	.43
−329	Almost always dream	.43
100	Indecision when problems complex	.43
345	Often feel things not real	.42
342	Forget what people say to me	.40
374	Mind works too slowly	.40
459	Can't fight some bad habits	.39
297	Bothered by sex thoughts	.39
33	Peculiar, strange experiences	.37

Other:

No.	359	389	356	40	31	134
F_c	.56	.53	.49	.38	.36	.28

Reliability: For $n = 23$, .856; $n = 17^a$, .810

C7, T, tension, worry, fears

No.	Content	F_c
555	About to go to pieces	.72
431	Worry over possible misfortunes	.70
337	Anxious most of time	.68
217	Frequently worry over something	.67
238	So restless, can't sit	.63
506	High-strung	.62
543	Something dreadful might happen	.56
442	Lose sleep over worry	.55
43	Sleep fitful, disturbed	.55
−242	More nervous than others	.54
340	So excited can't sleep	.53
−152	Can't sleep for recurrent thoughts	.52
448	Bothered by being watched	.49
186	Hand shakes when do something	.48
499	Worry without reason	.46
−66	Afraid of high places	.45
338	More than my share of worries	.44

Other:

No.	−407	182	32	439	335	102	473	158	303	13
F_c	.68	.56	.54	.52	.50	.48	.46	.43	.43	.42
No.	388	322	360	22	351	−131	365	494	492	
F_c	.41	.40	.40	.38	.38	.37	.36	.34	.32	

Reliability: For $n = 36$, .923; $n = 17^a$, .882

[a] Denotes abridged set of best defining items.

derived from residuals), the three clusters of variables that define the dimensions are themselves not completely independent. They are slightly oblique to each other. Note also that they encompass the other form clusters, D, R, A, T, which are therefore called "dependent" clusters. This means that actual composite scores on these four dependent clusters can be predicted with fair accuracy from cluster scores on the three pivotal clusters, I, B, S. Table 11.3 presents in metric form most of the facts represented pictorially in the geometry of Fig. 11.1. The actual magnitudes of the intercorrelations between the clusters are given in the correlation matrix of Sec. A of Table 11.3. In Sec. B of Table 11.3 is shown the index of overall generality of each of the seven clusters, i.e., the proportion of the total squares of raw correlations over the whole original correlation matrix that is reproduced by the sums of squares of correlations by the given dimension taken singly. It was the fact that the tension cluster had the highest reproducibility ratio of .88 that partly disposed us to include T as a fourth cluster in O-analysis, below. The very high α reliabilities computed by

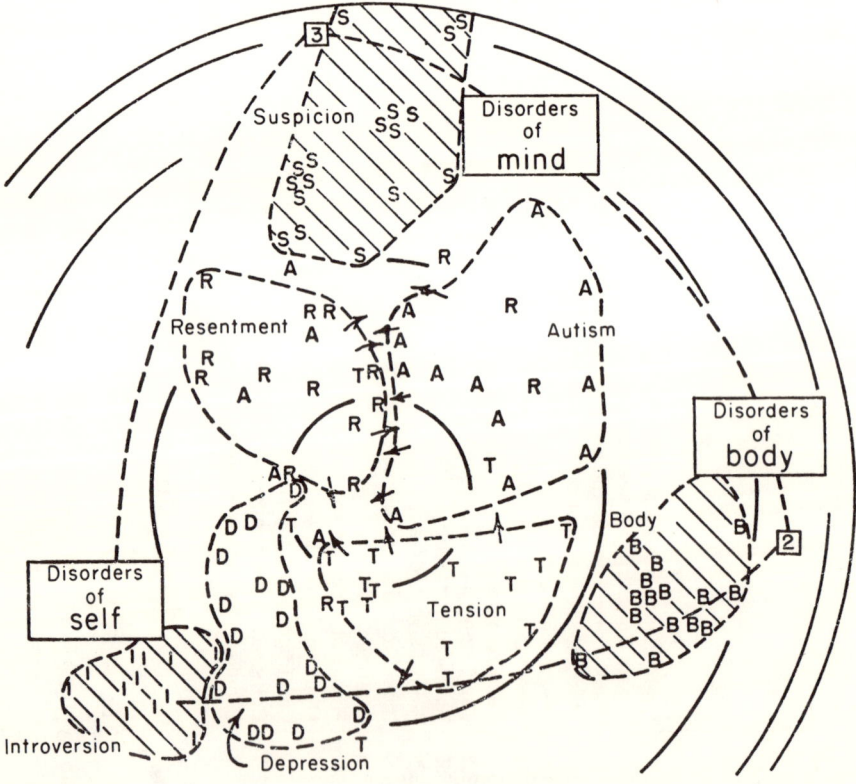

FIGURE 11.1
Representative cluster structure by SPAN of the 566 MMPI items.

the CSA component and mentioned earlier in Table 11.2 are also given in Sec. C of the table.

One significant finding from this cluster structure is that psychiatric symptoms as measured by the MMPI compose an oblique structure. That is, psychiatric symptoms are generally positively correlated. Furthermore, it is found that the most independent pivotal three clusters, introversion, body symptoms, suspicion, refer respectively to disorders of self, disorders of body, and disorders of mind or thinking.

It may legitimately be questioned whether the estimation of the cluster structure of the full supply from random samples is reliable. For evidence on this point, the first three stages of BIGNV were repeated on new sets of random drawings. The composite structure on stage 3 of this second trial was virtually identical to that of the first trial for the two dimensions of introversion and body, but the third dimension was a fuzzy mixture of suspicion and resentment items. It was known from the first trial that S and R were quite oblique and could be statistically combined, so there is in fact no inconsistency in the results of the two trials. The other three dependent clusters, D, A, and T, showed about the same differentiation

TABLE 11.3 INTERCORRELATIONS, GENERALITY, AND RELIABILITY
OF THE SEVEN OBLIQUE CLUSTERS

	A. Intercorrelations[a]						
	I	**B**	**S**	**D**	**R**	**A**	**T**
I		.47	.27	.73	.49	.41	.68
B	.36		.32	.48	.44	.51	.75
S	.28	.35		.37	.54	.54	.48
D	.80	.54	.37		.64	.57	.78
R	.57	.51	.66	.74		.59	.69
A	.48	.61	.66	.67	.72		.66
T	.61	.66	.41	.87	.81	.78	

	B. Generality						
Reproducibility of the 13,806 raw correlations	.65	.55	.41	.83	.72	.67	.88

	C. Reliability						
$s = 17$	.91	.89	.83	.91	.82	.81	.88
$s > 17$ (*including "other"*)	.93	.92	.85	.94	.87	.86	.92

[a] Above diagonal, raw correlations; below, interdomain (common factor) correlations ($s = 17$). $s =$ number of items per cluster (cluster R has $s = 16$).

as in the first trial. Since the results of the second trial confirm those of the first, we can accept the seven clusters of Table 11.2 as the final cluster structure of the MMPI.

BIGNV in O-analysis

BIGNV applied to the isolating of profile types of persons in O-analysis is identical in broad design to BIGNV in V-analysis. The similarity is clearly shown in Table 11.4, where the same four stages are given as they apply to two separate O-analyses. At the left side of the table is the O-typing of the MMPI profiles of 310 psychotics, neurotics, and normal adults; at the right is the O-analysis of the achievement profiles of the 145 children in the Holzinger problem.

The MMPI person analysis is considered first. Each person is scored on the three pivotal clusters, I, B, and S, and on one of the dependent clusters. The reason for choosing an additional (dependent) cluster is that the three pivotal dimensions are not completely sufficient. The fourth cluster T, measuring tension, is selected because it is the largest of the dependent clusters, and, as shown in Table 11.3, it is the most general of all seven clusters. It also comes nearest conceptually to what is generally known as "anxiety," believed to be a central attribute in many psychiatric disorders.

In stage 1, a standard EUCO analysis, as described earlier, is performed on random samples of 120 subjects from the full supply of N. In stage 2 the full supply of N subjects is divided into random samples (three in the MMPI study, two in the Holzinger study). The cluster structures of these samples are determined on the preset dimensions found in stage 1 and tested for sufficiency. In stage 2, dummy or marker individuals are projected into the analyses and thereby the cluster score axes can be set into the configuration of persons so that the scores of any one of them can be read off to a close approximation.

In stage 3, MERGER procedures synthesize the cluster structure of the full supply from the samples, as in V-analysis. The results of MERGER are shown in Figs. 11.2 and 11.3. Plotted in Fig. 11.2 as points on the surface of a sphere are the persons whose intercorrelations (or interspace distances) can be accounted for by the first three pivotal person dimensions. The defining persons of the first three pivotal dimensions are circled as the three groups labeled C_1, C_2, and C_3 in Fig. 11.2. In order to see concretely what this figure means, the reader should look ahead at Fig. 11.4, where he will find the score profiles of these three groups of persons drawn in the orthodox way on scales with a mean of 50 and standard

TABLE 11.4 BIGNV PROCEDURES OF BC TRY APPLIED TO O–ANALYSIS
(MMPI AND HOLZINGER)

Preliminary

MMPI, 310 Adults	Holzinger, 145 Children
Calculation of cluster scores on I, B, S, T	Calculation of cluster scores on F, V, S, M
Calculation of six scattergrams between the four cluster scores	Same as MMPI

Stage 1. Sampler: Breakdown of the Supply of N Subjects into Samples of 120, and Determining the Pivotal O-dimensions from Them

Selection of a random sample of 120 subjects	Same as MMPI
Discovery of four pivotal O-cluster dimensions on the sample	Same as MMPI

Stage 2. Preset Cluster Analyses on the Samples: on Stratified Random Samples of Objects

Selection of three stratified samples of persons; no rejections	Selection of two stratified samples; no rejections
Preset factoring and structure analysis of the three samples on the four pivots	Same as MMPI
Projection of dummy model markers and rational markers into the structure (OMARK)	Same as MMPI
Sufficiency tests of the pivots in the samples	Same as MMPI

Stage 3. Merger: Synthesis of the Cluster Structure of the Full Supply of N Persons from the Samples

Formation of a composite structure and configuration of the supply from the samples	Same as MMPI

Stage 4. Description: Nature of the Cluster Structure of the Full Supply of N Persons

(To be reported in a monograph by Stein and Chu)	Decision on 14 Core O-types O-clusters: allocation of all 145 persons to the Core O-types (EUFIT) Level and homogeneity of the 14 O-cluster profile-patterns (OSTAT)

deviation of 10. It should be noted that groups C_1, C_2, and C_3 form very tight profile groups. In short, the SPAN diagram of Fig. 11.2 provides a total map of the similarities and differences in profiles among all the persons plotted on the sphere.

Some of the person points have negative signs in front of them, meaning that their profiles are mirror images, a "reflection," of those of persons next to them who do not carry negative signs. This can be

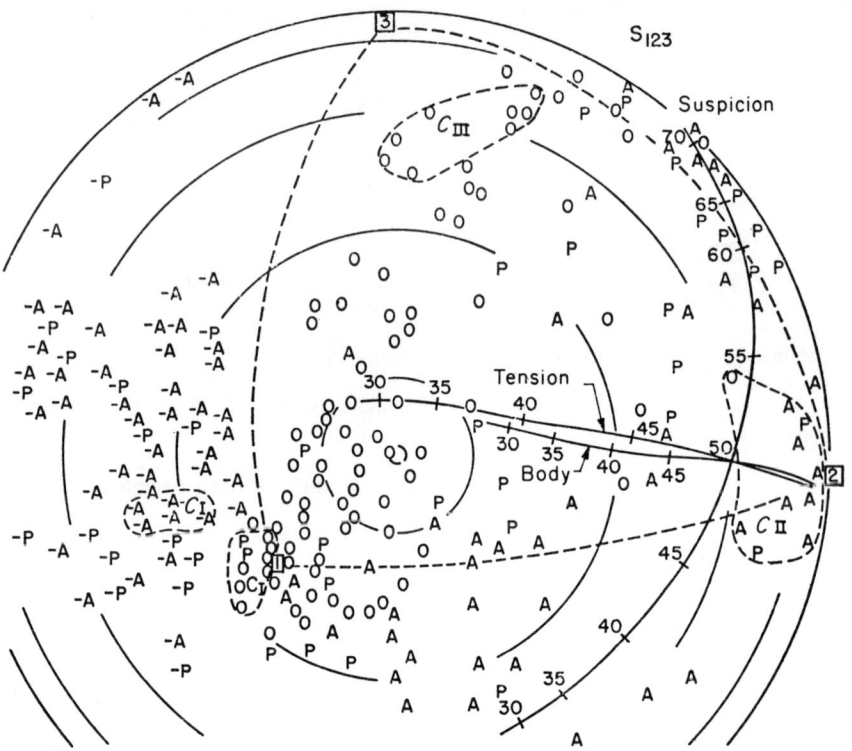

FIGURE 11.2
Cluster structure of individuals whose score profiles are described at more than 90 percent sufficiency on sphere S_{123}.

seen concretely in the profiles of the reflected group labeled $-C_1$ in Fig. 11.4. Just as those of C_1 are uniformly below average on the four cluster scores, those of $-C_1$ are uniformly above average, though more extreme.

The scaled cluster score axes have been drawn on the surface of the sphere so that the scores of the persons can be observed. In Fig. 11.2 they are labeled Body, Tension, and Suspicion. Because of the curvature of the surface, the projections of persons on these axes are somewhat distorted in the plane of the paper. However, it will be observed that cluster C_1 is below average on B, S, and T and that cluster $-C_1$ is above average on B, S, and T and farther from the average but in the reflected, i.e., above average, position.

The significant finding of the O-analysis goes directly to the matter of validity of the MMPI clusters and of the O-types produced on them. The normal officers are abbreviated as O's in Fig. 11.2, the psychotics as P's, and the anxieties as A's. The officers generally cluster around C_1, whereas not a single officer lies among the psychiatric cases in the

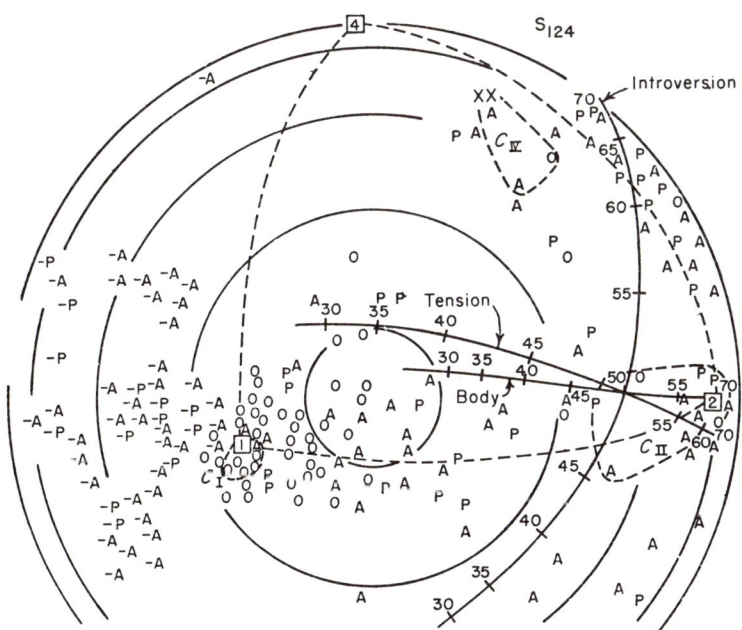

FIGURE 11.3
Cluster structure of individuals where score profiles are described at more than 90 percent sufficiency on sphere S_{124}.

reflected zone of $-C_1$. Furthermore, some of the officers and none of the psychiatric cases form cluster C_3, high on the suspicion and mistrust dimensions (and aggression) though still low on the body and tension dimensions. Perhaps it is "normal" for the Armed Services to include some officers who have such a belligerent, healthy, and relaxed pattern.

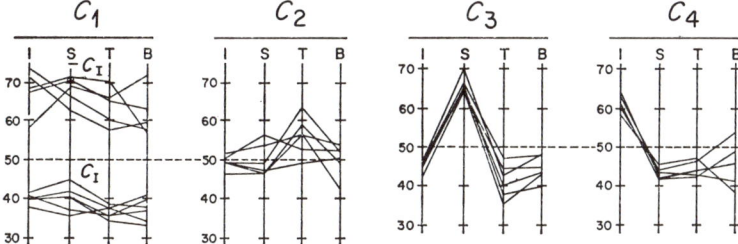

FIGURE 11.4
Cluster score profiles of the individuals shown in person-clusters, C_1, C_2, C_3, and C_4 of the SPAN spheres.

The next sphere, Fig. 11.3, brings in the introversion, body, and tension dimensions. Unlike the preceding figure, except for the few persons in cluster C_4, there are virtually no persons high on the introversion dimension and also low on the body and tension dimensions. There are, however, quite a fair number high on all three clusters, and these are composed almost exclusively of psychiatric cases.

In the description phase, stage 4, the objective is taxonomic, namely, to study the configuration of the full supply, such as that shown in Figs. 11.3 and 11.4, and to select a set of contrasting collinear core O-types that is representative of the complete structure. In choosing the final core types, the analyst will naturally orient to loci in the configuration where there are heavy concentrations of individuals. These are empirical or "natural" clusterings, i.e., types of individuals whose higher frequencies of special score patterns on the dimensions ensue from selective biosocial forces. Lacunae in the configuration signify "missing types," biosocially incompatible, disjunctive patterns of scores on the attributes. This search for "natural clusters" is precisely the same objective as that of the recently emerging field of biology called "numerical taxonomy" (Sokal and Sneath, 1963).

Sharp concentrations or lacunae may not appear, in which case the analyst may segment the configuration into any arbitrary classes of core O-types that are convenient for his purposes of taxonomic analysis. When the selection of core types appears to be unduly arbitrary, nothing, indeed, should prevent the analyst from selecting alternative sets of core types for study and for prediction if that is his purpose.

The next step in stage 4 is to allocate every individual in the full supply to that particular core type that has a score pattern that fits closest to his own. The procedure for doing so is EUFIT, by which the euclidean distances in score space between each individual and each of the different core types are computed, followed by his allocation to that core type with which he has the smallest distance. The final groupings of all individuals by this allocation process are termed the "final O-clusters." Not all individuals fit in O-clusters; there usually remains an interesting subset of mavericks whose smallest euclidean distance is too large to admit them to any O-cluster.

In stage 4, the objective description of the final set of O-clusters, made by the program called OSTAT, consists of determining the level of the mean profiles of each O-cluster and its homogeneity H. The core O-types in the MMPI study and the O-type analysis in the Holzinger problem have already been described extensively in Chap. 8.

Chapter 12

STATISTICAL THEORY AND COMPONENT
PROGRAMS OF BC TRY

This chapter is a brief description of the statistical and logical basis of the BC TRY System and of the programs themselves. We do not attempt to develop here the entire theory and practice of cluster and factor analysis, our aim being the more modest one of illustrating the major theoretical and practical aspects of the statistical and logical components of the System. A number of general texts contain descriptions of the procedures and mathematics of some of the orthodox factor analysis components of the System. These subjects are given relatively scant attention here, deference being paid to other authors on these topics.

The plan of this chapter is to describe each major statistical component of the System, one at a time. Each component is described in terms of the general statistical and logical theory behind the component. Where it is not obvious, the programming of the component is described at the level of a general algorithm that is machine independent. It is not intended to exhibit the details of the component programs of the System here. The actual programs (at the time this was written) make up some 60,000 Fortran statements. The program listings themselves are equivalent to several volumes the size of this book. As a consequence, it must be obvious

TABLE 12.1   DST INPUT–OUTPUT TABLE

Programs Which Write DST
Data processor program, DAP2
Euclidean distance program, EUCO2

Programs Which Read DST
Correlation program, COR2
Correlation/covariance program, COR3
Cluster and factor scoring program, FACS
Object typology program, OTYPE (optional)
Typological prediction program, 4CAST (optional)

that this chapter and the next are sufficient to give only a general idea of the operation of the components and the logic behind them. The detailed descriptions of the data, input and output, component programs, and details regarding the control of the System by the user constitutes the BC TRY User's Manual and will not be repeated here in any but the most general form when it is useful in explaining the logic and implementation of a program. Since the input and output for BC TRY components are fundamental in discussing the components, we reproduce here, as Tables 12.1 and 12.2, the DST and IST I/O tables. Chapter 13 is an abridged User's Manual, showing the detail of BC TRY applications in standard problems.

Cluster structure analysis

A cluster of variables is merely a composite or a grouping of variables. In BC TRY the clusters are selected in an analysis of multivariate data to satisfy certain criteria. However, in general, any group of variables is logically a cluster. Clearly, some clusters represent domains of variation in the population of variables or objects better than other clusters. Cluster structure analysis is the means whereby the statistical properties of clusters are studied and statistical relationships of clusters and domains can be studied. The main correlational properties of clusters, particularly linear composites of scores of variables in clusters, have long been known (Spearman, 1913; Guilford, 1950).

The important statistics one wants to know about cluster composites (cluster scores) are (1) their reliability, (2) the validity with which they represent some (unspecified) domain, (3) the correlations between different cluster scores, (4) the correlations between estimates of the domains the composites represent, (5) the correlations of each of the observed variables with the cluster composites, (6) the correlations of each of the observed variables with the domains represented by the cluster composites, (7) the generality of each oblique dimension, and (8) data on expansion of clusters with respect to reliability.

All that is required for a cluster structure analysis is a set of NV variables, the correlations (Pearson product-moment) among them, and the clusters. The nature of the clusters is immaterial, as is the method of selecting them. The clusters may be selected to satisfy rational-theoretical considerations or certain statistical considerations as in factoring, etc., or they can be arbitrarily selected. The analysis is applicable to any collection of subsets of the NV variables, all the variables, or only a portion of them.

TABLE 12.2 IST INPUT–OUTPUT TABLE

Program	IDFILE	VNAMS1	VSUMS1	MEANS1	VAREN1	STDEV1	CORRM1	ONAMS1
DAP2	O	O	O	O	(O)	O		O
COR2	I	I	I	I		O	O	
COR3	I	I		I	(I)(O)		(O)	
REDE	I	I					I	
RLIST	I	I			I		I	
SLEP1	I	I,O					I,O	
DVP	I						I	
FALS	I						I	
NC2	I						I	
CC5	I						I	
SLEP2	I	I,O					I,O	
GYRO	I							
NCSA2	I	(I)					I	
CSA2	I	(I)					(I)	
FAST	I	(I)					(I)(O)	
SPAN2	I	(I)						
FACS	I	(I)		I	(I)	I	I	
EUCO2	I	O	O	O		O		
COMP1	I	(I)						
COMP2	O	O					O	
RSCAT	I	(I)						(I)
OTYPE	I	(I)		(I)		(I)		I
OSTAT	I							I
4CAST	I	(I)		(I)		(I)		

Notes: I = input for all control options; (I) = input for some control options; O = output for all control options; (O) = output for some control options. All files can be input and/or output by GIST and GIVE.

TABLE 12.2 IST INPUT–OUTPUT TABLE (Continued)

Program	2MEAN1 2SENS1 2STDEV1	MEANS1	STDEV2	COVAR1	REORD1	CORRM2	DIAGV1
DAP2							
COR2							
COR3	(O)	(O)	(O)	(O)			
REDE							
RLIST	I	I	I				
SLEP1					O	O	
DVP							O
FALS							I,O
NC2							I,O
CC5							I,O
SLEP2					I	I	O
GYRO							
NCSA2							
CSA2							(I)
FAST							
SPAN2							
FACS		O	O				(I)
EUCO2		I	I				
COMP1							
COMP2							
RSCAT							
OTYPE		(I)	(I)				
OSTAT			I				
4CAST		(I)	(I)				

Also, one may apply the methods of cluster structure analysis to the "reproduced" correlations from a factor analysis published in the literature even though the correlation matrix is not available.

The basic model for cluster structure analysis is the domain sampling model that is utilized throughout BC TRY. In the model, a cluster composite is defined as a sum of values on the variables in the cluster. Imagine that the sample of variables in a study is identified as

$$X_1, X_2, X_3, \ldots, X_{NV}$$

A subset of these observed variables is selected as a cluster, say the ith cluster out of K clusters, and designated as

$$V_{i1}, V_{i2}, \ldots, V_{iS_i}$$

TABLE 12.2 IST INPUT–OUTPUT TABLE (Continued)

Program	UFACT1	CLUST1	REFLX1	RFACT1	BASIS1	FSCOR1	WEIGH1
DAP2							
COR2							
COR3							
REDE							
RLIST							
SLEP1							
DVP							
FALS	O						
NC2	O	O	O				
CC5	O	O	(I),O				
SLEP2	I,O	(I)(O)	(I)(O)				
GYRO	I			O	O		
NCSA2	(I)		(I)	O	O		
CSA2	(I)		I	O	O		
FAST	I						
SPAN2	(I)		(I)	(I)			
FACS	(I)		(I)	(I)	(I)	O	(I)O
EUCO2						I	
COMP1		(I)	(I)	I			
COMP2							
RSCAT						(I)	
OTYPE						(I)	
OSTAT						I	
4CAST						(I)	

where S_i indicates the number of variables in the ith cluster. The several clusters are defined by sets of variables out of the total set of observed variables. The clusters may have common variables, but no special procedures are involved when a variable is in two or more clusters and we need not make special note of the fact. A complete list of the K clusters might appear as follows:

$$\text{Cluster 1: } V_{11}, V_{12}, \ldots, V_{1S_1}$$
$$\text{Cluster 2: } V_{21}, V_{22}, \ldots, V_{2S_2}$$
$$\cdots \cdots \cdots \cdots \cdots \cdots \cdots \cdots$$
$$\text{Cluster } i: V_{i1}, V_{i2}, \ldots, V_{iS_i} \tag{12.1}$$
$$\cdots \cdots \cdots \cdots \cdots \cdots \cdots \cdots$$
$$\text{Cluster } K: V_{K1}, V_{K2}, \ldots, V_{KS_K}$$

A cluster composite is simply the sum of the values of the cluster variables

$$C_i = V_{i1} + V_{i2} + \cdots + V_{iS_i} \qquad (12.2)$$

It is assumed that the variables are in standard form, i.e., with mean of 0 and standard deviation of 1. This assumption is not necessary, but it simplifies the notation and equations. Since the desired results are in terms of correlations, the standardization of the variables is implied from the first even though the equations could be stated in terms of the obtained means and standard deviations.

Corresponding to cluster C_i there is a cluster domain and a set of variable domains from which the observed variables were sampled. A variable domain is defined as a very large set of variables that are theoretically observable, but unobserved, and precisely "parallel," or collinear, with the corresponding observed variables. Thus, for variable X_i the variable domain is defined by the set

$$X_i, X_i^{(1)}, X_i^{(2)}, X_i^{(3)}, \ldots, X_i^{(\infty)} \qquad (12.3)$$

The domain composite for the variable domain is defined as

$$D_{X_i} = X_i + \sum_{g=1}^{\infty} X_i^{(g)} \qquad (12.4)$$

The domain composite for a cluster domain is defined as the sum of the variable domain composites corresponding to the variables defining the cluster. Thus, for cluster i

$$D_i = C_i + \sum_{j=1}^{S_i} \sum_{g=1}^{\infty} V_{ij}^{(g)} \qquad (12.5)$$

The definitions of cluster composite, domain, domain composite, etc., just outlined, permit some rather far-reaching statements about the correlation characteristics of clusters and domains. These characteristics can be made quantitative from knowledge only of the clusters, the intercorrelations of the observed variables, and the "communality" of the variables. For the time being assume that the diagonal values of the correlation matrix are filled with the communalities of the variables

$$h_1^2, h_2^2, h_3^2, \ldots, h_{NV}^2$$

or, in an alternative notation,

$$r_{X_1 X_1}, r_{X_2 X_2}, \ldots, r_{X_{NV} X_{NV}}$$

When reference is made to the communality of some variable in a cluster, an alternative notation is

$$h_{V_{ij}}^2 = r_{V_{ij} V_{ij}}$$

for the ith cluster and the jth variable in the cluster. The domain sampling interpretation of these values and means whereby they are obtained are presented later in this chapter.

Domain validity

The domain validity of a cluster composite can be interpreted as the accuracy of the cluster score in estimating the cluster domain composite. The domain validity of the cluster composite C_i in estimating the cluster domain composite D_i is the correlation of D_i and C_i

$$r_{D_i C_i} = \sqrt{\frac{\displaystyle\sum_{j=1}^{S_i}\sum_{k=1}^{S_i} r_{V_{ij}V_{ik}}}{\displaystyle\sum_{j=1}^{S_i}\sum_{k=1}^{S_i} r_{V_{ij}V_{ik}} + \sum_{j=1}^{S_i}(1 - h^2_{V_{ij}})}} \qquad (12.6)$$

Reliability of the cluster composite

The Spearman-Brown reliability of a cluster composite is given by the square of the domain validity

$$r_{C_i C_i'} = r^2_{D_i C_i} \qquad (12.7)$$

where C_i' is a parallel domain (see Tryon, 1957a, and Ghiselli, 1964).

Cluster composite intercorrelation

The intercorrelations among cluster composites are given by

$$r_{C_i C_j} = \frac{\dfrac{1}{S_i S_j}\displaystyle\sum_{g=1}^{S_i}\sum_{k=1}^{S_j} r_{V_{ig}V_{jk}}}{\sqrt{\left(\dfrac{1}{S_i^2}\displaystyle\sum_{g=1}^{S_i}\sum_{k=1}^{S_i} r_{V_{ig}V_{ik}}\right)\left(\dfrac{1}{S_j^2}\displaystyle\sum_{g=1}^{S_j}\sum_{k=1}^{S_j} r_{V_{jg}V_{jk}}\right)}} \qquad (12.8)$$

The diagonal elements $r_{V_{ig}V_{ik}}$ and $r_{V_{jg}V_{jk}}$, where $g = k$, are assumed to be unities in this equation rather than communalities. In orthodox factor analysis this coefficient is known as the "correlation between oblique factor estimates."

Cluster domain intercorrelations

The intercorrelations among cluster domains are the intercorrelations among cluster composites, Eq. (12.8), except that the diagonal elements are the communalities of the respective observed variables. Equation

(12.9) is the same as (12.8), except that the diagonal elements are unities in (12.8) and communalities in (12.9). The interdomain correlations are denoted

$$r_{D_iD_j} \tag{12.9}$$

A rondo of relationships exists among the correlations already indicated. This is given in

$$r_{C_iC_j} = r_{D_iD_j}r_{C_iD_i}r_{C_jD_j} \tag{12.10}$$

The intercorrelation of cluster composites is equal to the cluster domain intercorrelation reduced proportionally by the invalidity of the cluster composites as measures of the respective domains.

Correlation of cluster composites and cluster domains

These correlations are simply the interdomain correlations reduced proportionally by the invalidity of the cluster composite

$$r_{C_iD_j} = r_{D_iD_j}r_{D_iC_i} \tag{12.11}$$

Correlation of variables with cluster domains

The correlation of an observed variable with a cluster domain is known as the "oblique factor coefficient" of that variable with regard to the domain. All diagonal elements are communalities

$$r_{X_iD_j} = \frac{\sum_{g=1}^{S_j} r_{X_iV_{jg}}}{\sqrt{\sum_{g=1}^{S_j}\sum_{k=1}^{S_j} r_{V_{jg}V_{jk}}}} \tag{12.12}$$

Augmented oblique factor coefficient

This coefficient is the correlation between the cluster domain and the domain represented by a single variable D_{X_i}

$$r_{Dx_iD_j} = \frac{r_{X_iD_j}}{h_{X_i}} \tag{12.13}$$

The augmented factor coefficient is the correlation between the variable and the domain, adjusted in accordance with the communality of the variable.

Generality of each oblique dimension

One of the major points made in discussing factoring is the degree to which the orthogonal factors account for the intercorrelations among the NV variables or for the communality of the NV variables. In cluster structure analysis, since the clusters are not orthogonal, in general, these considerations must be made separately for each of the cluster dimensions. The generality across orthogonal factors is cumulative, but the generality across nonorthogonal cluster dimensions is not.

The generality of an oblique cluster can be represented by either of two indexes: (1) the degree (proportion) to which the cluster dimension accounts for the intercorrelations among the NV variables or (2) the degree to which it accounts for their common variances, their communalities. "Account for" means the degree to which the intercorrelations or communalities can be reproduced by the given cluster dimension. The formulas are given below.

The cluster domain D_i can be represented as a variable or dimension. If the domain D_i represents all that is general to the relationship between two observed variables, say X_j and X_k, the partial correlation

$$r_{X_j X_k \cdot D_i} = \frac{r_{X_j X_k} - r_{X_j D_i} r_{X_k D_i}}{r_{X_j D_i} r_{X_k D_i}} \tag{12.14}$$

will be exactly zero. When the partial correlation is zero, the correlation between the two variables is equal to the product of the correlation of the variables and the domain. That is,

$$r_{X_j X_k} = r_{X_j D_i} r_{X_k D_i} \quad \text{when } r_{X_j X_k \cdot D_i} = 0 \tag{12.15}$$

In general, the right-hand term of (12.15) will not be equal to the left-hand term. The right-hand term, however, is a measure of the degree to which the domain correlation of the variables with D_i accounts for the correlation of X_j and X_k. The quantity is known as the "reproduced correlation"

$$r'_{X_j X_k} = r_{X_j D_i} r_{X_k D_i} \tag{12.16}$$

The more nearly $r'_{X_j X_k}$ equals $r_{X_j X_k}$, the more nearly does D_i account for the correlation of the two variables. In taking a general index of the degree to which D_i accounts for the correlation among the NV variables the respective coefficients are first squared and then the squares are summed over the entire set of variables and a ratio of the sums of squares formed

$$T_i^2 = \frac{\displaystyle\sum_{j=1}^{NV}\sum_{k=1}^{NV} (r'_{X_j X_k})^2}{\displaystyle\sum_{j=1}^{NV}\sum_{k=1}^{NV} r_{X_j X_k}^2} = \frac{\displaystyle\sum_{j=1}^{NV}\sum_{k=1}^{NV} (r_{X_j D_i} r_{X_k D_i})^2}{\displaystyle\sum_{j=1}^{NV}\sum_{k=1}^{NV} r_{X_j X_k}^2} \tag{12.17}$$

To the degree that T_i^2 approaches unity, the domain D_i has generality across the entire set of observed variables, with respect to their inter-correlations. The diagonal values in the correlation matrix in (12.17) are the communalities obtained from a BC TRY factoring component, such as CC5.

The communality $h_{X_j}^2$ of the variable X_j is the portion of the variance of X_j predictable from the other $NV - 1$ variables in the study. In the degree to which the squared domain correlation $r_{X_j D_i}^2$ of a variable on a domain is equal to the variable's communality, that domain accounts for the generality of the variable in the set of NV variables. In applying this to all NV variables for a given domain, the total generality of the NV variables is the sum of the communalities, and the degree to which this is accounted for by a domain is the sum of the squared domain correlations

$$U_i^2 = \frac{\sum_{j=1}^{NV} r_{X_j D_i}^2}{\sum_{j=1}^{NV} h_{X_j}^2} \qquad (12.18)$$

The degree that U_i^2 approaches 1.00 is an indicator of the generality of the domain D_i over the set of NV variables with respect to their communality. This ratio is adversely affected as an adequate measure of generality to the extent that the estimates of communality used are inadequate.

Expanding a cluster by adding variables that maximize the reliability coefficient of the new cluster composite

If one adds to the definers of a cluster additional variables with high factor coefficients on the cluster dimension, the reliability of the expanded composite may be increased. When such an increase can be made, however, there is a maximal value of the reliability coefficients of the expanded composite beyond which further additions of variables can give reliabilities less than the maximal value. The problem is to discover what additional variables will contribute to the maximizing of reliability and what the optimal value will be after including them in the composite.

The reliability of a cluster composite is given by Eq. (12.7). If the cluster is expanded by a single new variable, say V_q, the equation can be stated (C_i' is the expanded ith cluster and $C_{i'}'$ is its parallel)

$$r_{C_i' C_{i'}'} = \frac{\sum_{j=1}^{S_i} \sum_{k=1}^{S_i} r_{V_{ij} V_{ik}} + \left[h_q^2 + 2 \sum_{j=1}^{S_i} r_{V_{ij} V_{qi}} \right]}{\sum_{j=1}^{S_i} \sum_{k=1}^{S_i} r_{V_{ij} V_{ik}} + \sum_{j=1}^{S_i} (1 - h_{V_{ij}}^2) + \left[1 + 2 \sum_{j=1}^{S_i} r_{V_{ij} V_{qi}} \right]} \qquad (12.19)$$

The reliability of the extended cluster C_i' depends on the relationship of the bracketed parts in the numerator and denominator. If the ratio

$$W_{iq} = \frac{h_q^2 + 2\sum_{j=1}^{S_i} r_{V_{ij}V_{qj}}}{1.0 + 2\sum_{j=1}^{S_i} r_{V_{ij}V_{qj}}} \tag{12.20}$$

has a value exceeding the reliability of the unexpanded cluster, the reliability of the expanded cluster will be greater than the reliability of the unexpanded cluster. Whenever the ratio (12.20) is less than the reliability of the unexpanded cluster, expansion of the cluster will not increase the reliability. Thus W_{iq} is the critical value of the variable V_q with respect to the expansion of the cluster, by adding V_q.

CSA uses this development in "expanding" clusters. Each variable which is not in the cluster but which has its highest oblique factor coefficient on the cluster is called a "potential expander." The potential expanders are "added" to the cluster one at a time in the order of the magnitude of the (12.19) as evaluated for the single variable expansion. Equation (12.19) is evaluated next for the progressively expanded cluster. These statistics, and that given as a lower bound (12.22), are used by the analyst to determine a reclustering if needed. The reclustering is not done by CSA.

Equations (12.19) and (12.20) are written in terms of the correlations among the observed variables. For computational purposes the best equation is based on the factor coefficients

$$r_{C_i'C_{i'}'} = \frac{h_q^2 + 2r_{X_qD_i}\left(\sum_{j=1}^{S_i} r_{V_{ij}D_i}\right) + \left(\sum_{j=1}^{S_i} r_{V_{ij}D_i}\right)^2}{S_i - \sum_{j=1}^{S_i} h_{V_{ij}}^2 + 1.0 + 2r_{X_qD_i}\left(\sum_{j=1}^{S_i} r_{V_{ij}D_i}\right) + \left(\sum_{j=1}^{S_i} r_{V_{ij}D_i}\right)^2} \tag{12.21}$$

Theoretically, the critical value of a variable as a potential expander of a cluster is related to the factor coefficient of the variable with respect to that cluster dimension. Re-expressing (12.20) in terms of the factor coefficient, setting (12.20) equal to the reliability coefficient of the cluster, and solving for the root of the quadratic equation implied by the expression gives a lower bound for the factor coefficient of a variable that can be added to the cluster and increase the reliability of the cluster. This lower bound of the factor coefficient of a variable X_q can be calculated by

$$\text{LBFC} = r_{C_iC_i'}\left(\sum_{j=1}^{S_i} r_{V_{ij}D_i}\right) - \sum_{j=1}^{S_i} r_{V_{ij}D_i}$$
$$+ \sqrt{\left(\sum_{j=1}^{S_i} r_{V_{ij}D_i}\right)^2(1 - r_{C_iC_i'})^2 + r_{C_iC_i'}} \tag{12.22}$$

If the variable X_q has a factor coefficient $r_{X_q D_i}$ equal to or greater than the LBFC, it should enhance the reliability of C_i when C_i is expanded to include it. This lower bound is only theoretical, and the empirical method of trying the potential expanders is used in the program, although the LBFC is output for each cluster.

When an added variable carries a negative sign on its factor coefficient with the defining cluster, it must be reflected. This means that its factor coefficient becomes positive and that all its correlations with definers and other added variables must have signs changed. Reflection of variables is discussed in more detail in the section below on cluster analysis procedures.

Miscellaneous statistics and analyses within cluster structure analysis

A number of statistical and logical analyses of secondary interest are performed by the BC TRY component. They are listed here with brief explanations.

The augmented correlation between observed variables is the correlation coefficient divided by the product of the square roots of the communalities of the respective variables

$$r_{Dx_{,i}Dx_{,i}} = \frac{r_{X_i X_j}}{h_{X_i} h_{X_j}} \tag{12.23}$$

These coefficients represent the correlations of the variable domains for all NV observed variables.

All-objective, nontheoretical description of the cluster structure in a sample of observations can be achieved by reorganizing the raw correlation matrix. The rows and columns of the matrix are simply arranged so that variables in a cluster appear together in the matrix, placing nonclustered variables next to the cluster with which they correlate most highly $r_{X_j D_i}$. Variables are reflected to make intracluster submatrices positive, and variables with communalities below .20 are given positions in the rightmost columns and bottommost rows of the matrix.

The mean correlation of each variable with the definers of each cluster is also calculated by CSA. These values can be calculated without resort to the raw correlation matrix by the use of

$$\bar{r}_{X_i C_i} = r_{X_i D_i} \frac{\sum\limits_{k=1}^{S_i} r_{V_{ik} D_i}}{S_i} \tag{12.24}$$

The BC TRY component CSA

The purpose of CSA is to apply the equations given above to clusters of variables. The clusters may come from a form of cluster analysis, or they may be input through the GIST option of the system. The limits of the component are similar to the limits on the cluster analysis components of the system: 15 clusters, each consisting of up to 20 variables.

In addition to the correlation information about the cluster composites, cluster domains, variables, and variable domains, other information is also generated by the component. A complete list of the IST input and output of the component is given in Table 12.2. The output to the printer is as follows:

1 Defining variables of the clusters
2 Correlations between cluster composites and between cluster domains
3 Domain validities of cluster composites
4 Correlations of variables with cluster domains
5 Generally of each cluster: the factor contribution to the communalities of the variables and to the correlation among variables
6 Expanded cluster characteristics: reliability of each cluster, reliability when each additional variable is included in the cluster, reliability when the additional variables are included incrementally
7 The clustered correlation matrix (optional)
8 The augmented correlation matrix or correlations corrected for uniqueness (optional)

Noncommunality cluster structure analysis

The domain sampling model for CSA depends on the exact collinearity of the variables in the domain. This assumption leads to the requirement that the communality of each variable be available, or at least estimated. A second domain sampling model avoids the necessity of having communalities. The second model is called the "noncommunality" or "NC model." In the NC model, a cluster composite is defined in the same way that it was defined in CSA. However, the cluster domain composites are defined in slightly different ways. The domain is composed of the observed variables in the cluster plus a very large number of other variables having the same pattern of correlations that exists in the sample of variables in the observed sample

$$C_i = V_{i1} + V_{i2} + \cdots + V_{iS_i} \tag{12.25}$$

$$D_i = C_i + \sum_{j=k}^{\infty} V_{ij} \quad \text{for } k = S_i + 1 \tag{12.26}$$

The variables in the last term of (12.26), that is, $V_{i(S_i+1)}$, $V_{i(S_i+2)}$, . . . , V_{i_∞}, are not in the observed set of variables but in the theoretical domain of the cluster. This definition is not unlike that of the domain in CSA except that in CSA the variables in the domain were collinear replicas of the observed variables. In an NC cluster domain the variables are not necessarily collinear but satisfy a pattern condition. The pattern of intercorrelations in the domain must be the same as the pattern in the sample to the extent that the mean intercorrelation in the domain must be equal in the limit to the mean intercorrelation in the sample cluster.

The cluster domain correlation properties of a cluster are essentially the same in NCSA and CSA. The equations in NCSA generally are the same as those in CSA except that the mean observed intercorrelation takes the place of communalities or the diagonal term is dropped in many of the expressions. These specific results will not be repeated here since they are similar to the CSA results and are also presented in Tryon (1957b and 1958a).

In practice, the values of the correlation properties of clusters computed in NCSA are usually very close to those of CSA.

It is possible to define clusters with mean intercorrelations that are negative, particularly when the clusters are determined by some psychological theory. When this happens, the cluster structure analysis becomes indeterminant and the BC TRY component NCSA suffers a program halt. Consequently, except in special need, CSA is recommended over NCSA in cluster structure analysis.

Diagonal values

The first major decision to be made in performing a cluster or factor analysis is what portion of the observed variance among individuals in each of the NV attributes should be described by scores on the different clusters or factors. This portion is called the "diagonal element" to be inserted in the correlation matrix when those forms of dimensional analysis are performed that require the presence of such an element (CC5 and FALS).

The usual decision is any one of three: (1) all the variance, hence the diagonal elements are 1.00; (2) the portion that is reliable, hence reliability coefficients are used; or (3) the portion that is general, hence communalities are used, denoting the amount of variance of each variable that is common to it and to all the other $NV - 1$ variables that sample different attributes.

Since the general objective of cluster and factor analysis is to replace the NV variables by K composites that describe all that is general among individual differences across all NV attributes, communalities are usually chosen as the diagonal values (though in some special studies one may wish to use 1.00 as diagonal values). We focus here on the methods of estimating the communalities of the NV variables by various subprograms of DVP.

On domain sampling principles, communality has a definition similar to that of the reliability coefficient of a variable. The reliability coefficient of a sample variable X_j is the square of the correlation between X_j and a hypothetical domain score that is a composite of many scores that sample different attributes, sample variables whose correlations with the remaining $NV - 1$ variables are collinear, i.e., proportional with those of X_j. Specifically

$$h^2_{X_j} = r^2_{X_j D_{X,j}} \tag{12.27}$$

where D_{X_j} is defined as the domain dimension of the variable with a form similar to that of (12.3) and (12.4). This model states that D_{X_j} is a collinear domain composed of scores on different attributes whose patterns of correlations with the other $NV - 1$ variables of the study are identical. A second basis for (12.27) is the model that defines the communality as the correlation between X_j and another collinear variable that has exactly the same magnitudes of correlation with the other $NV - 1$ variables of the study.

The use of communalities

Communalities are used in a number of ways in cluster and factor analysis. In the BC TRY System they have two important uses: (1) as an index of generality and (2) as a criterion to determine the number of cluster or factor dimensions. The most important use of communality of a variable is as an index of individual differences in the variable across the other $NV - 1$ variables of the study. This is a purely objective use of the definition of communality given in (12.27). Communalities are used in a number of ways in the process of factoring a set of variables. The first use in factoring is to estimate the communalities before factoring and then to factor until the estimates are "essentially" reproduced by the factors. In most analyses it is sufficient to factor up to the point where some portion, say 98 percent, of the grand total of the communalities is reproduced. Initial estimates of communalities here form a terminating criterion of factoring. The second use of communalities is to insert them in diagonal cells of the correlation matrix during factoring. In this use, communalities perform a double role, their necessary role as diagonal elements and their optional role in forming a saliency criterion for terminating factoring.

Solving for values of the communalities

In a given problem involving NV variables there is, in principle, a unique value defined by (12.27). In practice, it may be difficult to solve for the exact value, but a close approximation can always be found. Each type of solution implemented in the BC TRY component DVP is a special operational design formulated to secure approximations to (12.27). There are three such designs, namely, the communality $h_{X_j}^2$, estimated from

  1 The correlations of X_j with the other $NV - 1$ variables
  2 A subset of variables most collinear with X_j
  3 A salient dimensional analysis performed by factoring

Some of the estimates are direct solutions in which the communalities are unknown in equations and are solved for directly. Others are indirect or iterated solutions in which "trial" values of the communalities are set in equations that permit one to solve for the communalities; then iterated solutions are performed until all communalities converge at some specified degree of precision or until a specified number of iterations are made. Both direct and reiterated solutions are available under the three main computational designs.

The quadratic formula, QF

The variable X_j is predicted by the regression of X_j on the remaining $NV - 1$ variable domains in the QF procedure. It can be shown that D_{X_j} is contained in, i.e., linearly dependent on, the configuration of the $NV - 1$ domains representing the other $NV - 1$ variables and that the locus of D_{X_j} in the domain space can be determined precisely from the locus of the other $NV - 1$ points and their geometric relationships. Thus, we can replace D_{X_j} with a suitably weighted composite of the $NV - 1$ remaining variables. The squared correlation of X_j and this composite is an estimate of the communality of X_j. Let

$$D_s = \sum_{i=1}^{NV} w_i D_{X_i} \qquad i \neq j$$

where the w_i's are selected to maximize the squared multiple correlation $R_{X_j D_s}^2$. Then

$$h_{X_j}^2 = R_{X_j D_s}^2 \tag{12.28}$$

Calculating $R_{X_j D_s}^2$ for $i = 1, \ldots, NV$ using $NV - 1$ predictors for each variable is impractical where NV is large. Hence the DVP routine QF uses a most collinear subset of the $NV - 1$ variables (the number of variables in the subset is selected by the user, with a maximum of 10).

In order to calculate (12.28), initial values must be inserted into the diagonals of the subset of variables used to define D_s. The highest correlation of X_j with the remaining $NV - 1$ variables is used as this initial value ($j = 1, \ldots, NV$). Using these initial diagonal values, the submatrix for calculating (12.28) for X_j is augmented by dividing each correlation by the square root of the product of the corresponding initial diagonal values. This augmented matrix is used to evaluate (12.28) as a straightforward, squared multiple correlation from the symmetric augmented matrix with the one adjoined column of semiaugmented correlations of X_j, that is, divided by the square root of the initial diagonal values of the respective variables in the subset. Equation (12.28) is evaluated for $j = 1$, $\ldots, NV$, the values just obtained are placed into the diagonal of the correlation matrix, and the process is repeated until the communalities converge to stable values.

In empirical problems where the predictor variables are a subset of the NV variables in the problem, converged communalities are usually quickly secured, but a few may give somewhat erratic values. If all $NV - 1$ variables were used in the regression, the communalities should all converge rapidly.

Approximation B and modified approximation B

These approximations are based on the fact that a cluster composite of collinear variables, say V_{i1}, V_{i2}, V_{i3}, also collinear with X_i, can be used as a surrogate for the domain of the variable X_i. Hence the correlation of X_i and the cluster composite is approximately equal to the correlation of X_i and its variable domain D_{X_i}. Let the cluster composite be

$$C_i = V_{i1} + V_{i2} + V_{i3}$$

If the definers of this composite are perfectly collinear, it can be shown that

$$h_{X_i}^2 = r_{X_i C_i} = \frac{\displaystyle\sum_{j=1}^{3}\sum_{k=1}^{3} |r_{X_i V_{ij}} r_{X_i V_{ik}}|}{\displaystyle\sum_{\substack{j=1 \\ k \neq j}}^{3}\sum_{k=1}^{3} |r_{V_{ij} V_{ik}}|} \tag{12.29}$$

This approximation tends to give values biased downward because the cluster composite is based on only approximately collinear variables. To get a less biased estimate this approximation is modified by averaging the value obtained from (12.29) and the value of the highest correlation for any of the $NV - 1$ variables correlated with X_i. This average is called the "modified approximation B."

Triads

The triads method of calculating an approximation to the communality is equivalent to approximation B except that only two variables, say V_{i1}, and V_{i2}, are used to form the cluster composite C_i in finding the communality of the variable X_i.

Proportional fit, PF

This method of calculating communalities is similar to the QF method and approximation B with respect to the use of a subset of variables that are relatively collinear with the variable for which the communality is to be calculated. A set of variables most collinear with X_i (up to nine, under control of the user of BC TRY) are selected; these variables are called the "reference variables" for X_i. The submatrix of intercorrelations among these variables is formed, and the diagonal is filled with the highest correlation that the respective variables exhibit with the other $NV - 1$ variables in the sample. This matrix is adjointed with the vector of correlations of X_i with the variables in the matrix. The linear regression equation of the correlations in the adjoint vector on the correlations in the matrix is solved, giving rise to regression weights for each of the reference variables. These regression weights are applied to the respective correlations between X_i and the reference variables. The resulting value is the best estimate of the diagonal element for the variable X_i, with respect to the degree that the intercorrelation matrix of the reference variables predicts the intercorrelations between the reference variables and variable X_i. The fact that the reference variables are the most collinear with X_i tends to ensure that the regression-predicted value is proportionally the best estimate of the diagonal element. Although this method is not in an obvious way a direct calculation of communalities, it produces diagonal elements that are consistent with the pattern of intercorrelations in the rest of the correlation matrix.

The PF method is an indirect solution. First the diagonal elements in the matrix are filled with preliminary values. Then, new values are solved for, for each variable in the matrix, using the PF algorithm. The new values are entered as the diagonal elements in the matrix and the process repeated until the diagonal elements fail to differ on successive iterations or until a fixed number (determined by the user) of iterations are achieved. A maximum of 10 iterations is permitted by the PF routine in DVP.

Independent dimensional analysis

The independent successive dimensions from factoring give rise to communalities. When the factoring procedure is effective, K dimensions will represent the variability shared by the behavior properties of the observed variables. If these K independent dimensions are held constant, the correlations of the observed variables will be essentially zero. The correlation of a variable X_i with the K independent dimensions or domains, when squared and summed, is equal to the correlation of X_i and another hypothetical variable with the same magnitudes of correlations, that is, X_i

$$r_{X_iX_i'} = r^2_{X_iD_1} + r^2_{X_iD_2} + \cdots + r^2_{X_iD_K} \tag{12.30}$$

A definition of communality alternate to the definition given in (12.27) is implied in (12.30). It can be shown that the communality is the correlation between a variable X_i and another collinear variable that has exactly the same magnitudes of correlation with the other $NV - 1$ variables in the sample, i.e.,

$$h^2_{X_i} = r_{X_iX_i'} \tag{12.31}$$

where X_i' has the same magnitudes of correlations with the other variables as those exhibited by X_i.

The roles of factoring in calculating communalities and of communalities in factoring are encountered again in the next section.

Ad hoc diagonal values

An ad hoc estimate of communality is one that is not directly formulated on a basic psychometric model of communality. Several such definitions of diagonal values are in wide use because of empirical success with the definition in factoring or because of some rational bearing on the definition of communality. The PF method of calculating communalities is an example of an ad hoc method. One other ad hoc definition of diagonal values is in wide use in factor analysis: each diagonal value is set to 1.00. The practice of setting diagonal values to 1.00 is widely accepted in orthodox factor analysis. For principal-axes factor analysis the 1.00 values in diagonals may produce the most adequate results. In cluster analysis, as implemented in BC TRY, the diagonal values should not be set to 1.00 as this will generally result in an attempt to define too many clusters. The advantages and disadvantages and special implications of diagonal values of 1.00 in factor analysis are discussed extensively in other sources. Another frequently used ad hoc estimate of the communalities of a variable is the highest correlation of the variable with the other $NV - 1$ variables.

Collinearity in diagonal values routines

In several of the methods of calculating communalities, subsets of collinear variables must be selected out of the NV variables sampled. In general, collinearity is defined in BC TRY by large values of an index of proportionality defined and discussed in the following section. In the cluster analysis program CC5 of BC TRY a complicated algorithm is used to select collinear subsets. In DVP, only the collinearity, defined by the index of proportionality, between the variable X_i and the other $NV - 1$ variables is used to select the subset.

The BC TRY component DVP

The purpose of DVP is to apply the equations and algorithms discussed above to the matrix of correlations. In addition to the communality estimates, the component prints the 10 highest correlations of each variable. A complete list of the IST input and output of the component is given in Table 12.2.

Cluster analysis

The theory and techniques of key-cluster analysis are discussed in extensive detail in Chap. 6. This section is a more summary presentation of the step-by-step procedures, equations, and algorithms of the BC TRY procedures of key-cluster analysis, as expressed in the component programs CC5 and NC2. The options under which the programs are run are also spelled out here in a general way. The cluster structure analysis procedures and models are applicable in general to the theory of cluster analysis. The definitions of a domain, a cluster, correlations of variables with clusters, etc., all pertain to cluster analysis as well as cluster structure analysis. In the BC TRY System the general practice, in empirical research, is to cluster-analyze the multivariate data of the research by an application of the cluster analysis programs before doing a cluster structure analysis. In BC TRY the cluster analysis programs are not designed to evaluate cluster structure but more simply to select clusters on the basis of a matrix of successively reduced correlations, each successive matrix being uncorrelated with the previous clusters. In order to do this, each successively selected cluster is used to define a fictional "dimension" or factor. The correlations of each variable with the fictional dimensions are calculated at each step of the clustering process. The properties of these factor dimension correlations permit evaluation of the degree to which the clusters account for the mean squared correlation coefficient and the initial estimate of communality. Cluster selection is the primary purpose of the BC TRY cluster analysis programs. As a secondary goal, CC5 iterates the calculation of the correlations of variables with the factor dimensions in order to reestimate the communalities of the variables until the estimates converge. During cluster selection and after convergence, the program calculates and prints a variety of statistics giving the correlational structure of the factor dimensions.

Steps in selecting a collinear set of variables

Cluster selection in BC TRY components CC5 and NC2 is either by inputting the cluster members (preset) as part of the control and data cards or by collinear subset analysis. In selecting a collinear subset of variables to define a key-cluster dimension, the programs use the initial correlation matrix for the first cluster residual correlation matrices for each successive cluster. The description in this section uses notation relevant to the initial correlation matrix.

First, a pivot variable is selected. The criterion for pivot variable selection is intended to ensure that the pivot variable be most likely to appear at the "edge" of the multivariate space and be surrounded by a cluster of other variables. Such a variable would have relatively high correlations with its most collinear subset and low correlations with variables at remote points in the configuration. That is, it would tend to have a large variance of its correlations with the other $NV - 1$ variables. In order to avoid the effects of negative signs on the correlations, the squares of the correlations are used; the variance of the squared correlations of a variable in the entire set of variables is the index of pivotness used

$$\sigma_i^2 = \frac{1}{NV - 1} \sum_{j=1}^{NV} (r^2_{X_i X_j} - \overline{r^2_{X_i X_i}})^2 \qquad i \neq j \qquad (12.32)$$

The selected pivot variable for cluster j is the variable with the largest σ^2. This variable becomes V_{j1} in the cluster.

Next, at least one additional definer is added to the cluster, as follows. The second variable in the cluster V_{j2} is that variable with the highest index of proportionality $P^2_{V_{j2} X_i}$. The variable X_i, the third variable in the cluster V_{j3}, is added when the proportionality indexes of it with the previous two, that is, $P^2_{V_{j1} X_i}$ and $P^2_{V_{j2} X_i}$, are greater than .40 and it has the highest mean index of proportionality with the two variables already included. The same criterion is applied to the fourth variable of the cluster except that the average index is found for the previous three variables. Additional definers are included only if their mean index of proportionality is within twice the range of the indexes of proportionality among the four first-selected variables and if all of the indexes of proportionality of the variable and the previously selected variables are greater than .81 (an arbitrary criterion that can be modified by the program user). This combination of criteria is intended to select a larger number of variables for a cluster if the cluster is extended only by adding definers relatively collinear with the definers already selected. Also, when four variables are selected, the criterion for membership is tightened to permit extension of the cluster membership only when the additional variables are highly collinear with the already selected cluster definers: this permits definition of larger clusters only when the cluster is very "tightly" defined in the configuration.

If a pivot variable fails to pick up an additional definer, a trial-and-error procedure is utilized. A second pivot variable is tried, the variable with the second highest variance of squared correlations. Four (or another number specified by the user) such pivots are tried. If none picks up the additional definer, factoring is terminated.

The collinearity criterion of cluster selection is based on an index of proportionality P^2. The justification of this is as follows. We seek a subset of variables to define a cluster, such that the correlations of the variables with the cluster dimension can be used to reproduce the intercorrelations of the variables in the clusters. This would mean that the cluster dimension C_i, treated as a simple variable, would interact with two definers of the cluster so as to make the partial correlation of the definers, conditional on the cluster dimension, have a value of zero

$$r_{X_j X_g \cdot C_i} = \frac{r_{X_j X_g} - r_{X_j C_i} r_{X_g C_i}}{r_{X_j C_i} r_{X_j C_i}} = .00$$

where

$$r_{X_j X_g} = r_{X_j C_i} r_{X_g C_i} \tag{12.33}$$

Any two members of the subset of variables defining the collinear cluster have proportional correlations. This is given by taking the ratios of the correlations expressed in (12.33) for two variables in the cluster C_i, say V_{ij}, V_{ik}

$$\frac{r_{V_{ij} X_g}}{r_{V_{ik} X_g}} = \frac{r_{V_{ij} C_i} r_{X_g C_i}}{r_{V_{ik} C_i} r_{X_g C_i}} = B_{V_{ij} V_{ik}} \qquad g = 1, \ldots, NV \tag{12.34}$$

where B is the coefficient of proportionality of the two collinear variables. The degree to which B varies as a function of g in (12.34) is a measure of the degree to which the two variables V_{ij} and V_{ik} are not collinear. This can be measured by the departure from 1.00 by an index called the "index of proportionality" (see Burt, Tucker; 1951; and Wrigley and Neuhaus, 1955):

$$P^2_{X_i X_j} = \frac{\left(\sum\limits_{k=1}^{NV} r_{X_i X_k} r_{X_j X_k}\right)^2}{\sum\limits_{k=1}^{NV} r^2_{X_i X_k} \sum\limits_{k=1}^{NV} r^2_{X_j X_k}} \tag{12.35}$$

For any pair of variables, the closer P^2 is to 1.00, the more nearly collinear they are. The mutual collinearity criterion used in CC5 and NC2 is simply a procedure of selecting a subset of definers whose values of P^2 are both as near to 1.00 as possible, beginning with the pivot variable.

The procedures for calculating the index of pivotness and the index of proportionality in NC2 differ from those in CC5 in that there are no diagonal elements in the correlation matrix for NC2. Equations (12.32) and (12.35) are written for the CC5 procedures and must be modified to represent the NC2 procedures.

In CC5 it is possible to indicate the number of variables that will compose each cluster. A special application of this capability is to set this number S_i for the ith cluster to 1. Each cluster is thus defined by the pivot variable. The analysis defined in this way is known as the "square root method" or the "pivot variable method" of factoring the correlation matrix. It is possible to select the pivot variable in this application by other methods than the index of pivotness method described above. The communality, the sum of squared correlations, or the pivotness index are all possible selections.

When the user of the cluster analysis programs selects the number of variables to be in each cluster as the number of variables in the entire study, the resulting analysis is a centroid factor analysis. This option is applicable only where NV is no greater than 20. If NV is greater than 20, a special option permits subsets of 20 to be selected on the basis of the sum of squared correlations or the variance of the squared correlations; the 20 variables with the largest sums or largest variance are used to define a salient centroid in the entire set of variables.

Reflecting the definers of a cluster

The metric by which individuals are ordered on all the definers of a cluster may not be unidirectional, causing the definers to intercorrelate negatively. To ensure unidirectionality of the definers they must be reflected until the submatrix of correlations between definers is maximally positive. A reflection is accomplished simply by multiplying the row and column for a reflected variable in the correlation matrix by -1. For small clusters (10 or fewer members) it is practical to inspect the 2^{S-1} (S is the number of variables in the cluster) different patterns of positive and negative signs that define the possible reflections of the matrix. The optimal reflection is obtained in this way, giving the maximal sum or correlation coefficients among the definers of the cluster. Where S is too large to make this procedure attractive, the variables are reflected successively until the sum of correlations in the matrix fails to increase with additional reflection. The reflected variable at each step is that variable having the largest negative sum of correlations in the submatrix of the cluster definers.

Factor coefficients

For each cluster in the cluster analysis a fictional dimension is defined such that it correlates very highly with the cluster domain with the restriction that all the fictional dimensions are uncorrelated. These fictional dimensions are called factors. The factors are defined in the same succession that the clusters are defined. The first factor is defined by the first cluster, the second factor by the second cluster, etc. Each successive factor is used to determine the correlation among the NV variables that is independent of the factor. This process results in a residual matrix of correlations after the first factor, independent of the first factor; a second residual matrix of correlations after the second factor, independent of the first two factors; etc. The successive factor dimensions are defined in a cluster analysis by applying Eq. (12.36) to the correlations in the successive residual matrices. The superscript on the correlation coefficients in the equation denotes the factor; $r_{X_iX_j}^{(k)}$ denotes the correlation matrix used in determining the kth factor. In the instance of $k = 1$ the original correlations are involved; for $k = 2$ the correlations are those obtained by calculating the residual correlations after the first factor, i.e., independent of the first factor; etc. Denoting the K factors by F_1, F_2, . . . , F_K, the equation for the correlation between a variable X_i and the jth factor is

$$r_{X_iF_j} = \frac{\sum\limits_{g=1}^{S_j} r_{X_iV_{jg}}^{(j)}}{\sqrt{\sum\limits_{g=1}^{S_j} \sum\limits_{k=1}^{S_j} r_{V_{jg}V_{jk}}^{(j)}}} \tag{12.36}$$

Since the K factors are independent, the factor correlation variances are additive and the sum of the squared factor correlations of a variable, summing across all factors, is the communality of the variable

$$h^2_{X_i} = \sum_{j=1}^{K} r^2_{X_i F_j} \tag{12.37}$$

Each of the terms in this sum is known as the "partial communality" of the variable on the factor, that is, $r^2_{X_i F_j}$ is the partial communality of the ith variable on the jth factor.

For the purpose of stating the percentage of the communality of a variable determined by the successive dimensions, each of the squared factor correlations (12.36) is divided by the total communality (12.37). The square root of these values is called an "augmented or normalized factor coefficient." They are calculated directly by dividing each factor coefficient (12.36) by the square root of the total communality (12.37)

$$r_{Dx_i F_j} = \frac{r_{X_i F_j}}{h_{X_i}} \tag{12.38}$$

The sum of squares of these coefficients, for each variable $i = 1, \ldots,$ NV, is equal to 1.00. The implication is that each variable is mapped onto the surface of a unit hypersphere of K dimensions. The coordinates of these surface points are used in the geometric description of the cluster structure in the BC TRY component SPAN.

In noncommunality cluster analysis, component NC2, the basic calculations are similar to those of CC5 with the exception that there is no diagonal element in the correlation matrix. The average intercorrelation instead of the sum of intercorrelations is used in (12.37).

Residual and reproduced correlations

The so-called "fundamental factor theorem" states that the correlation of two variables can be reproduced by the inner product of factor coefficients for that variable if sufficient factors are defined in the factor analysis of the variables. Since, in general, only K factors are defined, only a certain portion of all the intercorrelations among the observed variables is reproduced. The reproduced correlation of X_i and X_j is given by

$$r'_{X_i X_j} = \sum_{g=1}^{K} r_{X_i F_g} r_{X_j F_g} \tag{12.39}$$

In the general case these coefficients are not equal to the original coefficients. The differences between the original correlations and the reproduced correlations are called the "residual correlations." In the process of factoring the correlation matrix, i.e., of calculating the successive factor dimensions, $F_1, F_2, \ldots, F_K$, the amount of the original correlations reproduced increases with each successive factor. Thus, there is a succession of residual and reproduced correlation matrices. The residual correlation between X_i and X_j after F_k has been accounted for in the factoring is

$$r_{X_iX_j}^{(k)} = r_{X_iX_j} - \sum_{g=1}^{k} r_{X_iF_g}r_{X_jF_g} \tag{12.40}$$

It is the matrix of these residuals that is used to define the $k + 1$ cluster and factor F_{k+1}. When the residuals are sufficiently small to satisfy some criterion of saliency of correlation, the factoring process is stopped and $K = k$ is accepted as the salient dimensionality of the multivariate data.

Terminating factoring

Two criteria for termination of factoring are built into the cluster programs of BC TRY. When the total squared correlations among the variables are approximated by the total squared reproduced correlations based on k factors, factoring is terminated. The degree of approximation is selected by the user of the system to represent what he thinks a salient amount of the correlation matrix is. The successive factors also account for a successively larger amount of the initial estimates of communality. When the total communality accounted for by k factors approximates total initial communality by some salient degree, the factoring process is terminated. When one or another of the following quantities is close enough to zero to satisfy the saliency criterion, factoring is terminated

$$T_k = \sqrt{\frac{\sum_{i=1}^{NV} \sum_{j=1}^{NV} (r_{X_iX_j}^{(k)})^2}{\sum_{i=1}^{NV} \sum_{j=1}^{NV} (r_{X_iX_j})^2}} \tag{12.41}$$

$$U_k = 1 - \frac{\sum_{i=1}^{NV} \sum_{g=1}^{k} r_{X_iF_g}^2}{\sum_{i=1}^{NV} h_i^2} \tag{12.42}$$

Iteration for convergence on communalities

In communality cluster analysis the communality estimated on the basis of the factoring is generally quite different from the communality initially estimated. If the initial estimates are replaced by the estimates from factoring and new factor coefficients calculated, the communalities estimated from the second factoring will be still different. In order to have estimates of communalities entirely consistent, the process of calculating factor correlations is repeated a number of times until the difference between communality estimates on successive calculations is essentially zero. The process is called "iteration to convergence on communalities." BC TRY component CC5 permits convergence to a desired criterion of convergence or as a standard option four iterations, after which the differences are quite likely to be small.

The BC TRY component CC5

Options in the component permit the application of a wide variety of analyses, among which are:

1 Empirical key-cluster analysis. The program selects the definers of the K clusters and determines the number of clusters.

2 Preset key-cluster analysis. This form of cluster analysis has been known by factor analysts as "multiple group factoring." The user of the program inputs the definers, selected on rational grounds or on the basis of a previous empirical key-cluster analysis.

3 Preset dimension analysis. The user lets the component select the clusters but indicates on the control cards the number of clusters desired.

4 Pivot variable analysis. Each dimension is defined by one variable, selected in the way pivot variables for empirical cluster analysis are selected. This mode of factoring is called "square root factoring" in the literature of factor analysis.

5 Centroid factor analysis. In standard centroid factor analysis the factors are defined on all NV variables. In centroid factor analysis in the CC5 component, the factors are defined on the 20 variables with the largest sum of squared correlations. If NV is equal to or less than 20, the analysis is a standard centroid analysis; otherwise it is referred to as a salient centroid analysis.

6 Bifactor analysis. The first dimension is defined by a general or salient centroid factor analysis, and the remaining dimensions are defined by key-cluster analysis, either empirical or preset. The use of CC5 in this analysis is repeated. The first use of CC5 involves the use of the centroid factor analysis option and a preset of the number of dimensions at one. The program FAST in the system is then used to replace the original correlation matrix with the residual matrix, communalities are reestimated by using the component DVP, and then CC5 is used to perform the key-cluster analysis.

7 Sleeper analysis. Before using CC5 the analyst uses the program SLEP1 to delete temporarily from the correlation matrix a number of variables he does not wish to be involved in the cluster factoring procedures. Once the cluster analysis is completed, the user calls SLEP2, which recalculates the factor coefficients and other statistics from CC5 on all the original set of variables, including those that were sleepers in the CC5 analysis. This design requires a sequential use of four components, SLEP1, DVP, CC5, and SLEP2.

8 Ordered and designed analyses. By sequencing the use of components in the BC TRY System before and after CC5, the cluster analysis program can be used to achieve a very wide spectrum of analytic designs.

The input and output of the component program to and from the IST are listed completely in Table 12.2. In addition to these files a large number of statistics and lists are output on the printer. At the beginning of the analysis the following information is output: descriptions of the methods and parameters selected by the user, mean squares of the original correlations, initial communality estimates of all the variables, and the sum of initial estimates of communalities. At the end of the last iteration of factoring the following are output as a summary of the analysis: lists of cluster definers, the factor coefficients of all variables on each factor dimension, the partial communalities, the cumulative communalities, the augmented factor coefficients, the squared factor coefficients, the proportion of the sum of squared correlations accounted for by the K dimensions, the proportion of the sum of communalities exhausted by the K dimensions, the mean of the squares of residuals after each dimension, and the matrix of residual correlations after the last dimension. For the first iteration of factoring the printout includes: sum of squared correlations or residual correlations for each variable, the variance of squared correlations, the index of proportionality P^2 of each definer of the dimension with each of the NV variables, the submatrix of correlations between the defining variables, the list of variables defining the cluster, the factor coefficients of all of the variables, the partial communality of the variables, the cumulative communality of the variables, the proportion of the communality and sum of squared correlations exhausted by the factors so far defined, and the sum of squares of the residual correlations.

The BC TRY component NC2

The NC2 component of BC TRY is so similar to CC5 that it suffices to point out only those details which differ significantly. The lack of diagonal elements implies that a pivot variable analysis is not possible. The communality exhaustion criterion is not a particularly meaningful criterion although it can be used. The solution is not iterated since the communality values are not involved in defining the factor dimensions.

The spherical configuration

The factored dimensions, being derived from residuals that are values obtained by holding prior dimensions constant, are independent (orthogonal) axes. The factor coefficients of a variable are coordinates of the variable on these axes. Plotting of the variables by their coordinates describes the configuration of points (variables), the structure of the relationships among all the variables.

If the variables fall in independent, uncorrelated clusters, the dimensions derived by key-cluster factoring usually will pass through them. But clusters of variables are usually correlated (oblique); the key-cluster dimensions usually come near to the oblique clusters, subject to the condition that the dimensions are orthogonal.

Efforts to interpret orthogonal dimensions of any sort are usually ill directed since they are merely an orthogonal "reference frame" that "holds" the configuration. To rotate the reference frame does not change the configuration; the axes, lying in new positions, e.g., varimax or quartimax axes, are not likely to take on meaning by virtue of being rotated because they are still orthogonal whereas the structure is usually oblique. In short, the configuration is the thing; the independent cartesian dimensions that frame it are rarely of real interest.

We need to represent visually the structural pattern as a whole, the total configuration. To do so is the object of spherical analysis, programmed by the component SPAN2. The total configuration would be invariant with respect to the method of factoring if, by whatever method, we factored through to NV dimensions. But we do not: we factor on $k < NV$ dimensions. But a distinctive feature of factoring is that, as factoring proceeds, additional dimensions describe a configuration that departs less and less from the total configuration. Thus a set of K salient dimensions describes the salient features of the configuration; additional $K + 1$, $K + 2$, etc., dimensions trivially alter the configuration. Key-cluster factoring shares with other factoring methods this attribute, but the configuration is stable even if we add more key-cluster dimensions. Rotated centroid or principal-axes (FALS) dimensions may present quite different aspects of the configuration as additional trivial dimensions are added and rotated, particularly if the varimax rotation method is used.

The main point here, however, is that graphic presentation of the salient configuration is extremely important in structure analysis. For when the configuration is visually examined, arbitrary features in cluster selection by the factoring method become immediately apparent. The investigator will study the configuration, revise the cluster selection if necessary, and rerun the whole analysis with preset definers of the dimension.

We can summarize the objectives of configuration analysis as follows: (1) to select the particular properties of a geometric model that provide perceptually the best layout of the total salient configuration; (2) to select the minimal subspaces of the K dimensions by which to present the salient configuration (SPAN2); (3) to select the particular type of factored dimension that presents aspects of the configuration most invariant to the adding of dimensions (GYRO); (4) to select from the salient configuration the final set of least oblique, most collinear clusters by which to score individuals (objects).

Selection of spatial properties to

depict the salient configuration

The geometric model on which the configuration is presented is the generalized unit sphere. This model relies on the fact that any variable X_i can be plotted as a point on such a sphere since the sum of its squared augmented factor coefficients (taken as coordinates) equals 1.00, that is,

$$r^2_{Dx,iF_1} + \cdots + r^2_{Dx,iF_K} = 1.00 \qquad (12.43)$$

As a surface point, variable X_i has an augmented communality of 1.00, (12.38), whence the point refers to the domain score for the variable X_i.

To clarify this logic, take an example of two variables, X_i and X_j, all of whose geometric properties in the spherical model are depicted in Fig. 12.1. Within this two-dimensional subspace of the full solution the reproduced correlation is .429, from Eq. (12.39). The interdomain correlation (augmented correlation) for the two variable domains is .507, from

$$r_{Dx,iDx,j} = r_{Dx,iF_1}r_{Dx,jF_1} + \cdots + r_{Dx,iF_K}r_{Dx,jF_K} \qquad (12.44)$$

By definition of the correlation (12.44) and analytic geometry of a sphere, $r_{Dx,iDx,j}$ is the cosine of the central angle θ of the variables. Here $\theta = .60$.

The communalities, from factoring, have the values $h_i^2 = .88$ and $h_j^2 = .81$. Since the observed variable X_i is by definition collinear with D_{X_i}, the raw score variable X_i is located at distance h_i from the origin and D_{X_i} at distance 1.00 on the surface of the sphere. The correlation $r_{X_iD_{X,i}}$ is the same as the square root of the communality h_i.

Thus, within a three-dimensional sphere defined by factored dimensions F_1, F_2, F_3, the plane in Fig. 12.1 is a slice through the origin, on the surface of which lie domains D_{X_i} and D_{X_j} subtended by a 60° angle from the origin and with the fallible (observed) variables on their vectors as shown. Any two variables can be represented geometrically in this manner whatever the dimensionality of the solution.

Note the other geometric properties of the relations between X_i and X_j in Fig. 12.1. Most important is the distance between D_{X_i} and D_{X_j}

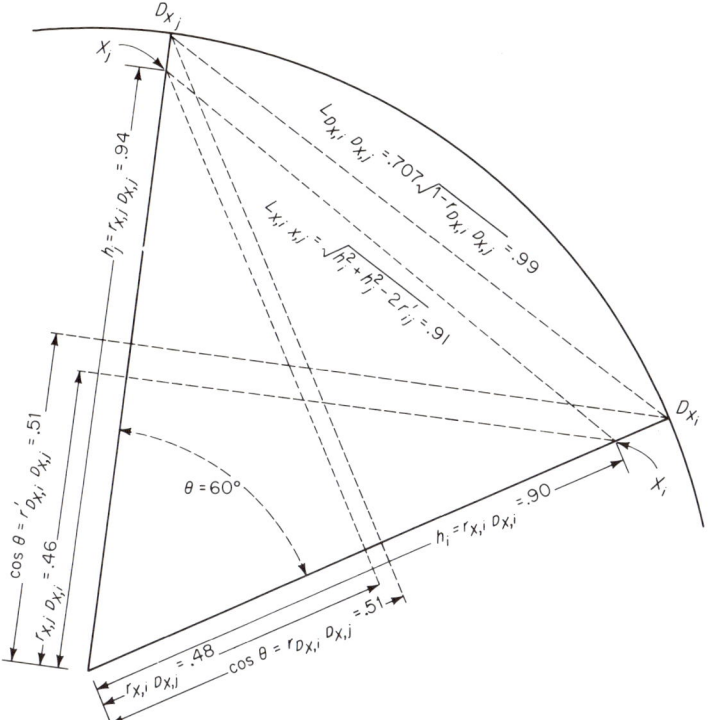

FIGURE 12.1
Spherical model of two variables and their domains.

and the distance between X_i and X_j. The actual value of $r_{D_{X,i}D_{X,j}}$ is represented twice, from the perpendicular projections of D_{X_i} and D_{X_j} on the two vectors, as shown. The correlations between the variables X_i and X_j and the domains D_{X_i} and D_{X_j} are perpendicular projections from their loci. These are the oblique factor coefficients, of special interest when the domains are oblique clusters.

Selection of minimally sufficient subspaces in which to describe the configuration (SPAN)

When the salient factored dimensionality K exceeds three, the total configuration cannot, as such, be visually presented, since it lies in a hypersphere. But the hyperspherical configuration can be fractionated into subspaces in which only those variables are plotted whose augmented communalities, i.e., the proportion of their communalities, approach 1.00 or any other high optional criterion of sufficiency S. If we let S be less than 1.00, there usually exists a set of three-dimensional subspaces in which all or parts of the total configuration can be described. For example, it is not uncommon to be able to describe the total configuration on a hypersphere as a configuration in one three-dimensional sphere at a sufficiency criterion of, say, .80.

The logic of decomposing the hypersphere into subspaces is as follows. At a criterion $S < 1.00$, all variables are allocated to one-dimensional subspaces (single dimensions). The dimension with a maximal number of variables that meet the criterion is selected first. The dimension that includes the maximal number of variables *not* included in the first is selected second, and so on. The dimensions are thus ordered by saliency in their accounting for the variables by one-dimensional sufficiency at level S. Then two-dimensional sufficiency (planes) is tested, in which all variables are allocated to pairs of dimensions. The plane that includes the most variables is selected first. Then the plane that includes the maximal number of variables *not* included in the first plane is selected, and so on. Some variables may carry over to two planes. The minimally sufficient two-dimensional subspaces are thus ordered by saliency. The same principle is followed for three-dimensional subspaces of larger spaces.

When the dimensionality is three, all the variables can be printed as loci on the surface of the sphere. When the dimensionality is greater than three, each sphere includes only variables that meet or exceed S, but any variable with $S < 1.00$ lies below the surface, i.e., projects onto dimensions not in the sphere to the degree $1 - S$. Many variables may carry over from one sphere to another. By simultaneously looking at the configuration on different spheres, "perceptually superimposing" the common variables, the investigator can often "see" the total configuration on the hypersphere.

On any sphere there is an octant formed by the spherical triangle bounded by the three planes that intersect at the three factored dimensions that define the sphere. If all variables included in the sphere lie within this octant, they define a "positive manifold"; i.e., their factor coefficients on all three dimensions are positive. To maximize the positiveness of the manifold, it may be necessary to reflect some variables. Procedurally, one selects as the "formal center" of the octant a point whose factor coefficients on the three dimensions are equal and the squares of which sum to 1.00. Then all variables are reflected if by so doing they are nearer the formal center. Such a procedure concentrates the configuration at one place on the sphere, thus facilitating the perceptualization of the configuration.

The configuration can best be visualized if the centroid of all plotted variables is in the line of the eye. If we call the factored dimensions at the vertexes of the spherical triangle dimensions F_a, F_b, F_c, then "centering" the configuration for visual presentation is achieved by "rigid rotation" of F_a, F_b, and F_c, as follows. Calling the horizontal, vertical, and the inferred third dimension in the plane of the paper F'_a, F'_b, F'_c, respectively, the centroid of the points is rotated to F'_c, and F_a and F_b are rotated to positions F'_b, F'_a, respectively, in such a fashion that the angle of rotation of F_a to F'_b equals the angle of rotation F_b to F'_a. In the BC TRY component the F'_a, F'_b, F'_c coordinates of each included variable are given in the printout and designated by a check (V or R). The transformation is performed by a matrix multiplication. The matrix of augmented factor coefficients, r_{Dx,iF_j} (NV by 3), for the three dimensions defining the sphere is postmultiplied by a matrix T (3 by 3):

$$T = \begin{bmatrix} E(F_a) & -E(F_a) & -E(F_c) \\ E(F_b) & 1 - \dfrac{E(F_b)^2}{1+E(F_a)} & -\dfrac{E(F_b)E(F_c)}{1+E(F_a)} \\ E(F_c) & -\dfrac{E(F_b)E(F_c)}{1+E(F_a)} & 1 - \dfrac{E(F_c)^2}{1+E(F_a)} \end{bmatrix} \tag{12.45}$$

The elements in T are

$$E(F_j) = \frac{\dfrac{1}{NV}\displaystyle\sum_{i=1}^{NV} r_{Dx,iF_j}}{\sqrt{\displaystyle\sum_{j}\frac{1}{NV}\displaystyle\sum_{i=1}^{NV} r^2_{Dx,iF_j}}} \qquad j = a, b, c \tag{12.46}$$

The configuration in the three-dimensional space can be represented by a two-dimensional graph. Each variable meeting the sufficiency criterion S in the space is projected perpendicularly onto the plane defined by F'_a and F'_b, simply plotted as cartesian coordinates. Since each point in the configuration is very close to the surface of the unit sphere (depending on S), the location of the surface point is implied by the location in the F'_a and F'_b plane.

Selecting factored dimensions that make the salient configuration most nearly invariant to added dimensionality, and the rotation of axes (GYRO)

Unaugmented factor coefficients on K dimensions of any variable X_i determined by key-cluster factoring do not materially change if one continues to factor to $K + 1$, $K + 2$, etc. dimensions. Augmenting the factor coefficients under added dimensionality alters only those variables that pick up added communality from the added dimensions, but this added communality is usually small on dimensions added to the salient set K. Therefore, the configuration described by K dimensions is usually not materially changed from that described by dimensions greater than K.

The problem is important because the decision on K is arbitrary, depending on the predilections of the investigator. Thus, if the configuration is changed radically because of the arbitrary decision to add more dimensions than K, then the presented configuration is sensitively a function of an arbitrary decision and becomes untrustworthy.

Since centroid and principal-axes dimensions are defined by all variables, variously weighted by their factor coefficients on the different dimensions, they fall not in centers of collinear clusters but centrally to all variables. Usually impossible to conceptualize, the dimensions are rotated "to meaning." This means placing them in the configuration so that they may be better defined by variables, hence taking on meaning derived from nearby variables.

Quartimax rotation (Neuhaus and Wrigley, 1954; see Harman, 1967, pp. 294ff) is a rotation that tends to concentrate the factor coefficients of each variable on a few dimensions. The net effect is to pass the quartimax dimensions into or near clusters of variables, as key-cluster dimensions do. Therefore, when the configuration is described by quartimax dimensions, it tends, like that described by key-cluster dimensions, to be immaterially affected by added, rotated dimensions.

Varimax rotation (Kaiser, 1958; see Harman, 1967, p. 301) is otherwise. This rotation tends to equalize the factor coefficients of all rotated dimensions. The consequence is to bound the configuration, i.e., generate a positive manifold, with each dimension given about equal weight as a bounding intersection of planes. Therefore, by arbitrarily adding additional dimensions the factor coefficients on earlier dimensions may radically change. The main difficulty is that if the investigator is interested in interpreting the dimensions, then by adding dimensions the interpretation of earlier dimensions may radically change. For this reason, describing the configuration by varimax dimensions seems less desirable than doing so by quartimax.

Selection of the final set of oblique clusters from the configuration

With the salient configuration visually presented in SPAN the investigator can finally select the oblique clusters of variables by which to score individuals by FACS. It is usually found that the key-cluster dimensions pass as near to collinear clusters as is possible subject to the condition of their being orthogonal. Usually the defining variables of key-cluster dimensions are retained as defining variables of the final set of oblique clusters selected by the analyst. But not always. Normally, the investigator wishes to add some variables to an oblique cluster that had not been accepted as definers by the somewhat arbitrary mutual collinearity criterion. The "unifactorial" allocation of variables in CSA2 is helpful here. Some definers at the edge of a defining collinear set may be rejected. Having decided on the final most independent, most collinear oblique clusters, the analyst usually reruns the analysis—especially CSA and SPAN—with preset definers.

The BC TRY component SPAN

The component SPAN is a rather straightforward application of the techniques just described. Each successive three-space is printed out as a map of the cartesian coordinates of F_a and F_b, along with suitable markings in the plot to indicate the original factor dimensions, F_a, F_b, and F_c, the surface arc connecting these points, and a scaling. Each variable plotted in a three-space is located by a spotter, showing the relative location of the variable on the page of the plot. The basic statistics of each of the three-spaces needed to account for the variables are printed. These are the coordinate values in F_a, F_b, and F_c, the augmented coordinate values, and the actual communalities of the variables. The user of the component has a wide variety of options regarding the operation of the component. For example, the user can indicate on control cards those three-spaces he wishes to have printed, or he can have printed certain selected subspaces along with the subspaces that the program would print as a matter of the logic of the program. The spaces can be defined in terms of the unrotated factor coefficients or the factor coefficients obtained by the use of GYRO. Subspaces not containing a certain minimal number of previously undescribed variables can be suppressed in the printout, saving the listing of subspaces with little information beyond that contained in other subspaces already printed. The criterion S can be set to some nonstandard value (standard is .80), variables with communality below a given amount can be deleted from consideration, and the transformation can be defined on the cluster definers alone instead of on all NV variables. The IST summary in Table 12.2 indicates all the files read by the component. SPAN does not output files to the IST or DST.

The BC TRY component GYRO

This component is a direct application of the rotation procedures of the quartimax method or the varimax method as described by their originators.

The rotated factor coefficients and related statistics are printed by the component. The user specifies which of the methods and how many iterations or what convergence accuracy he wishes.

V-analysis of many variables

Some investigators face the task of performing a cluster or factor analysis of data on a very large number of variables. A seeming difficulty is that the BC TRY System has a limitation of NV_{max} (120 or 90 on systems in operation) variables on many of its components. BIGNV surmounts this difficulty. The object is to provide systematic procedures by which the BC TRY System can be employed to perform a V-analysis on any problem, however large the number of its variables may be.

The objective can be quite simply stated. Imagine that we had a super BC TRY System that could perform a V-analysis of an indefinitely large number of variables, a system with no limit to the number of variables it could handle. Suppose, further, one was actually confronted with a problem with 1,000 variables. How could one use the system having only a single-run limit of NV_{max} variables to discover the cluster structure that would have been found in one run by the imaginary super BC TRY System?

The general answer to such a question is as old as statistics. One proceeds by the use of sampling procedures. There is, however, a new wrinkle to BIGNV beyond mere random sampling. Successive samples of NV_{max} variables are taken in such a fashion that V-analysis on these samples increasingly converge upon the results of a hypothetical super single V-analysis performed on all the variables. The V-analysis on the final sample of NV_{max} variables is designed, in short, to depict the structure of all NV variables, however large NV may be; i.e., it is the solution, we would expect, that would have been discovered in one run on all NV variables by a hypothetical super BC TRY System.

Strictly speaking, BIGNV is not a single component program of the System; it is a design of analysis that involves the use of a sequence of components to achieve the statistical objective. Nevertheless, since the sequence stands as a unit it is treated as a component. The principle of sample convergence embodied in BIGNV is applied also in O-analysis.

There are four major stages in BIGNV analysis.

1 *Stage* 1: Tentative selection of a stable set of pivotal dimension-defining clusters. The objective in this first stage is to run V-analyses on a small number of variables drawn randomly from the NV variables. The samples chosen are sufficient to establish the dimensionality K of the NV variables and to select a stable pivotal set

of defining variables of these dimensions. To do this, a full-cycle CC analysis is performed on a first random sample. Next, a second random sample of the variables is chosen that includes as markers the most collinear sets of definers observed in the first sample plus other variables having high communalities. Then, a CC analysis is run on this second sample. Its most collinear pivotal sets of dimension definers may or may not include the first pivotal sets; the second run provides an opportunity to add to the first pivotal sets, to change them, even to change the dimensionality. Now, a third random sample is chosen, including the best pivotal sets plus other high communality variables from the first and second run. It is a matter of judgment how many such runs with carry-over variables are necessary to select a final stable set of pivotal dimension definers; perhaps two or three should be sufficient even in problems with a very large NV.

The carry-over of variables into successive random samples may introduce some undesirable cumulative biasing effects. An alternative method of selecting dimension definers that seems to reduce, perhaps eliminate, such possible cumulative bias is to confine the selection to the most salient variables drawn from fully random samples, i.e., samples not augmented by any carry-over variables. By most salient here is meant the most general variables, as objectively indexed by the value of their communalities derived from CC factoring (or from a DVP program) or as measured by the mean squares of their correlations or by the variances of their squared correlations.

2 *Stage* 2: Construction of a composite configuration of all NV variables. In this stage the sets of pivotal clusters from stage 1 are used as common anchor dimension definers of a series of preset V-analyses on subsamples of the whole pool of all NV variables divided into lots having sizes manageable by the BC TRY components, i.e., into lots of $NV_{max} - S$, where S is the number of common variables that compose the dimension-defining clusters. These lots do not have to be random samples; they can be any grouping of the NV variables. The printouts of CSA in these lots are strictly comparable and can be concentrated into one composite CSA listing, because it is a matter of indifference what lot any particular nondefining variable is in. Findings from CSA are governed entirely by relations with the definers and not at all by the context of nondefiners.

The same is true of the spherical configurations on the SPAN spheres of all these lots. Across all lots, the configurations on a sphere of a given set of three dimensions are strictly comparable. With a little juggling, one can draw a single composite sphere that includes all the variables appearing on it in all lots. The composite spheres of the problem give an overall composite configuration of all NV variables. In drawing the composites one eliminates from consideration variables with low communality.

3 *Stage* 3: Sector analysis for the final selection of M core V-clusters. With all NV variables projected into the same configurations, the final core V-clusters can be drawn from all NV variables. They consist, first, of K pivotal dimension-defining clusters and, second, of any dependent clusters that appear in the configuration, i.e., "natural" clusters that may lie in between the K pivotal clusters. Together the two groups form the final M clusters that describe the total configuration.

In stage 3, therefore, one chooses what appears in the composite configuration to be the best sets of most nearly independent pivotal dimension definers and, using them as new pivots, one reruns the analyses only on samples of variables that are most likely to be chosen to be either in final clusters of dimension definers or independent clusters. For example, for the first rerun in stage 3 preset on the new revised sets of K pivots, one would probably include, in addition to the pivots, only variables in the zone in SPAN around the terminus of the first dimension. The second rerun might include, along with the common pivots, those in the larger sector embracing variables lying in the plane of the first and second dimensions and those also around the terminus of the second dimension. Guidance on what sectors to select in the runs comes from a study of the general structure of the configuration assessed in the light of the objectives to select the most collinear, meaningful, and salient core V-clusters.

At the end of sector analysis the best set of M core V-clusters composed of the most nearly independent set of pivotal clusters will have been selected. These core V-clusters could be the ones on which scores would be computed later by FACS.

4 *Stage* 4: The final representative configuration. In most studies there will be hardly more than NV_{max} variables in the final M core V-clusters. These can all be run in one single V-analysis. In principle, the results should be about what would be found if all NV variables had been run together in a one-run super V-analysis.

In a really massive problem, it may be that more than NV_{max} variables survive to stage 4. If so, then more than one run may be necessary, but since they would be preset on the same final pivotal clusters, the results would be strictly comparable and hence projected into one composite.

The BC TRY components for BIGNV

There are several programs that are used in various combinations for the execution of a BIGNV analysis. They are quite complex, and their description here would serve no useful purpose. The components are available only on the IBM 7094 in Fortran II. The procedures described above can be approximated, with difficulty, without any special component programs.

Comparative cluster and factor analysis

The BC TRY component COMP is designed primarily to permit a direct comparison between the dimensions discovered in one group of individuals and those discovered in other groups. The data on which the comparisons are based are the dimensions of the various groups derived individually, by group, from a V-analysis. There are no requirements in comparative analysis that the different groups have the same number of dimensions.

The dimensions to be compared across the groups may be termed the comparison entities, the columns to be compared. The row entities on which the dimensions are "observed" are the NV variables, termed the common referents of the dimensions. In general, there must always be a set of common referents in relation to which entities are observed if we are to compare them. The observations in the matrix of entities vs. dimensions are the oblique factor coefficients of the dimensions on the

NV variables. The degree to which two dimensions are alike is measured by the index of similarity of their two columns of entities. The index of similarity is optional in the BC TRY component, the index of proportionality, the correlation coefficient, or, best, the cosine of the interangular separation $\cos \theta$ of the comparison entities in the solution space.

There are four main steps in a comparative analysis. The first step is to obtain the matrix of factor coefficients for each of the separate groups to be entered into the comparative analysis. This job is executed by the component program COMP1, which punches the oblique factor coefficients from the results of a cluster structure analysis, i.e., from CSA. When COMP1 is preceded by FALS and GYRO, it outputs the rotated orthogonal factor coefficients. The second step is to adjoin these factor coefficient matrices so that there are NV rows and $K_1 + K_2 + \cdots + K_G$ columns, where K_i is the number of dimensions in the ith group, and where there are G groups. This is the first step of the component COMP2; the G separate decks of cards output from each of the G V-analyses by COMP1 in each of the analyses are stacked together and entered in the data deck for the component COMP2. The third step is to delete selected columns and rows from this adjoined matrix, as indicated by the user in COMP2 control cards. The fourth step is to compute the similarity of each pair of the dimensions, within and between the separate groups.

There has been considerable uncertainty about how to measure the similarity of two dimensions when the observations are not scores on individuals but are factor coefficients. What should be the index of similarity? General concurrence is that it should not be the correlation between their two columns of factor coefficients. Some writers have proposed P, the proportionality index, the square of which is our index of mutual collinearity used in key-cluster factoring (see Burt, 1948; Tucker, 1951; Wrigley and Neuhaus, 1955). The answer, however, is quite straightforward. Clearly within a group, the formula for the index of similarity between two dimensions computed from factor coefficients should give the same value as the correlation between those dimensions (factors) already computed in CSA. We already have those values actually calculated in CSA as correlations between domains. Therefore, all we need is the formula by which these same values can be computed from the columns of oblique factor coefficients also listed in CSA and set up by COMP2 as columns adjoined to those of the other groups.

Within a group, the correlation between two dimensions, the value of which has been computed in CSA from the original correlation matrix, turns out to be a simple quadradic function of the P value, the collinearity measure used in CC factoring but here computed from the two columns of oblique factor coefficients of the two dimensions. Since the correlation between two dimensions as given in CSA is the cosine of their central angle in SPAN, it is perforce an index of their collinearity; hence the index computed in COMP2 from P is called the "collinearity index" or simply cos θ, symbolized as L.

The value of L between any two dimensions within a group can be directly compared with their interdomain correlation given in CSA for that group. These are in correspondence only if the definers of each dimension are themselves exactly collinear, usually only approximately the case. In sum, L is a general index of similarity between dimensions both within and across groups since there is no logical restriction that the dimensions be determined on the same group of individuals.

A full-cycle V-analysis is finally executed on the similarity matrix. This matrix is accepted as an r matrix by the CC factoring component. Structure analysis by CSA and SPAN describes all the relations among the dimensions within and across groups. From the results one can thus draw conclusions about the similarities and differences among the dimensions in the various groups.

The BC TRY component COMP1

This component of the BC TRY System, like the first part of COMP, is designed to facilitate the preparation of data cards for COMP2. COMP1 simply prepares a deck of data cards required by COMP2 on each of the comparison groups from information available on the intermediate storage tape, put there as the result of a prior factoring procedure (NC and NCSA, CC and CSA, or FALS and GYRO). These cards are saved and later combined with similar data decks from the other groups for input into COMP2. The use of COMP1 is *not* a prerequisite for the use of COMP2. If the analyst likes, he may punch all or any part of the data cards required by COMP2, or he may allow COMP1 to do the work.

The BC TRY component COMP2

COMP2 performs the comparative cluster or factor analysis. This program is designed to calculate a matrix of indexes of similarity among the adjoined sets of dimensions and to store this matrix on the intermediate storage tape for the use of other programs in the BC TRY System.

COMP2 is quite general and can perform comparative analyses of column entities in any kind of matrix in different groups so long as, in all groups, the rows refer to common referents (variables or objects) and the entries in the matrix are relations between the row and column entities.

COMP2 executes three major steps: (1) It inputs the COMP1 cards and adjoins the factor matrices of the different groups, thus forming one large rectangular observation matrix. (2) At the user's option, it deletes any column dimensions (comparison entities) in any groups that are not to be compared with the other dimensions, and it also deletes any row variables not wanted as common reference entities. (3) It calculates indexes of similarity among all column dimensions to be compared and stores them as a correlation matrix (in IST file CORRM1), ready for full-cycle key-cluster analysis.

Following these steps of COMP2 it is, of course, the full-cycle key-cluster analysis of the similarity matrix (CORRM1) that reveals the degrees of identity of the dimensions of the different groups.

Cluster or factor scores on individual objects

The psychometric principles of forming composite scores on individuals have been known for a long time. These principles are basically grounded on domain sampling doctrine, namely, defining a domain of behavior, taking a number of test samples of it, and forming a composite score on them, usually additively, as a better description of the attribute than one dealing with a single-test-sample score. In cluster or factor analysis, forming composites occurs at two stages: (1) Each of the NV specific attributes is usually a composite score itself consisting either of the sum of observed item samples or of an integrated judgment or rating composite of a complex multifaceted behavior domain. (2) After performing a V-analysis of NV such composites, one reduces NV by forming K supercomposites, each designed to measure a general attribute defined either by a cluster of S highly collinear variables or by all NV variables appropriately weighted to measure an independent or oblique dimension derived by factoring.

When the composites are defined by clusters of variables, we refer to the scores as "cluster scores." When the composites are defined on all NV variables, we refer to the scores as "factor scores." The general formula for any composite may be written

$$C_{ip} = \sum_{j=1}^{NV} w_{ij}X_{jp} \qquad (12.47)$$

where w_{ij} is a weight associated with the variable. Each object is evaluated by this equation, with the value of X_{jp} being the score of the pth object on the jth observed variable. How the weights are defined determines the type of composite obtained. It is assumed in this equation that the scores are standardized scores.

Cluster scores as a simple sum

The Tryon method of scoring clusters is simply to sum the scores of variables in each cluster. This corresponds in (12.47) to substituting 1.00 as the weight for variables in the cluster and .00 as the weight for variables not in the cluster:

$$C_{ip} = \sum_{j=1}^{S_i} V_{ijp} \qquad (12.48)$$

When these cluster composites are rescaled to standard scores, the simple sum is transformed to a differentially weighted sum that depends on the intercorrelation of the variable with the variables in the cluster. A definer with higher average correlations among the other definers will generate more variance in the composite score than those with lower average intercorrelations do.

Regression estimates as composites

A rather large number of methods and techniques for obtaining the weights in (12.47) have been developed. The major procedures among these are described by Harman (1967) and will not be discussed here. Two of these methods are included in the BC TRY component FACS.

The BC TRY components FACS and FACS3

Two separate components are available in the BC TRY System to calculate composites. Both FACS and FACS3 calculate cluster composites, but FACS3 does not calculate regression estimate composites. The simple-sum, or Tryon, method is straightforward in its application of (12.48). Each of the score dimensions is standardized in accordance with parameters input on data cards or standardized to a mean of 50 and a standard deviation of 10 if no parameters are input by the user. When FACS3 is used with missing data, the mean observed score on the other variables in the cluster is inserted in the missing data cells and the cluster scores are standardized.

The IST and DST input-output summaries are in Tables 12.1 and 12.2. The printed output of the components is the basic data defining the composites, the standard deviations of the composites, the effective weight matrices, the defining variables of clusters for the Tryon method, the input or regression-determined weight matrix, the intercorrelations between the composite scores, and the composite score matrix. The composite score matrix is punched out at the option of the user, giving the title of the job, the format in which the scores are punched (perhaps more than one card per object), and the matrix of scores punched in accordance with the format.

O-analysis with EUCO

Because the logic of EUCO for O-analysis was extensively discussed in Chap. 8, we do not describe it in very much detail here. Two primary options are available in the BC TRY component EUCO2; the first compares each of the NS subjects with each other, and the second compares each of the NS subjects with a set of specially defined "subjects" called OMARKs. Each of the NS subjects is defined by a set of K scores. These scores compose a vector indicating the location of the subjects as points in a K-dimensional space. By applying the simple geometry of Euclid, the distance, or euclidean distance, of each point from the other points can be calculated. For the pth and the qth object the distance in the K-dimensional space is given by

$$L_{pq} = \sqrt{\sum_{j=1}^{K} (C_{jp} - C_{jq})^2} \qquad (12.49)$$

The C values in (12.49) are defined by (12.48). This matrix of distances is output to the DST as a raw score matrix, and the O-analysis proceeds as though it were a standard cluster analysis. In the second option the NS subjects are compared against specially defined points in the score space called OMARKs. Each OMARK is defined as a point having all but one dimension score equal to the mean value of the scores. A set of six OMARKs is defined to represent a given score dimension, the scores being equal to $-3, -2, -1, +1, +2$, and $+3$ standard deviations from the mean on the score dimension, respectively. Thus for 11 dimensions there are a total of 6 times 11 OMARKs defined in this way. In addition, the average point is defined by an OMARK having average scores on all dimensions. Each of the NS objects is compared with each of the $M = 6K + 1$ OMARKs to give a score matrix of M by NS with the NS playing the role of the number of variables for the following O-analysis. The distance of each object from each of the M OMARKs is calculated by (12.49) for the pth object and the qth OMARK, where C_{jp} is the score of the pth object on the jth score dimension and C_{jq} is the score of the qth OMARK on the jth score dimension. This matrix of distances is stored on the DST as a raw score matrix with NS variables and M observations. The O-analysis proceeds as a standard V-analysis.

O-analysis with OTYPE

The procedures of the BC TRY component OTYPE were described in some detail in Chap. 8, and this section is therefore limited to the bare essentials of the statistical formulation of the component. The NS objects are represented in a K-dimensional space defined by the K cluster or factor scores from FACS. The K-dimensional space is divided into sectors by sectioning each of the K dimensions as specified by input parameters (either the number of sectors on a dimension or scores on the dimension specifying the sectors). Sectors in the K-dimensional space that have more objects falling within their boundaries than specified by a parameter established by the user (or set to a standard value by the component) define the first approximation to O-types. For each of these O-types, a mean score on each dimension is computed; these scores define a new "subject" that is the centroid of the trial core O-type. The distance (12.49) between each of these O-types is computed, and the closest O-types are "merged," their scores are averaged, and the merged O-type takes the place of the two O-types just merged. This process is repeated until all the O-types are merged into a single O-type. The distances of the successive O-types with the remaining O-types are printed in the process. The merging is strictly an arithmetic process, the actual O-type groups remaining distinct in the program.

Each of the NS objects is compared with each of the trial core O-types. The objects are assigned to the core O-type with which they have smallest euclidean distance. This results in the exclusion of some objects from a core O-type, because the objects are closer to another core O-type in an adjacent sector, and the inclusion of some objects from adjacent sectors, some of which may have been in other core O-types or in sectors that were too sparsely filled to define a core O-type. An object is not included in an O-type if it is at a distance greater than c times the RMS standard deviations from the O-type centroid (see Chap. 8).

Each of the O-clusters now occupies a different locus than that of its progenitor. The changes in the O-clusters are revealed by (1) the drift of its locus as measured by the distance between its new locus and its old locus, (2) the changes in mean values on each of the K dimensions, (3) changes in the membership of the O-cluster. These data are printed by the component. If two O-clusters are closer than specified by a user-controlled lower bound, the two O-clusters are condensed, thus eliminating one O-cluster from further consideration.

At this point in the process the O-clusters that remain are treated in the same way as the initial trial O-types, and the process is repeated.

The process is reiterated until no O-type changes membership or until a user-determined number of iterations are performed.

O-cluster homogeneities are calculated and printed. These values are described in detail in the section below on OSTAT. The means of scores on the K dimensions of all O-clusters are punched at the option of the user.

The formalisms of OTYPE are as follows. Let the matrix of cluster or factor scores be C_{ij} for the ith dimension and the jth object. Let the mean cluster score within the gth O-cluster be denoted $\bar{C}_{ig}$ for the ith dimension. The distance between the jth object and the gth O-cluster is defined by

$$L_{ig} = \sqrt{\sum_{i=1}^{K} (C_{ij} - \bar{C}_{ig})^2} \qquad (12.50)$$

Each object is reassigned to that O-cluster for which (12.50) is smallest, contingent on the restriction that the value L_{ig} is smaller than

$$c \sqrt{\sum_{i=1}^{K} \sigma_i^2} \qquad (12.51)$$

where c is a constant and σ_i^2 is the variance of the ith score dimension. The mean dimension score of an O-cluster will change when the membership of the O-cluster changes. Let $\bar{C}_{ig}^{(n-1)}$ be the mean of the gth O-cluster on the ith dimension for the $(n-1)$st iteration. On the nth iteration the mean is $\bar{C}_{ig}^{(n)}$, and the difference in means is

$$\Delta_n \bar{C}_{ig} = \bar{C}_{ig}^{(n-1)} - \bar{C}_{ig}^{(n)} \qquad (12.52)$$

The drift can be defined in terms of the total euclidean distance covered by

$$\sqrt{\sum_{i=1}^{K} (\Delta_n \bar{C}_{ig})^2} \qquad (12.53)$$

The intercluster distances are computed from the centroids of the O-clusters

$$L_{gh} = \sqrt{\sum_{i=1}^{K} (\bar{C}_{ig} - \bar{C}_{ih})^2} \qquad (12.54)$$

L_{gh} is computed at each iteration, and the two O-clusters with the smallest L value are merged if their distance is less than a quantity formed like (12.51) with the appropriate constant for cluster condensation. The new mean of the merged cluster is calculated by

$$\frac{S_g \bar{C}_{ig} + S_h \bar{C}_{ih}}{S_g + S_h} \qquad (12.55)$$

where the S's are the number of objects in the respective clusters. The intercluster distances are recalculated for the merged cluster and the condensation procedure is repeated until no distance is smaller than the condensation criterion.

The component program of BC TRY, OSTAT, is used to calculate a variety of statistics on the O-clusters from the dimension scores used to determine the O-clusters. OSTAT lists the K cluster or factor scores of the members of each of the O-clusters produced on the last iteration of OTYPE; for each O-type OSTAT computes the means, standard deviations, and homogeneities of scores on each dimension, correlation ratios η for each dimension on the discontinuous series of O-clusters, and the overall homogeneities of each O-type across all dimensions. Two values of η are calculated, one for the actual scores of objects in O-clusters using only the objects in O-clusters, and a Tryon η, which uses the variance of scores in the entire sample of objects whether or not they fall in O-clusters. The K cluster or factor scores are now differentiated according to the O-type in which a subject is located. Letting the third subscript designate the O-type, we have for the ith dimension and the jth subject in the gth O-type the cluster score C_{ijg}. Let M stand for the number of O-types. The O-type means are simply

$$\bar{C}_{ig} = \frac{1}{S_g} \sum_{j=1}^{S_g} C_{ijg} \tag{12.56}$$

where there are S_g objects in the gth O-type. The variance within an O-type for a given cluster score dimension is given by

$$\sigma_{ig}^2 = \frac{1}{S_g} \sum_{j=1}^{S_g} C_{ijg}^2 - \bar{C}_{ig}^2 \tag{12.57}$$

The variances of the cluster scores initially, for the entire sample of objects, are σ_i^2 for $i = 1, \ldots, K$. The squared homogeneity for the gth O-cluster on the ith cluster score dimension is

$$H_{ig}^2 = 1 - \frac{\sigma_{ig}^2}{\sigma_i^2} \tag{12.58}$$

The square root of the quantity given in (12.58) is the homogeneity of the O-type on the dimension when the quantity in (12.58) is positive. When the quantity (12.58) is negative, the homogeneity is defined as the negative of the square root of the absolute value of (12.58). The overall squared homogeneity of an O-cluster, across all K cluster dimensions, is given by

$$\bar{H}_g^2 = \frac{\sum_{i=1}^{K} \sigma_i^2 H_{ig}^2}{\sum_{i=1}^{K} \sigma_i^2} \tag{12.59}$$

The homogeneity is the square root of this quantity, with the same provision for negative values as above. The Tryon η is given by the square root of

$$\tau_i^2 = \frac{\sum\limits_{g=1}^{M} S_g H_{ig}^2}{\sum\limits_{g=1}^{M} S_g} \tag{12.60}$$

The standard η is given by the square root of

$$\eta_i^2 = 1 - \frac{\sum\limits_{g=1}^{M} S_g \sigma_{ig}^2}{\sum\limits_{g=1}^{M} \sum\limits_{j=1}^{S_g} (C_{ijg} - \bar{C}_i)^2} \tag{12.61}$$

where

$$\bar{C}_i = \frac{\sum\limits_{g=1}^{M} \sum\limits_{j=1}^{S_g} C_{ijg}}{\sum\limits_{g=1}^{M} S_g} \tag{12.62}$$

Typological prediction

The BC TRY component 4CAST applies the basic principles of computer simulation or Monte Carlo estimation to the problem of determining the degree to which a dependent variable is differentially distributed within the O-clusters. 4CAST computes the mean, standard deviation, and homogeneity of scores on a variable, generally a dependent variable in the data set on the DST or on a dependent cluster score in the factor score file on IST. The full set of NS scores on the dependent variable, the predicted variable, are brought into memory of the computer, and the equations of OSTAT pertinent to a single variable $(K = 1)$ are applied to the variable for the O-clusters from OTYPE. In addition, for each O-type, many random samples of S_g scores from the full set of NS scores are selected, and the mean, standard deviation, and homogeneity of each sample are computed. These statistics on the random samples are tallied against the empirical statistics. The relative frequency with which the random samples give statistics more extreme than the empirical statistics is printed out. The distribution of sample means, standard deviations, and homogeneities is printed in histogram form.

The workings of the program are straightforward, parallel with the prose algorithm and the statistical procedures of OSTAT. However, the random sampling procedures require comment. If a very large number of random samples is called for by the user, the computer time involved can be very large. Consequently it is necessary to take care in constructing the computer method of selecting the random samples. The random number generator and the method of selecting the scores should be efficient, and the random number generator should have certain statistical properties. The reader is referred to Lehman and Bailey (1968) for an introductory discussion of these issues and references to more advanced discussions.

Miscellaneous BC TRY components and procedures

The BC TRY System contains a number of component programs or specially named procedures performed by component programs. This section reviews these components or procedures in brief. The intent is to give a thumbnail sketch of the peripherally important components of the system but not a detailed statistical description of the components. Many of these components are simply housekeeping programs to assist in the management of data, or they are programs to perform standard statistical calculations, e.g., DAP to input data and COR2 to calculate a correlation matrix.

Access

Since BC TRY is a very large system, it must be maintained in the library of the computer center, available to users through submission of job decks containing only cards that make a request to the computer for the use of the System and control and data cards for the system itself. The mechanism to establish accessibility of the System on a computer varies with the computer installation. In the next chapter this is referred to as part of the local conventions of a job deck makeup.

General executive program, GEP

The GEP was described in some detail in Chap. 3. Some of the procedures described below are actually a part of the GEP.

Binary tape dump, BTD

If, for any reason, the user of the System becomes confused about what he has on the IST, he can obtain a "picture" of the contents of the IST by including the EC card calling for a binary tape dump, /BTD. This procedure is particularly useful in determining the status of the IST after it is defined by a TAKE deck.

COMMENT

The user can have as much explanatory material printed in his output as he wishes by including EC cards (at places where EC cards are expected by the System) with /COMMENT followed by cards containing the material he wishes to have printed. The material is printed, line by line, as it is encountered by the GEP in its attempt to find an EC card after the /COMMENT card.

Correlation, COR2 and COR3

These two components simply calculate the Pearson product moment correlation matrix for the data input through DAP. COR2 assumes that all the data are defined. COR3 assumes that there are missing data and calculates the correlation coefficients on objects for whom data are mutually defined. A variety of other statistics are output by COR3, including the matched sample sizes, covariances, standard deviations, etc.

Data processor, DAP

The data processor program simply inputs to the DST the basic raw data of a study. DAP permits one to identify each variable by name and each object by name if so desired. DAP is capable of reflecting variables, of reordering variables, and of accommodating to missing data. Means and standard deviations of variables are calculated and output.

Data printing, DPRINT

This EC card causes all the data on the DST to be printed, clearly labeled as to variable and observation name and sequence number. Problems caused by faulty input, either because of keypunch errors, faulty format statements, or computer failure, are spotted easily by scanning this output.

Least squares factoring, FALS

Whereas we described the processes and statistical procedures of cluster analysis in some detail, we refer the interested reader to other sources, e.g., Harman, for a discussion of least squares factor analysis. The BC

TRY component FALS computes factors by three different methods, the principal-axes method, the augmented factor analysis method, and the canonical factor analysis method.

Factor statistics, FAST

The factoring program FALS does not print out the residual matrix, and the program FAST is included in the system to do this. FAST also can print out the successive residual matrices or the successive reproduced matrices. The IST is modified by FAST so that the reproduced or the residual matrix takes the place of the correlation matrix in file CORRM1. This permits a cluster or factor analysis of the residual matrix.

Input and output to the IST, GIST

The component GIST (Generate IST) enables users of the System to input to or output from (on punched cards) the IST. This component permits recovery of data in keypunch form from a BC TRY run. It also permits the user to set up nonstandard uses of the System by defining IST files in special ways, independently of the use of certain BC TRY components.

Restart, GIVE-TAKE

The components GIVE and TAKE permit the user to capture the IST at any point in a run where an EC card is expected by GEP. If at a later date the user wishes to restart the analysis at that point, he uses the deck of cards containing the IST files to reestablish the IST precisely as it was defined when he used the GIVE EC card.

Missing data statistics, RLIST

The missing data correlation program COR3 can output all the statistics describing the matched samples, etc. However, these are output one matrix at a time. RLIST outputs the statistics for each cell of all the matrices appearing together in the printout.

Correlation matrix organization, REDE

For purposes of publication, a BC TRY user may want to reorganize a correlation matrix. In order to reorder the variables, delete variables, and reflect variables in a printed matrix the BC TRY user would use the component REDE. This component does nothing but print out the reorganized matrix.

Scatter diagrams, RSCAT

The scatter diagram of the joint distribution of two variables is printed out by the component RSCAT. Also included in the output are the frequency distribution of each of the variables, the Pearson product moment correlation coefficient, and the coefficient of nonlinear correlation η.

Temporary suppression of variables, SLEP1 and SLEP2

There are two principal reasons why an investigator would want to suppress variables in a cluster or factor analysis until after factoring is completed: (1) To suppress experimentally dependent variables, i.e., those which include other of the NV variables as physical components; such variables are likely to generate communalities over 1.00 in factoring. (2) To implement an experimental design in which some variables are suppressed from factoring in order to see the portion of their variance accounted for by the dimensions factored from the nonsuppressed variables. SLEP1 is the component that suppresses variables from factoring in order to see the portion of their variance accounted for by the dimensions factored from the nonsuppressed variables. SLEP1 is the component that suppresses the designated variables. SLEP2 completely reactivates the suppressed variables after factoring is completed and recomputes factor statistics on all NV variables in the customary summary form.

Matrix-algebraic operations, SMIS

This component performs general matrix-algebraic operations and provides the user with almost unlimited capacity for matrix and vector operations with access to the IST. Each operation defined under SMIS is initiated by the user by control cards. By combining the operations provided in SMIS the user can perform virtually any calculation desired, even though the calculation is not an integral part of the other components of BC TRY. Hence, with creative use of SMIS pioneering work can be done in multivariate analysis within the setting of BC TRY.

Chapter 13

ABRIDGED USER'S MANUAL OF THE BC TRY SYSTEM[1]

This chapter is an abridgement of the User's Manual of the BC TRY System. The unabridged manual, which contains detailed descriptions of all control options in the System and extensive discussions of the purpose and methods of the components of the System, makes up approximately 200 pages of computer-printed material. Most of the general discussion in the Manual is covered in this book. In most applications of BC TRY the control options are standard. In standard options the control cards of BC TRY are generally blank or have stereotyped punches. As a consequence, for most applications the card decks required are quite simple. This chapter presents those segments of the Manual needed for standard use of the System.

General conventions

Three general kinds of cards are required in each BC TRY run: local computer center cards, executive control cards, and component control

[1] This chapter was written with the collaboration of John Bauer and Jin-Yu Yen.

TABLE 13.1 STANDARD SEQUENCES AND DECKS BY
TYPES OF ANALYSES[a]

V-analysis

Empirical CC5		Principal axes–varimax		Preset CC5	
Program	ID	Program	ID	Program	ID
Local	a	Local	a	Local	a
START	b:1	START	b:1	START	b:1
COMMENT	c:2+	COMMENT	c:2+	COMMENT	c:2+
DAP2	d:5+	DAP2	d:5+	TAKE	d:deck
DPRINT	e:1	DPRINT	e:1	DVP41	e:2
COR2[b]	f:1	COR2[b]	f:1	CC5	f:3
DVP41	g:2	DVP41	g:2	CSA2	g:3
CC5	h:3	FALS	h:5	SPAN2	h:2
CSA2	i:3	FAST	i:2	GIVE	i:1
SPAN2	j:2	GYRO	j:4	COMP1	j:1
GIVE	k:1	GIST	k:3	END	k:1
END	l:1	SPAN2	l:2	Local	l:1
Local	m:1	GIVE	m:1		
		END	n:1		
		Local	o:1		
		Then Preset CC5			

O-analysis

Condensation Method		Convergence Method		Typological Prediction	
Program	ID	Program	ID	Program	ID
Local	a	Local	a	Local	a
START	b:1	START	b:1	START	b:1
COMMENT	c:2+	COMMENT	c:2+	COMMENT	c:2+
DAP2	d:5+	DAP2	d:6+	DAP2	d:5+
TAKE	e:deck	DPRINT	e:1	TAKE	e:deck
FACS	f:4	GIST	f:9+	4CAST	f:2+
RSCAT	(opt.)	EUCO2	g:2	END	g:1
OTYPE	g:2	COR2	h:1	Local	h:1
OSTAT	h:2	DVP71	i:2		
GIVE	i:1	CC5	j:3		
END	j:1	CSA2	k:3		
Local	k:1	SPAN2	l:2		
		GIVE	m:1		
		END	n:1		
		Local	o:1		

[a] Under ID are the sequence identifications of the cards (following the sequential order of the captions given in the text); after the colons are the minimal numbers of required control cards. The table can be used to check the card and sequence numbers of a standard deck.
[b] COR3 is used if there are missing data.

cards (including data cards). Because of the wide variety of requirements imposed by local computer centers the local computer center cards are not described in detail here. Although there may be different requirements in various installations, there are generally three types of local cards: job cards, monitor control cards to access BC TRY in the computer center library, and end of job cards to signal the end of the BC TRY run. Wherever local cards are likely to be required, they are indicated in this chapter by the phrase "local cards" or simply "local."

Instructions for punching cards are stated in detail, or the columns in which literal information is to be punched are indicated along with the literal information in quotation marks. For example, on the first executive control card of a BC TRY job the literal "/START" is punched in columns 1 through 6. This is indicated in the following by

Executive card

1–6 "/START"

which means "punch a card with the characters/START in columns 1 through 6." If other columns are to be punched, they are indicated in a like manner. The phrase "pack right" in instructions below mean simply to punch the indicated number to the right of the field indicated. The value 41 punched in columns 5 through 8 appears in columns 7 and 8 when packed right.

Table 13.1 gives six standard sequences of BC TRY executive and component cards. These sequences cover the most common applications of BC TRY.

Empirical key-cluster analysis

a. Local cards.
b. *START* card, activating the BC TRY System. *One* card.
   1) *Executive card.*
      1–6 "/START"
c. *COMMENT* cards, for comments about the job. Optional. *Two* or more cards.
   1) *Executive card.*
      1–8 "/COMMENT"
   2) *Component cards.*
      1–72 Punch messages on as many such cards as desired.
d. *DAP2* program: prepares data, reads and error-checks raw scores for input to DST (Data Storage Tape). *Five* or more cards plus data deck.
   1) *Executive card.*
      1–5 "/DAP2"

2) *Component cards.*

Card A. Title card.

1–72 Title of problem to be printed on all output.

Card B. Parameters and controls. Pack right.

1–4   Number of variables, NV.

5–8   Number of subjects, NS.

9–12   Length of field ($LFIELD$), i.e., number of columns used per variable on raw-score data cards. If $LFIELD$ exceeds 9, or is not the same for all variables, or if data are in decimal format, or if NV is only part of the total number of variables on data cards, or if ONAMS are not on columns 73 to 78, $LFIELD$ must be left blank and card C (format card) below must be prepared.

13–16 A "1" punch means data are complete; a blank means missing data.

Card C. Format card, necessary only if columns 9–12 of card B above are blank. For details of preparing card C, see section headed "Format statement in DAP2 and GIST" below.

Card D. Variable names, called V-names, providing for the printing of user-assigned names of variables on output. The program automatically assigns V-names if user does not.

    Card 1. Directs program to input V-names.

    1–14 "VARIABLE NAMES"

    Card 2. Detail cards of V-names.

    1–72 Punch variable names in six or less alphameric characters, separate each name by a comma, and punch a period after the last name. Use as many cards as necessary. Do not split a name between cards.

Cards E, F, G. For optional reordering or reflecting variables and for supplying O-names.

Data cards. The deck of raw scores comes last, preceded and followed respectively by control card 1 and control card 2.

    Card 1. Directs input of data to DAP2.

    1–4 "DATA"

    Data Deck is inserted here. It is prepared as follows. The NV raw scores of each individual are punched in fields of constant length, $LFIELD$ (number of columns). $LFIELD$ cannot exceed nine. A missing score is left blank in its field. The O-names are punched in columns 73–78 on each data card of each individual. A score must not be split between two cards. The sequence number of a given individual's cards is punched in columns 79 and 80, packed right.

    Card 2. Signals end of data input.

    73–75 "END"

e. DPRINT program, printing out the raw scores as written on DST by the DAP2 program. Optional, but desirable as a check to be sure data are correctly input to the computer. *One* card.
  1) *Executive card.*
     1–7 "/DPRINT"

f. CORRELATION (COR2) program for complete data: computes correlations among the variables and other statistical constants of them. *One* card.
  1) *Executive card.*
     1–5 "/COR2"

f'. CORRELATION (COR3) program for missing data. *Two* cards. See below in section headed "Optional programs in V-analysis" for controls.

g. DVP program: estimates communalities for input to the diagonal cells of the correlation matrix. (It can also insert unities.) *Two* cards.
  1) *Executive card.*
     1–4 "/DVP"
  2) *Component card.*
     3–4 "41"

h. CC5 program: determines K, the number of dimensions sufficient to reproduce the correlation matrix; uses diagonal values, reiterates and computes residuals; defines dimensions by key clusters (subsets) of variables. *Three* cards.
  1) *Executive card.*
     1–4 "/CC5"
  2) *Component cards.*
     Card A. Blank.
     Card B. Blank.

i. CSA2 program: computes factor coefficients on oblique cluster domains (factors), interdomain (common factor) correlations, reliability of cluster scores (accuracy of factor estimates), and other aspects of cluster or factor structure. *Three* cards.
  1) *Executive card.*
     1–5 "/CSA2"
  2) *Component cards.*
     Card A. Blank.
     Card B. Blank.

j. SPAN2 program: based on the results of factoring, the program allocates the variables to a minimal set of subspaces and plots the variables as points on the surface of a minimal set of salient spheres, physically rotated for visually centering the configuration. *Two* cards.
  1) *Executive card.*
     1–6 "/SPAN2"

2) *Component card.*
    Card A. Blank.
k. GIVE restart card, producing an output deck of binary cards that contain IST files produced by the above programs. *One* card.
    1) *Executive card.*
        1–5 "/GIVE"
The printout from GIVE gives the title of an output TAKE card and a list of the IST files punched on the output binary GIVE deck. These are:

VNAMS1 (names of variables)
VSUMS1 (sums of variables)
MEANS1 (means of variables)
ONAMS1 (names of individuals)
IPFILE (master identification)
IDFILE (title)
STDEV1 (standard deviations of variables)
CORRM1 (correlation matrix)
CLUST1 (cluster definers, i.e., indices)
REFLX1 (cluster definers with signs)
DIAGV1 (diagonal values)
UFACT1 (orthogonal factor coefficients)
RFACT1 (rotated factor coefficients)
BASIS1 (correlations between cluster domains, factors)

Along with this deck are various local cards, which must be present in the restart of the job through the TAKE component. These local cards vary from installation to installation.
l. END of job. *One* card.
    1) *Executive card.*
        1–4 "/END"
m. Local cards.

Preset key-cluster analysis

After a study of the output of the empirical key-cluster solution, one usually wishes to revise the defining variables of one or more of the dimensions, perhaps even to eliminate some dimensions entirely. This is done by restarting the analysis with the TAKE deck that was output by the GIVE restart card.

The revised run is identical with the empirical run except that (1) the segment that includes DAP2, DPRINT, and COR2 (or COR3) is now replaced by the TAKE deck input and (2) the preset features of the CC5 program are now utilized. Here is the full sequence:

a to *c*: Local, START, COMMENT.

d. TAKE program: the Intermediate Storage Tape, IST, is restored with files that are needed by the CC5 program, namely, IDFILE, CORRM1, and DIAGV1. A TAKE is composed as follows.

1) *Executive card.*
   1–5 "/TAKE" This card is included in the Hollerith (BCD) part of the punch output from GIVE.

2) *Binary data cards.*
   These cards of the GIVE deck contain the data to be restored on the IST.

e. *DVP* program: same as the empirical run. *Two* cards.

f. *CC*5 program: same as the empirical run, except for punching several parameter values on component card A, and detail cards informing the program of the revised defining variables of the dimensions. *Three* cards plus deck of detail cards C.

1) *Executive card.*
   1–4 "/CC5"

2) *Component cards.*
   Card A. Parameters. Pack right.
   1–4 "4" informing program that the number of dimensions is preset in columns 17–20 below.
   9–12 "3" informing program that the defining variables of each dimension are provided in detail card C below. *If only* the number of dimensions is to be preset, leave blank.
   17–20 Punch K, the number of dimensions in this preset run.
   Card B. Blank.
   Card C. Detail cards specifying definers of preset dimensions. Punch a detail card for each of the K dimensions, giving the ordinal numbers of not more than 20 definers of the dimension in *integer card format* (successive four column fields, beginning in column 1 and ending in column 80; use no punctuation marks; pack numbers to the right in each field), without reflection signs (indicators) attached to the definers. Skip this card if only the number of dimensions is to be preset.

g. *CSA2* program: same as empirical run. *Three* cards.

h. *SPAN2* program: same as empirical run. *Two* cards.

i. *GIVE* restart card: same as empirical run. When the GIVE deck is received from this preset run, destroy the GIVE deck received from the empirical run. *One* card.

j. Optional *COMP1* program, for a later comparison of the dimensions found here with those discovered in other groups, as described in section 3 in "Optional programs in V-analysis" for program COMP2. *One* card.

   1) *Executive card.*
     1–6 "/COMP1"

k. *END* of job: same as empirical run.

l. *Local cards:* same as empirical run.

Optional programs in V-analysis

1. *Correlation program for missing data*, COR3 (in place of COR2): computes correlations and covariances among the variables and other statistical constants. *Two* cards.

   1) *Executive card.*
     1–5 "/COR3"

   2) *Component card.*
     Punching a "1" in each of the columns 4, 8, 16, 20, 40, will give the following: correlation matrix, standard deviations, matched N's.

2. *Correlation scattergram program*, RSCAT: plots the correlation scatter between any or all of the NV variables or K cluster scores, computes for each scatter the value of r, both η's, the regression constants, frequency tables, and other statistical values. *Two* cards, unless selected scattergrams are specified, which requires a detail card.

   1) *Executive card.*
     1–6 "/RSCAT"

   2) *Component card:* one or more cards.
     Card A.
     1–4 Blank. Compute the scattergrams between all the variables or clusters. (NV or K not greater than 18.)
     "2" Compute the scattergrams between all variables or clusters and a common variable or cluster. Designate the ordinal number of the common variable/cluster in columns 17-20.
     "3" Compute only scattergrams specified on card B.
     5–8 Blank. Scattergram between clusters. FACS required before this program.
     "2" Scattergram between variables. DAP2 required before this program.
     17–20 Necessary only when punched "2" in column 4. Punch the ordinal number of common variable or cluster.
     Card B. Necessary only when punched "3" on column 4, card A. Punch "1+3, 2+4, 10+20." for the combinations of 1 and 3, 2 and 4, 10 and 20. Only 24 scattergrams can be specified.

3. *Comparative cluster analysis* of the dimensions found in different groups: performs a comparative analysis by COMP2 of the key-cluster dimensions discovered in separate groups of individuals that are measured on the same variables. COMP1 prepares the decks to be input to COMP2. COMP2 adjoins the input factor matrixes of the separate groups, computes similarity indexes between all dimensions in the form of a correlation matrix on which a full cycle empirical key-cluster analysis is then performed. *Two* cards plus the COMP1 data cards following cards *a* through *c* of an empirical key-cluster analysis.
   1) *Executive card.*
      1–6 "/COMP2"
   2) *Component cards.*
      Card A. Title card.
      1–72 Punch title card describing the analysis.
      Card B.
      1–4 Punch the number of the groups to be compared.
      Data Cards from COMP1 that were output during the preset key-cluster analysis, step *j*: these COMP1 cards from each preset run should be interpreted, then stacked together and inserted after card B. Use the whole COMP1 punch output as data cards. Do not remove any cards.
   3) Full cycle CC5 analysis: follow the cards for the COMP2 program by exactly the same control cards of steps *g* to *m* of the empirical key-cluster solution, namely, by DVP, CC5, CSA2, SPAN2, GIVE, END, and local.
4. *Principal axes, varimax analysis:* computes an orthodox principal-axes solution with varimax or quartimax rotation, from which a cluster structure analysis can be derived, if desired. Required input files: the programs that must precede the principal-axes program are identical with those that must precede the CC5 program, namely, those from steps *a* through *g* of an empirical key-cluster analysis. Here is the full sequence:

   a through *g*: local, START, COMMENT, DAP2, DPRINT, COR2, DVP41. Same as above.
      h. FALS, using the principal factor analysis option. *Five* cards.
         1) *Executive card.*
            1–5 "/FALS"
         2) *Component cards.*
            Card A. Specifies the type of least squares solution desired.
                  1–3 "PFA"
            Card B. Blank.
            Card C. Blank, but if one wishes to preset the dimensionality, then in columns 1–4 punch the number of preset dimensions desired. (Pack right.)
            Card D. Blank.

i. *FAST:* computes the matrix of residuals after the last principal axis dimension. *Two* cards.
  1) *Executive card.*
     1–5 "/FAST"
  2) *Component card.*
     1–8 "RESIDUAL"

j. *GYRO* program: computes the varimax rotation of the principal axes calculated above in FALS. *Four* cards.
  1) *Executive card.*
     1–5 "/GYRO"
  2) *Component card.*
     Card A. Specifies method of rotation.
     1–6 "VARMAX" (for a quartimax rotation, punch "QRTMAX").
     Card B. Blank.
     Card C. Blank.

k. *GIST* program for the output of rotated varimax factor coefficients on a deck of cards (a GIST "output package"). Reason: the factor coefficients of some dimensions may be predominantly negative but can be made positive by duplicating the coefficients on a new card with reversed signs, using an ordinary key punch. *Three* cards.
  1) *Executive card.*
     1–5 "/GIST"
  2) *Component card.*
     Card A. Call output file package.
     1–6 "RFACT1"
     Card B. Blank.

l. *SPAN2* program: provides the spherical configuration of the variables on the varimax dimensions. *Two* cards.
  1) *Executive card.*
     1–6 "/SPAN2"
  2) *Component card.*
     5–8 "2"

m. *GIVE* restart card. *One* card.
  1) *Executive card.*
     1–5 "/GIVE"

n. *END* of job. *One* card.
  1) *Executive card.*
     1–4 "/END"

o. Local card.

NOTE 1: If the SPAN configuration is distorted because some dimensions are preponderantly negative in the signs of their factor coefficients, reflect the factor coefficients as indicated under GIST above and input by GIST the new RFACT1 file package.

NOTE 2:   It is always possible to get any desired sphere by calling for it.

NOTE 3:   From the SPAN configuration choose the defining variables of the K clusters that are most nearly independent and meaningful and run a preset CC5 analysis. The CSA2 and SPAN components provide an oblique solution to these varimax-defined clusters.

Condensation method of O-analysis (OTYPE analysis)

The condensation method is an iterated condensation of the NS individuals around a small number M of core O-types. The full sequence follows:

a through d: local, START, COMMENT, DAP2. Same as empirical run.

e. *TAKE*. Identical with preset key-cluster run but using the GIVE deck output by that run. Same as preset run.

f. *Cluster scoring* program, FACS: computes simple sum Z scores (mean = 50, standard deviation = 10) on each cluster whose definers are selected in the preset key-cluster analysis or in the initial empirical analysis if that is considered satisfactory. *Four* cards.

1) *Executive card.*
   1–5 "/FACS"
2) *Component cards.*
   Card A. Blank. If cluster scores are wanted on an output deck (the FSCOR1 deck), punch "1" in column 12.
   Card B. Blank.
   Card C. Punch in columns 1–72 an identifying title of this FACS run.

NOTE: The defining variables of the clusters can be preset differently from those in the REFLX1 file of the GIVE deck by GIST input of REFLX1 file package inserted after TAKE.

g. *OTYPE* program: assigns all the NS individuals to arbitrary sectors in the cluster score space of K dimensions; selects as core types those in sectors which include at least 2 percent of the group; reassigns all NS individuals to those core types; iterates the process to a final solution consisting of M O-types plus a set of unique individuals; and computes the hierarchical order of the M O-types. *Two* cards.

1) *Executive card.*
   1–6 "/OTYPE"

2) *Component card.*
Card A. Blank.

h. *OSTAT* program: lists the K cluster scores of the members of each of the M O-types produced by the last iteration of program OTYPE; for each O-type, computes its means, sigmas, and homogeneities of scores on each dimension; and calculates the correlation ratio η of each dimension on the discontinuous series of M O-types. *Two* cards.
  1) *Executive card.*
     1–6 "/OSTAT"
  2) *Component card.*
     1–4 "2"
     13–16 "1" which outputs OMEAN1 cards of the M O-types for the convergence method below.

i. *GIVE*, the restart card: reestablishes the IST files. To do so is important for two reasons: (1) by inputting this GIVE deck (via TAKE) and calling OTYPE again, if more iterations are desired after looking at the results, they can be continued from where they left off at the end of the last OTYPE run, and (2) if one wishes to perform typological prediction (see below), this GIVE deck reestablishes the OTYPE1 file after the last iteration, a necessary input file for the use of the 4CAST program in prediction (see below). *One* card.
  1) *Executive card.*
     1–5 "/GIVE"

j. *END* card. *One* card.
  1) *Executive card.*
     1–4 "/END"

k. Local card.

Convergence method (EUCO analysis) of O-analysis

The convergence method is called EUCO analysis. As a general method of projecting all individuals into a single spherical configuration from which a core set of M O-types is selected as descriptive of the O-cluster structure in the full supply of subjects, EUCO analysis is expensive in computer time and lacks the objectivity of the condensation method, which now supersedes it as the means of discovering the typology of a group.

But EUCO analysis is nevertheless critically important as a means of revealing the spherical configuration of the O-type structure. Hence, it is now included as an essential step *following* the description of the typology by the condensation method. Only the M O-types are carried into EUCO analysis, these being abstract individuals whose score profiles on the K cluster dimensions are output as the OMEAN1 file from the

OSTAT run in the section above. In addition, hypothetical marker "individuals" are also projected into the configuration by means of which the Z score axes are represented in the configuration so that it becomes possible to read off approximately the actual score patterns of the different O-types. The model markers are called the "OMARK objects," whose score profiles on the K dimensions are also output by the OSTAT program for use in the procedure described below. The full procedures follow:

a to c: Local, START and COMMENT: same as empirical run.

d. $DAP2$ program: used here to set up a problem title (IDFILE) and to provide the O-names (ONAMS1 file) of the O-types and OMARK objects that are projected into the spherical configuration. *Three* or more cards plus data deck.

1) *Executive card.*
   1–5 "/DAP2"

2) *Component cards.*
   Card A. Title card.
   1–72 Title of problem to be printed on output.
   Card B. Parameters. Pack right.
   1–4 Punch K, the number of Z score dimensions on the input data cards, below.
   5–8 Punch NS, the number of objects in this particular EUCO analysis, these being the M O-types whose OMEAN1 cards are in the data deck plus the OMARK objects whose scores are also output by OSTAT. (Of course, any other kinds of cases may be added.)
   Card C. Insert a format statement here. OMEANS punch output is in F8.4 format (see below).
   Card D. Variable names: in this case punch the names that have been assigned to the K dimensions. See the V-names procedure described above.
   Cards E, F, G. Usually not required.
   Data Cards. This is the data deck of OMEAN1 and OMARK cards, input thus:
   Card 1. Initiating input.
   1–4 "DATA"
   Data deck is inserted here.
   Card 2. Terminating input.
   73–75 "END"

e. $DPRINT$ program: provides a listing of the input cards. Same as above. Spot check this listing. *One* card.

f. $GIST$ input of IST files needed by program EUCO later.
   1) *Executive card. One* card.
      1–5 "/GIST"

2) *Component cards for MEANS2. Two* cards.
Card A. Input file package for MEANS2.
1–6 "MEANS2"
9–12 "1"
Card B. Value of MEANS2.
3–4 "50"
3) *Component cards for STDEV2. Three* cards.
Card A. Input file package for STDEV2.
1–6 "STDEV2"
9–12 "1"
Card B. Number of dimensions (factors) K.
1–4 Punch K, that is, the value of K.
Card C. Values of STDEV2 on the K dimensions.
3–4, 11–12, 19–20, 27–28, etc., for K such fields, punch "10" in each field.
4) Component cards for FSCOR1. *Two* or more cards plus NS cards of Data Deck.
Card A.
1–6 "FSCOR1"
9–12 "1"
13–16 Blank for standard format, i.e., successive eight-column fields with decimal point at fourth place from the right.
"1" for nonstandard format. Format statement, card B, is required.
Card B. Necessary only when column 16 of card A is punched "1". Punch format statement with open- and close-parentheses in column 1 through 72. No "INFORMAT" is required as in DAP2.
Card C. Number of column variables and row objects on data cards. Pack right.
1–4 Punch K, that is, the value of K.
5–8 Punch NS, the number of subjects in this particular EUCO analysis.
Data Deck of FSCOR1 cards.
5) A final blank card is *mandatory. One* card.

g. *EUCO* program: for COR2 below, computes the D values of each of the NS objects taken as column comparison entities with a common set of row referent objects that are a special set of OMARK objects that spans the K-dimensional score space; also computes D value between the NS objects. *Two* cards.
1) *Executive card.*
1–6 "/EUCO2"
2) *Component card.*
17–20 "1", which sets us the D matrix with the common row referents as OMARK objects.

h. *Correlation* (COR2) program: computes the r_{DD} values between the NS objects where the row referents are the OMARK objects that span the score space. *One* card.
1) *Executive card.*
   1–5 "/COR2"

i. *Diagonal values* (DVP) program: inserts unities in the diagonal cells of the correlation matrix. *Two* cards.
1) *Executive card.*
   1–4 "/DVP"
2) *Component card.*
   3–4 "71"

j to *o*: CC5, CSA2, SPAN2, GIVE, END, and local. Same as *h* to *m* of the empirical run. The spherical configuration of the objects with the OMARK score axes projected into the configuration is given on the SPAN diagrams.

NOTE: If one wants to converge onto the configuration of the full supply of objects by running different subsamples of individual objects on the same sequence of steps above, it can be done by superimposing their SPAN plots provided the CC5 analysis is preset on a common set of definer objects and dimensions, these being the set used in the CC5 analysis above.

Typological prediction

In the condensation method, program OTYPE describes the M O-types in an IST file called OTYPE1. Typological prediction by program 4CAST reveals for each of the M O-types described in OTYPE1 its prediction of scores on one or more variables. The sequence in typological prediction is as follows:

a to *d*: Local cards, START, COMMENT, DAP2. Same as in empirical run. DAP2 is not necessary if the predicted variables are taken from the FSCOR1 file on IST.

e. *TAKE* program: reestablishes the OTYPE1 file from the GIVE deck output by the condensation method. (If OTYPE1 comes from OSTAT, the GIVE deck used should, of course, have been output by OSTAT.)

f. *4CAST* (forecast of predicted variables from the O-types) program: for each of the M O-types, computes the mean, sigma, and homogeneity of scores on a predicted variable; deposits the NS scores of the full supply in the core of the computer and draws many random samples having the same number of cases as in the O-type and computes the mean, sigma, and homogeneity of each sample; prints the distribution of the sample values, spotting the observed value in the distribution; and states the probability of recovering sample values at least as deviant as the observed, for different levels of confidence. *Two* or more cards.

1) *Executive card.*
   1–6 "/4CAST"
2) *Component cards.*
   Card A. Parameters. Pack right.
       1–4 Blank means predicted variables are from IST file FSCOR1; punch "1" if they are to come from DST via DAP2.
       5–8 Blank takes *all* the variables in the file; punch "1" to select only some variables for 4CAST analysis via detail card B below.
      9–12 Blank draws 300 random samples; punch the number wanted if not 300.
    13–16 Blank lists the scores of the variables; punch "1" to suppress list.
    17–20 Blank plots distribution of sample means; punch "0" to suppress it.
    29–32 Blank leaves raw scores as is; punch "1" to convert them to Z scores.
    33–36 Blank means complete data; "1" *must be punched* if any scores are missing.
   Card B. Detail card used if variables are selected by a "1" punch in column 5–8 above. Punch the ordinal numbers of the selected variables (from DAP2) or clusters (from FSCOR1 on IST file) in free field integer format (commas between integers and a period at the end).

g. END of job card. *One* card.
1) *Executive card.*
   1–4 "/END"

h. Local.

Format statement in DAP2 and GIST

DAP2—Card C.
1–8 "INFORMAT"
9–80 In columns 9–80, punch F, X, and A symbols describing the input format, thus:
    ((open parenthesis): the beginning of the format statement.
) (close parenthesis): the end of the format statement.
    , a comma, separating each of the F, X, or A statements.

F the number of data fields.

X the number of columns to be skipped.

A the number of columns where O-names are punched.

/ skip the rest of the columns and go to next card.

For example,

9–80 "(1F1.0,11X,4F3.2,48X,A6/30X, 5F8.4//20X,F8.3)''

means that there are four data cards per subjects, as follows:

Card 1:

The first field is read from column 1 with the decimal point to the right of the column.

Skip the next 11 columns.

Read four fields from the next 12 columns with three columns per field and with decimal points at the second places from the right of all four fields.

Skip the next 48 columns.

Read the next six columns as O-names. (The ONAMS1 file will *not* be set up unless this "A6" statement is punched for this first card.)

Go to card 2 (skipping the last two columns of card 1).

Card 2.

Skip the first 30 columns.

Read five fields from next 40 columns with the decimal points at fourth places from the right from all five fields.

Go to card 3 (skipping the remaining columns of card 2).

Card 3:

Skip whole card and go to card 4.

Card 4.

Skip the first 20 columns.

Read the next eight columns as a field with decimal point at third place from the right of the field.

Data fields end.

GIST—Format statement in GIST is same as the DAP2 except no "INFORMAT" is necessary. Start "(", open parenthesis, from column 1.

Presetting defining variables of composites

There are three types of composites for which one may wish to preset the defining variables:

a. Preset collinear clusters for CC5-CSA2-SPAN2-FACS.

The definers are preset in CC5 and are thereafter carried in the REFLX1 file for CSA2, SPAN2, and FACS.

b. Composites for CSA2 and FACS.

The definers are preset by the GIST (REFLX1) input package. *Six or more* cards.

1) *Executive card.*

1–5 "/GIST"

2) *Component cards.*

Card A. File.

1–6 "REFLX1"

9–12 "1" for input file package.

Card B. Number of dimensions (clusters).

1–4 Punch K, the number of dimensions. Pack right.

Card C. Numbers of definers of the clusters.

1–4, 5–8, 9–12, etc. The numbers of definers in each of the clusters are punched in successive four column fields. Pack right.

Card D. (Deck).

Card D1. Ordinal numbers (indexes) of the defining variables of cluster 1.

1–4, 5–8, 9–12, etc. Punch the indexes of the definers of cluster 1.

Card D2, D3, etc., up to DK. Same as Card D1, but for clusters 2, 3, etc., up to cluster K.

Card E. Blank, if no other GIST cards follow. *Mandatory.*

c. Composites for FACS only.

Defining variables of clusters can be directly specified in the control cards of FACS, and furthermore, there can be up to NV definers of a single cluster. Specification is by inputting the "weight matrix." See any printout of FACS for the format of the weight matrix. A card in decimal card format, i.e., (9F8.4), corresponds to each column of the printed weight matrix.

REFERENCES

Anastasi, A., 1961: "Psychological Testing," The Macmillan Company, New York.

Bailey, D., 1960: "The Role of Psychological Distance in Human Discrimination Learning," Ph.D. thesis, University of California, Berkeley.

Broverman, D. M., 1961: Effects of Score Transformations in Q and R Factor Analysis Techniques, *Psychol. Rev.*, **68:**68–80.

Burt, C., 1915: General and Specific Factors Underlying the Primary Emotions, *Rept. Brit. Assoc. Advan. Sci.*, **85:**694–696.

————, 1937: Correlation Between Persons, *Brit. J. Psychol.*, **28:**59–96.

————, 1940: "The Factors of Mind," University of London Press, Ltd., London.

————, 1941: "The Factors of the Mind: An Introduction to Factor Analysis in Psychology," The Macmillan Company, New York.

————, 1948: The Factorial Study of Temperamental Traits, *Brit. J. Psychol. Statist. Sec.,* **1:**178–203.

Cartwright, D. S., and K. I. Howard, 1966: Multivariate Analysis of Gang Delinquency: I, Ecological Influences, *Multivariate Behav. Res.*, **1:**321–371.

Cattell, R. B., 1944: Psychological Measurement: Ipsative, Normative, and Interactive, *Psychol. Rev.*, **51:**292–303.

————, 1952: "Factor Analysis," Harper & Row, Publishers, Incorporated, New York.

Chu, Chen-Lin, 1966: "Object Cluster Analysis of the MMPI," Ph.D. thesis, University of California, Berkeley.

Cooley, W., and P. Lohnes, 1962: "Multivariate Procedures for the Behavioral Sciences," John Wiley & Sons, Inc., New York.

Etkin, W. (ed.), 1964: "Social Behavior and Organization Among Vertebrates," The University of Chicago Press, Chicago.

Ezekiel, M., and K. Fox, 1959: "Methods of Correlation and Regression Analysis," 3d ed., John Wiley & Sons, Inc., New York.

Fisher, R. A., 1921: Mathematical Foundations of Statistics, *Phil. Trans. Roy. Soc. London, Ser. A,* **222.**

Ghiselli, E. E., 1964: "Theory of Psychological Measurement," McGraw-Hill Book Company, New York.

Guilford, J. P., 1954: "Psychometric Methods," 2d ed., McGraw-Hill Book Company, New York.

————, 1965: "Fundamental Statistics in Psychology and Education," 4th ed., McGraw-Hill Book Company, New York.

Hadden, J. K., and E. F. Borgatta, 1965: "American Cities: Their Social Characteristics," Rand McNally & Company, Chicago.

Harman, H., 1941: On the Rectilinear Prediction of Oblique Factors, *Psychometrika,* **6:**29–35.

————, 1967: "Modern Factor Analysis," 2d ed., The University of Chicago Press, Chicago.

Holzinger, K., 1930: "Statistical Resume of the Spearman Two-factor Theory," The University of Chicago Press, Chicago.

———— and H. Harman, 1941: "Factor Analysis: A Synthesis of Factorial Methods," The University of Chicago Press, Chicago.

———— and F. Swineford, 1939: A Study in Factor Analysis: The Stability of a Bi-factor Solution, *Univ. Chicago Dept. Educ. Suppl. Educ. Monograph* 48.

Hotelling, H., 1933: Analyses of a Complex of Statistical Variables, *J. Educ. Psychol.,* **24:**417–441, 498–520.

————, 1936: Relations between Two Sets of Variates, *Biometrika,* **28:**321–377.

Kaiser, H., 1958: The Varimax Criterion for Analytic Rotation in Factor Analysis, *Psychometrika,* **23:**187–200.

—— and J. Caffrey, 1956: Alpha Factor Analysis, *Psychometrika*, **30:**1–14.

Kelley, T., 1928: "Crossroads in the Mind of Man," Stanford University Press, Stanford, Calif.

——, 1935: "Essential Traits of Mental Life," Harvard University Press, Cambridge, Mass.

——, 1947: "Fundamentals of Statistics," Harvard University Press, Cambridge, Mass.

Lawley, D., 1940: The Estimation of Factor Loadings by the Methods of Maximum Likelihood, *Proc. Roy. Soc. Edinburgh, ser. A*, **60:**64–82.

Lederman, W., 1939: On a Shortened Method of Estimation of Mental Factors by Regression, *Psychometrika*, **4:**109–116.

Lehman, R. S., and D. E. Bailey, 1968: "Digital Computing: Fortran IV and Its Applications in Behavioral Science," John Wiley & Sons, Inc., New York.

Lingoes, J. C., 1966: An IBM-7090 Program for Guttman-Lingoes Multidimensional Scalogram Analysis: I, *Behav. Sci.*, **11:**76–78.

McQuitty, L. L., 1960: Hierarchical Syndrome Analysis, *Educ. Psychol. Meas.*, **20:**293–304.

Meehl, P. E., 1950: Configural Scoring, *J. Consult. Psychol.*, **14:**165–171.

Meredith, W., 1964: Rotation to Achieve Factorial Invariance, *Psychometrika*, **29:**187–206.

Neuhaus, J., and C. Wrigley, 1954: The Quartimax Method: An Analytic Approach to Orthogonal Simple Structure, *Brit. J. Statist. Psychol.*, **7:**81–91.

Pearson, K., 1901: On Lines and Planes of Closest Fit to Systems of Points in Space, *Phil. Mag.*, 6th ser., 559–572.

Peters, C., and W. R. Van Voorhis, 1940: "Structural Procedures and Their Mathematical Bases," McGraw-Hill Book Company, New York.

Rao, C. R., 1955: Estimation and Tests of Significance in Factor Analysis, *Psychometrika*, **20:**93–111.

Shepard, R., 1962: The Analysis of Proximities: Multidimensional Scaling with an Unknown Distance Function," *Psychometrika*, **20:**125–140, 219–246.

Shevky, E., and W. Bell, 1954: "Social Area Analysis: Theory, Illustrative Application, and Computational Procedures," Stanford University Press, Stanford, Calif.

—— and M. Williams, 1948: "The Social Areas of Los Angeles," University of California Press, Berkeley.

Sokal, R., and P. Sneath, 1963: "Principles of Numerical Taxonomy," W. H. Freeman and Company, San Francisco.

Spearman, C., 1904: General Intelligence, Objectively Determined and Measured, *Am. J. Psychol.*, **15:**201–293.

——, 1913: Correlations of Sums and Differences, *Brit. J. Psychol.*, **5:**417–426.

——, 1927: "The Abilities of Man," The Macmillan Company, New York.

Stephenson, R. W., 1962: "Originality and Affect," Ph.D. thesis, University of California, Berkeley.

Stevenson, W., 1935: "Correlating Persons instead of Tests," *Char. Person.*, **4:**17–24.

Thomson, G. H., 1916: A Hierarchy without a General Factor, *Brit. J. Psychol.*, **8:**271–281.

——, 1949: On Estimating Oblique Factors, *Brit. J. Psychol. Statist. Sec.*, **2(1):**1–2.

——, 1951: "The Factorial Analysis of Human Ability," 5th ed., Houghton Mifflin Company, Boston.

Thurstone, L. L., 1931: Multiple Factor Analysis, *Psychol. Rev.*, **38:**406–427.

——, 1935: "The Vectors of Mind," The University of Chicago Press, Chicago.

——, 1947: "Multiple Factor Analysis," The University of Chicago Press, Chicago.

Tryon, R. C., 1932: Multiple Factors vs. Two Factors as Determiners of Abilities, *Psychol. Rev.*, **39**:324–351.

———, 1935: A Theory of Psychological Components: An Alternative to Mathematical Factors, *Psychol. Rev.*, **42**:425–454.

———, 1939: "Cluster Analysis," Edwards Brothers, Inc., Ann Arbor, Mich.

———, 1940: Studies in Individual Differences in Maze Ability: VIII, Prediction Validity of the Psychological Components of Maze Ability, *J. Compar. Physiol. Psychol.*, **30**:535–582.

———, 1955: "Identification of Social Areas by Cluster Analysis," *Univ. Calif. Publ. Psychol.*, **8**(1):1–100.

———, 1957*a*, Reliability and Behavior Domain Validity: Reformulation and Historical Critique, *Psychol. Bull.*, **54**:229–249.

———, 1957*b*: Communality of a Variable: Formulation from Cluster Analysis, *Psychometrika*, **22**:241–259.

———, 1958*a*: Cumulative Communality Cluster Analysis, *Educ. Psychol. Meas.*, **18**:3–35.

———, 1958*b*: General Dimensions of Individual Differences: Cluster Analysis vs. Factor Analysis, *Educ. Psychol. Meas.*, **18**:477–495.

———, 1959: Domain Sampling Formulation of Cluster and Factor Analysis, *Psychometrika*, **24**:113–135.

———, 1966: Unrestricted Cluster and Factor Analysis with Applications to the MMPI and Holzinger-Harman Problems, *Mult. Behav. Res.*, **1**:229–244.

———, 1967*a*: Person-clusters on Intellectual Abilities and MMPI Attributes, *Mult. Behav. Res.*, **2**:5–34.

———, 1967*b*: Predicting Individual Differences in Cluster Analysis: Holzinger Abilities and MMPI Attributes, *Mult. Behav. Res.*, **2**:325–348.

———, 1967*c*: Predicting Group Differences in Cluster Analysis: The Social Area Problem, *Mult. Behav. Res.*, **2**:453–475.

———, 1968*a*: Comparative Cluster Analysis of Variables and Individuals: Holzinger Abilities and the MMPI, *Mult. Behav. Res.*, **3**:115–144.

———, 1968*b*: Comparative Cluster Analysis of Groups: The Social Area Problem, *Mult. Behav. Res.*, **3**:213–232.

——— and D. Bailey (eds.), 1966*a*: "User's Manual of the BC TRY System of Cluster and Factor Analysis," Tryon-Bailey Associates, Inc., Boulder, Colo.

——— and ———, 1966*b*: The BC TRY Computer System of Cluster and Factor Analysis, *Mult. Behav. Res.*, **1**:95–111.

———, K. Stein, and C. Chu, 1965: Item-clusters and Profile Types Derived from Responses of Adults to the 556 MMPI Items, *Amer. Psychol.*, **20**:396.

Tucker, L. R., 1951: A Method for Synthesis of Factor Analysis Studies, *Dept. Army Person. Res. Sec. Rept. 984.*

Walker, H., and J. Lev, 1953: "Statistical Inference," Holt, Rinehart and Winston, Inc., New York.

Warner, W. L., M. Meeker, and K. Eels, 1949: "Social Class in America: A Manual for the Measurement of Social Class," Science Research Associates, Inc., Chicago.

Wolfe, John H., 1963: Object Cluster Analysis of Social Areas, M.A. thesis, University of California, Berkeley.

Wrigley, C., and J. Neuhaus, 1952: A Refactorization of the Burt-Pearson Matrix with the ORDVAC Electronic Computer, *Brit. J. Psychol. Statist. Sec.*, **5**:105–108.

——— and ———, 1955: The Matching of Two Sets of Factors, *Am. Psychol.*, **10**:418–419.

Index

Index